Advanced Research on Sprouts and Microgreens as a Source of Bioactive Compounds

Advanced Research on Sprouts and Microgreens as a Source of Bioactive Compounds

Editors

Beatrice Falcinelli
Angelica Galieni

MDPI • Basel • Beijing • Wuhan • Barcelona • Belgrade • Manchester • Tokyo • Cluj • Tianjin

Editors
Beatrice Falcinelli
Department of Agricultural,
Food and Environmental
Sciences
University of Perugia
Perugia
Italy

Angelica Galieni
Council for Agricultural
Research and Economics
Research Centre for Vegetable
and Ornamental Crops (CREA-OF)
Monsampolo del Tronto
Italy

Editorial Office
MDPI
St. Alban-Anlage 66
4052 Basel, Switzerland

This is a reprint of articles from the Special Issue published online in the open access journal *Plants* (ISSN 2223-7747) (available at: www.mdpi.com/journal/plants/special_issues/sprouts_microgreens_bioactive_compounds).

For citation purposes, cite each article independently as indicated on the article page online and as indicated below:

LastName, A.A.; LastName, B.B.; LastName, C.C. Article Title. *Journal Name* **Year**, *Volume Number*, Page Range.

ISBN 978-3-0365-7177-5 (Hbk)
ISBN 978-3-0365-7176-8 (PDF)

Contents

Review

Sprouts and Microgreens—Novel Food Sources for Healthy Diets

Andreas W. Ebert [†]

World Vegetable Center, 60 Yi-Min Liao, Shanhua, Tainan 74151, Taiwan; ebert.andreas6@gmail.com
† Current address: Oderstr. 3, D-73529 Schwäbisch Gmünd, Germany.

Abstract: With the growing interest of society in healthy eating, the interest in fresh, ready-to-eat, functional food, such as microscale vegetables (sprouted seeds and microgreens), has been on the rise in recent years globally. This review briefly describes the crops commonly used for microscale vegetable production, highlights *Brassica* vegetables because of their health-promoting secondary metabolites (polyphenols, glucosinolates), and looks at consumer acceptance of sprouts and microgreens. Apart from the main crops used for microscale vegetable production, landraces, wild food plants, and crops' wild relatives often have high phytonutrient density and exciting flavors and tastes, thus providing the scope to widen the range of crops and species used for this purpose. Moreover, the nutritional value and content of phytochemicals often vary with plant growth and development within the same crop. Sprouted seeds and microgreens are often more nutrient-dense than ungerminated seeds or mature vegetables. This review also describes the environmental and priming factors that may impact the nutritional value and content of phytochemicals of microscale vegetables. These factors include the growth environment, growing substrates, imposed environmental stresses, seed priming and biostimulants, biofortification, and the effect of light in controlled environments. This review also touches on microgreen market trends. Due to their short growth cycle, nutrient-dense sprouts and microgreens can be produced with minimal input; without pesticides, they can even be home-grown and harvested as needed, hence having low environmental impacts and a broad acceptance among health-conscious consumers.

Keywords: microscale vegetables; sprouts; microgreens; phytonutrients; functional foods; malnutrition; seed priming; biofortification; illumination; health-promoting compounds

Citation: Ebert, A.W. Sprouts and Microgreens—Novel Food Sources for Healthy Diets. *Plants* **2022**, *11*, 571. https://doi.org/10.3390/plants11040571

Academic Editors: Beatrice Falcinelli and Angelica Galieni

Received: 2 February 2022
Accepted: 18 February 2022
Published: 21 February 2022

Publisher's Note: MDPI stays neutral with regard to jurisdictional claims in published maps and institutional affiliations.

1. Introduction

Healthy diets are essential for nutrition and health [1]. As defined by Neufeld et al. [2], a healthy diet is "health-promoting and disease-preventing. It provides adequacy without excess, of nutrients and health-promoting substances from nutritious foods and avoids the consumption of health-harming substances." About three billion people cannot afford healthy diets around the globe. This figure includes most people living in sub-Saharan Africa and South Asia [3,4]. The Sustainable Development Goal 2 (SDG 2) 'Zero Hunger' of the United Nations calls for the eradication of hunger and all forms of malnutrition. All people ought to have access to safe, nutritious, and sufficient food all year round by 2030 [5]. The triple burden of malnutrition, i.e., undernutrition, micronutrient deficiency, and overnutrition, affects most nations around the globe. As incomes rise and food consumption patterns change, overnutrition from imbalanced diets increasingly becomes a concern in developed and developing countries.

Malnutrition is a high-risk factor for non-communicable diseases (NCDs), also known as chronic diseases. Diet-related NCDs, such as diabetes, cardiovascular disease, hypertension, stroke, cancer, and obesity, are escalating globally. Out of the estimated 40.5 million people killed by NCDs each year (71% of the annual deaths worldwide), approximately 32.2 million NCD deaths (80%) were attributable to cancers, cardiovascular diseases, chronic respiratory diseases, and diabetes [6,7]. The remaining 8.3 million NCD deaths (20%) have other root causes. These figures illustrate the seriousness of diet-related diseases for the healthcare sector. Under SDG 3—'Good Health and Well-Being'—SDG target 3.4 aims at reducing premature mortality from NCDs by one-third by 2030 [5]. The diversity and quality of food produced sustainably and made accessible to a wide range of consumers are decisive factors that enable substantial dietary shifts [8,9] and, in turn, help to address SDG targets 2.1 and 3.4. A truly transformed global food system may not only provide universal access to healthy diets but may also co-deliver on climate and environmental SDGs [10].

The nutrition community frequently highlights the importance of fruits, vegetables, and nuts in combating the triple burden of malnutrition [11]. The British Royal Society's strategy to eliminate hidden hunger involves measures that promote increased access to fruits and vegetables to enhance dietary diversity [12]. The World Health Organization (WHO) recommends a population-wide daily intake of 400 g of edible fruits and vegetables to prevent NCDs and alleviate several micronutrient deficiencies [13]. This WHO recommendation translates to roughly five portions of fruits and vegetables per day. People able to enjoy more diverse diets, in general, also have better nutrition and health. A recent study analyzing data of a health survey in Great Britain revealed a robust inverse association between fruit and vegetable consumption and mortality [14].

Despite this general acceptance that fruits and vegetables are essential for a healthy diet, the authors of several studies concluded that current and projected fruit and vegetable production levels would fail to meet healthy consumption levels [15,16]. Based on age-specific recommendations, only 40 countries representing 36% of the global population had adequate availability of fruits and vegetables in 2015 [17]. Although there was a sharp increase in vegetable consumption in sub-Saharan Africa over the last three decades, the combined fruit and vegetable intake (268 g) remains well below the WHO recommendation of 400 g [18].

With society's growing interest in healthy eating and lifestyles, e.g., the Slow Food movement and the promotion of novel and superfoods, the interest in fresh, ready-to-eat functional and nutraceutical food has been on the rise in recent decades [19,20]. In this context, microscale vegetables, i.e., sprouted seeds and microgreens, are becoming increasingly popular worldwide as fresh, ready-to-eat functional and nutraceutical food. They have great potential to diversify and enhance the human diet and address nutrient deficiencies due to their high content of phytochemicals [21–26].

Sprouts are commonly grown in the dark under high relative humidity. They are harvested when the cotyledons are still under-developed and true leaves have not begun to emerge, usually after 3–5 days from seed hydration. The entire plant (root, seed, and shoot) is consumed. Ancient Egyptians have already practiced sprouting seeds around 3000 B.C. [27]. During the germination process, the amount of antinutritive compounds (trypsin inhibitor, phytic acid, pentosan, tannin, and cyanides) decreases, while palatability and nutrient bioavailability, as well as the content of health-related phytochemicals (glucosinolates and natural antioxidants), are enhanced [28–30]. While sprouts usually take less than a week to mature, microgreens are harvested for consumption within 10–20 days of seedling emergence [31]. Microgreens, defined as tender, immature greens, are larger than sprouts, but smaller than baby vegetables or greens. They have a central stem with two fully developed, non-senescent cotyledon leaves and mostly one pair of small true leaves [21,32–35]. The stem, cotyledons, and first true leaves are consumed. Microgreens have been produced in Southern California since the 1980s [36,37] and have since gained popularity due to their vivid colors (like red and purple), delicate textures, and flavor-enhancing properties. They are used as garnishes in salads, sandwiches, soups, appetizers, desserts, and drinks [19] and are highly appreciated because of their nutritional benefits [21–26]. In vitro and in vivo research studies have demonstrated microgreens' anti-inflammatory, anti-cancer, anti-bacterial, and anti-hyperglycemic properties, further strengthening their attractiveness as a new functional food that is beneficial to human health (see, e.g., review by Zhang et al. [37]).

Commercial and home-grown microgreen production comprises several aspects: selecting appropriate species, growing systems, substrates, quality of seeds, seeding and germination, irrigation and fertilization, harvesting, phytosanitary quality, and post-harvest storage practices. Di Gioia et al. [32] provided a detailed insight into these aspects. For a recent review of microgreen product types and production practices, readers may also consult Verlinden [38].

This comprehensive review describes the main crops used in microscale vegetable production and the factors that impact sprouts' and microgreens' nutritional and bioactive profile and their consumer acceptance. It also reflects on underutilized species (landraces, wild food plants, and crops' wild relatives) that offer the scope to widen the range of crops used for this purpose. In addition, this paper reviews the effects of plant growth stages on the nutritional and bioactive composition of edible plant parts.

2. Crops Commonly Used for Microscale Vegetable Production

Sprouts and microgreens are grown from the seeds of many crops, such as legumes, cereals, pseudo-cereals, oilseeds, vegetables, and herbs [21,38]; Table 1. The significant traits of interest for consumers are the appearance, texture, flavor, phytochemical composition, and nutritional value of sprouts and microgreens [39]. Most crops are grown for sprouts and microgreens, except for beans and some oilseed tree species that are commonly grown as sprouts only (Table 1). Mungbean and soybean sprouts have long been an essential, year-round component of Asian and vegetarian dishes [21,40]. In recent decades, mungbean sprouts have become increasingly popular in the Americas, Europe, and Africa. They are commonly recognized as "bean sprouts," although this group comprises several different crops (see Table 1). Most food and forage legumes are known for their high nutritional value and an abundance of minerals and secondary metabolites. Sprouted seeds and microgreens often contain higher concentrations of bioactive compounds than raw seeds (Figures 1–3) [41,42]. Sprouting cereal grains enhances their nutritional value, especially when applying a sprouting duration of at least 3 to 5 days [43]. The sprouting process activates hydrolytic enzymes and releases nutrients from their phytate chelates, making them bioavailable; in addition, vitamins are synthesized and accumulate [43]. Sprouted grains are also used in many staple foods such as bread, pasta, noodles, and breakfast flakes, but food processing often compromises their nutritional value.

Table 1. The crop groups commonly used for sprouting and microgreen production.

Crop Group	Family	Species	Common Name	Main Use [1]
Legumes	Fabaceae	*Arachis hypogaea*	peanut	S
		Cicer arietinum	chickpea	S & M
		Glycine max	soybean	S
		Lens culinaris	lentil	S & M
		Medicago sativa	alfalfa	S & M
		Trifolium repens	clover	S & M
		Vigna angularis	adzuki bean	S (& M)
		Vigna mungo	black gram	S
		Vigna radiata	mungbean	S (& M)
		Vigna unguiculata	cowpea	S
Cereals	Poaceae	*Hordeum vulgare*	barley	S & M
		Zea mays	maize	S & M
		Avena sativa	oat	S &M
		Oryza sativa	rice	S & M
		Secale cereale	rye	S & M
		Triticum aestivum	wheat	S & M
		Zea mays	maize, popcorn	S & M
Pseudocereals	Amaranthaceae	*Amaranthus sp.*	amaranth	S & M
		Chenopodium quinoa	quinoa	S & M
	Polygonaceae	*Fagopyrum esculentum*	buckwheat	S & M

Table 1. *Cont.*

Crop Group	Family	Species	Common Name	Main Use [1]
Oilseeds	Asteraceae	*Helianthus annuus*	sunflower	S & M
	Betulaceae	*Corylus avellana*	hazelnut	S
	Linaceae	*Linum usitatissimum*	linseed, flax	S & M
	Pedaliaceae	*Sesamum indicum*	sesame	S & M
	Rosaceae	*Prunus amygdalus*	almond	S
Vegetables & herbs	Amaranthaceae	*Beta vulgaris*	beet	S & M
		Spinacia oleracea	spinach	S & M
	Amaryllidaceae	*Allium cepa*	onion	S & M
		Allium fistulosum	spring onion	S & M
		Allium porrum	leek	S & M
		Allium schoenoprasum	chives	S & M
	Apiaceae	*Apium graveolens*	celery	S & M
		Coriandrum sativum	coriander	S & M
		Daucus carota subsp. *sativus*	carrot	S & M
		Foeniculum vulgare	fennel	S & M
		Petroselinum crispum	parsley	S & M
	Asteraceae	*Lactuca sativa*	lettuce	S & M
	Brassicaceae	*Brassica juncea*	purple mustard	S & M
		Brassica oleracea, var. *alboglabra*	Chinese kale	S & M
		Brassica oleracea var. *capitata*	(red) cabbage	S & M
		Brassica oleracea var. *gongylodes*	purple kohlrabi	S & M
		Brassica oleracea var. *italica*	broccoli	S & M
		Brassica rapa var. *chinensis*	pak choi	S & M
		Brassica rapa var. *niposinica*	mizuna	S & M
		Brassica rapa var. *rapa*	turnip	S & M
		Brassica rapa var. *rosularis*	flat cabbage; tatsoi	S & M
		Eruca sativa	arugula, rocket	S & M
		Lepidium bonariense	peppercress	S & M
		Nasturtium officinale	watercress	S & M
		Raphanus raphanistrum subsp. *sativus*	daikon, small radish	S & M
	Fabaceae	*Trigonella foenum-graecum*	fenugreek	S & M
		Pisum sativum	garden pea	S & M
		Pisum sativum var. saccharatum	snow peas	S & M
	Lamiaceae	*Melissa officinalis*	lemon balm	S & M
		Ocimum basilicum	sweet basil	S & M
		Perilla frutescens	purple perilla	S & M

[1] Sprouts (S), Microgreens (M) or both (S & M).

Pseudocereals are underutilized food crops that are receiving increasing attention as highly nutritious and functional foods [44]. Among those, amaranth, quinoa, and buckwheat are increasingly becoming popular for sprout and microgreen production [45–48]. Apart from soybean, peanut (listed under legumes), and mustard (classified here as a vegetable), almond, hazelnut, linseed, sesame, and sunflower are other oilseed crops that one can use for sprouting or microgreen production (Table 1). Among the group of vegetables and herbs, members of the Brassicaceae family are widely used for sprouting and microgreen production, followed by crops of the Apiaceae, Fabaceae, and Amaranthaceae families (Table 1).

Figure 1. Dormant seed with stored reserves. A display of a seed mix for sprouting, consisting of quinoa (*Chenopodium quinoa*), lentil (*Lens culinaris*), and radish (*Raphanus sativus*) seed.

Figure 2. Home-grown 3-day old pea (*Pisum sativum*) sprouts. Seed germination process and nutritional benefits of sprouted seeds:

- Seed activation through imbibition, favourable temperature, oxygen, light, or darkness
- Enhanced respiration and metabolic activities
- Enzymes mobilize stored seed reserves and convert starch to sugar
- Hydrolysis of storage proteins, release of essential amino acids
- Accumulation of phenolic compounds with antioxidant ability
- Accumulation of vitamins (C, folate, thiamin, pyridoxin, tocopherols, niacin, etc.).

  Reduction of antinutritional factors:

- Phytate, oxalate, and tannin degradation, leading to enhanced palatability, improved bioaccessibility of iron and calcium, and enhanced digestibility of proteins.

Figure 3. Home-grown 9-day old pea (*Pisum sativum*) microgreens. Commonly recognized nutritional benefits of microgreens:

- Photosynthetic activity in microgreens further enhances vitamin C, phylloquinone, and tocopherol accumulation compared to sprouts
- Accumulation of carotenoids is often higher than in mature vegetables
- Increased accumulation of chlorophyll and phenolic compounds with antioxidant ability, compared to sprouts
- Often higher content of macro- and micronutrients and lower content of nitrate in microgreens compared to the adult growth stage
- Biofortification with specific elements (iodine, iron, zinc, selenium) made easy in hydroponic systems
- Microgreens are consumed raw, hence thermolabile ascorbic acid content can be fully utilized, unlike in cooked mature vegetables.

2.1. Bioactive Composition and Potential Health Effects of Brassica Microscale Vegetables

As evident from Table 1, the Brassicaceae family comprises a wide range of crops commonly used for microscale vegetable production. The intrinsic qualities of *Brassica* vegetables, including their color, aroma, taste, and health properties, are profoundly determined by secondary plant metabolite profiles and their concentrations in plant tissues [49]. *Brassica* vegetables are rich sources of bioactive compounds, such as glucosinolates (GSLs), polyphenols, anthocyanins, ascorbic acid, carotenoids, and tocopherols [50–53]. The biosynthesis of secondary plant metabolites is closely linked to plant protection and defense mechanisms and can be modulated by environmental and agronomical factors. Those factors may significantly change the concentration of secondary plant metabolites with up to 570-fold increases for specific compounds, such as isothiocyanates [49].

Among the bioactive compounds of *Brassica* vegetables, polyphenols and GSLs have been widely studied due to their known health-promoting effects [54,55], including the impact of cooking methods on the retention of these essential compounds [56]. In addition, polyphenols are good sources of natural antioxidants, which help decrease the risk of diseases associated with oxidative stress [57]. GSLs, defined as aliphatic, aromatic, or indolic based on their side chains, are important secondary metabolites that are predominantly found in *Brassica* crops [58].

The cancer-preventive potential of kale (*B. carinata*) has been demonstrated through in vitro studies which indicated the protection of human liver cells against aflatoxin in vitro [59]. Rose et al. [60] obtained similar results with broccoli (*Brassica oleracea* var. *italica*) and watercress (*Nasturtium officinale*). Isothiocyanates—hydrolysis products of GSLs—extracted from broccoli and watercress sprouts suppressed human MDA-MB-231 breast cancer cells in vitro. In addition, extracts of 3-day-old broccoli sprouts were highly effective in reducing the incidence, multiplicity, and rate of development of mammary tumors in rats treated with the carcinogen DMBA (7,12-dimethylbenz[*a*]anthracene) [61]. Therefore, diets high in *Brassica* vegetables may contribute to the suppression of carcinogenesis, and this effect is at least partly related to their relatively high content of GSLs [62].

Among five of the microgreen species of the Brassicaceae, namely broccoli (*Brassica oleracea* var. *italica*), daikon (*Raphanus raphanistrum* subsp. *sativus*), mustard (*Brassica juncea*), rocket salad (*Eruca vesicaria*), and watercress (*Nasturtium officinale*), broccoli had the highest polyphenol, carotenoid, and chlorophyll contents, as well as strong antioxidant power [53]. Mustard microgreens showed high ascorbic acid and total sugar contents. On the other hand, rocket salad exhibited the lowest antioxidant content and activity among the five evaluated microgreen crops [53].

Broccoli, curly kale, red mustard, and radish microgreens are good sources of minerals. They provide considerable amounts of vitamin C (31–56 mg/100 g fresh weight) and total carotenoids (162–224 mg β-carotene/100 g dry weight), the latter being higher than in adult plants [63]. In digestion studies, total soluble polyphenols and total isothiocyanates showed a bioaccessibility of 43–70% and 31–63%, respectively, while the bioaccessibility of macroelements ranged from 34–90% [63]. Among the four microgreen crops tested, radish and mustard presented the highest bioaccessibility of bioactive compounds and minerals.

2.2. Consumer Acceptance of Sprouts and Microgreens and Nutritional Profile of Microscale Vegetables

Six commonly grown and consumed microgreen species were tested by Michell et al. [64] for consumer acceptance, as follows: (a) Brassicaceae: arugula (*Eruca sativa*), broccoli (*Brassica oleracea* var. *italica*), and red cabbage (*B. oleracea* var. *capitata*); (b) Amaranthaceae: bull's blood beet (*Beta vulgaris*) and red garnet amaranth (*Amaranthus tricolor*); and (c) Fabaceae: tendril pea (*Pisum sativum*). All six microgreen crops received high ratings for appearance acceptability; hence they could easily be used to enhance the visual appearance of meals if they have the appropriate sensory attributes [64]. Among the six microgreen crops evaluated, broccoli, red cabbage, and tendril pea received the highest overall acceptability score with similar trends for taste and texture.

In a similar approach, Xiao et al. [39] evaluated six microgreen species for their sensory attributes and nutritional value. The six species consisted of (i) three Brassicaceae crops: Dijon mustard—*Brassica juncea*, peppercress—*Lepidium bonariense*, and China rose radish—*Raphanus sativus*; (ii) two representatives of the Amaranthaceae family: bull's blood beet—*Beta vulgaris* and red amaranth—*Amaranthus tricolor*; and (iii) one representative of the Lamiaceae family: opal basil—*Ocimum basilicum*. Overall, all six microgreen species included in the study received "good" to "excellent" consumer acceptance ratings and showed high nutritional quality. Among those six crops, bull's blood beet received the highest acceptability score regarding flavor and overall eating quality, while peppercress received the lowest score [39]. In addition, the authors detected the highest concentrations of total ascorbic acid and tocopherols in China rose radish, the highest contents of total phenolics and phylloquinone (vitamin K1) in opal basil, and the highest content of carotenoids in red amaranth.

In trials conducted at the World Vegetable Center, the consumer acceptance of amaranth (*Amaranthus tricolor*) landraces, conserved in the Genebank, were compared with commercially available cultivars [45]. A Genebank accession (VI044470) consistently received the highest ratings for appearance, texture, taste, and general acceptability at the sprout, microgreen, and fully grown stages.

A consumer acceptance study conducted in India comprised the following ten microgreens: carrot, fenugreek, mustard, onion, radish, red roselle, spinach, sunflower, fennel, and French basil [65]. The organoleptic acceptability of all ten microscale vegetables ranged from very good to excellent.

The high appreciation of microgreens compared to mature vegetables might also be related to their aroma profile. Recent research undertaken by Dimita et al. [66] has shown that the aroma profile of *Perilla frutescens* var. *frutescens* (Chinese basil or perilla; green leaves) and *P. frutescens* var. *crispa* (red leaves) is much higher at the microgreens stage than at the later adult stage. Both varieties have a clearly distinct aroma profile at the microscreen stage. The red variety emitted a citrusy, spicy, and woody aroma, while the green type produced a fruity, sweet, spicy, and herbaceous aroma at the microgreens stage [66]. After the microscreen stage, at the age of four weeks, green Chinese basil no longer emitted any aroma volatiles. Hence, the aroma profile of Chinese basil leaves at the microgreen stage is clearly variety-specific and not related to the content of total phenols or the antioxidant capacity of the leaves.

Attempting a nutritional determination among five Brassicaceae microgreen crops (broccoli, daikon, mustard, rocket salad, and watercress), broccoli excelled [53]. Broccoli microgreens had the highest content of isothiocyanates, known for their cancer-preventing abilities [61,62] and displayed the most potent antioxidant power. Broccoli microgreens exhibited the overall best nutritional profile and, therefore, are considered as one of the most promising functional food species [53].

Based on the determination of the contents of 11 nutrients and vitamins, as well as the anti-nutrient oxalic acid, and their relative contribution to the diet as per the estimated daily intake published in the United States Department of Agriculture (USDA) database for green leafy vegetables, Ghoora et al. [67] computed a nutrient quality score (NQS) to assess the nutritional quality of ten culinary microgreen species. The selected species included vegetable crops (spinach, carrot, mustard, radish, roselle, and onion); leguminous crops (fenugreek); oleaginous crops (sunflower); and aromatic species (French basil and fennel). All microgreen crops are moderate to good sources of protein, dietary fiber, and essential nutrients. Concerning their vitamin content, the studied microgreens are excellent sources of ascorbic acid, vitamin E, and beta-carotene (pro-vitamin A), meeting 28–116%, 28–332%, and 24–72% of reference daily intake of the respective vitamins [67]. In general, microgreens had low levels of oxalic acid, which is a predominant anti-nutrient in mature leafy vegetables. Based on the calculated NQS, radish microgreens showed the highest nutrient density, followed by French basil and roselle microgreens. On the other hand, fenugreek and onion microgreens are the least nutrient dense. Furthermore, the calculated NQS revealed that all microgreens were 2–3.5 times more nutrient dense than mature leaves of spinach cultivated under similar conditions.

While high nutrient density and high phytochemical content are considered a must in sprouts and microgreens, these microscale vegetables must also have high consumer acceptability in flavor attributes and visual appearance. Based on organoleptic and nutritional properties, Caracciolo et al. [68] assessed different microgreens species regarding consumer acceptance of appearance, texture, and flavor. The 12 microgreen species included in the studies were amaranth, coriander, cress, green basil, komatsuna, mibuna, mizuna, pak choi, purple basil, purslane, Swiss chard, and tatsoi. The results revealed that while the visual appearance of the microgreens played a role, the flavor and texture of microgreens were the main determining factors for consumer acceptance. In general, low astringency, sourness, and bitterness enhanced the consumer acceptability of microgreens [68]. Among the 12 examined microgreen species, mibuna (*Brassica rapa* subsp. *nipposinica*) and cress (*Lepidium sativum*) received the lowest consumer acceptance score, while Swiss chard (*Beta vulgaris* subsp. *vulgaris*) and coriander (*Coriandrum sativum*) were the most appreciated microscale vegetables.

Unfortunately, phenolic content strongly correlates with flavor attributes such as sourness, astringency, and bitterness. Therefore, microscale vegetables rich in phenolics, such as red cabbage (*Brassica oleracea* var. *capitata*), sorrel (*Rumex acetosa*), and pepper-cress (*Lepidium bonariense*), in general, receive a low consumer acceptability score [22,39]. However, rich content in minerals, vitamins, phenolics, and antioxidant activity can also be found in species of more acceptable tastes, such as amaranth, coriander, and Swiss chard [22,25,37,45,69,70]. As shown with the above examples, identifying microgreen species that satisfy both sensory and health attributes at a high degree remains a challenge since acrid taste's acceptability is subject to both inherited and acquired taste factors [22]. Providing concise, crop-specific information about the culinary uses and the outstanding nutritional and health benefits of microscale vegetables might increase consumer interest. Such information might convince them to try products of high nutritional value but less agreeable tastes, eventually broadening the overall consumer acceptability of such produce [68].

3. Underutilized Species with Potential for Microscale Vegetable Production to Enhance Nutrition Security

Breeding for high yield, appearance, etc., may sometimes unintentionally lead to a decline in essential nutrients and phytochemicals. This hypothesis is supported by a review study conducted on 43 garden crops that revealed a statistically reliable reduction in six nutritional factors (protein, Ca, P, Fe, riboflavin, and ascorbic acid) between 1950 and 1999 based on USDA food composition data for this period [71]. These changes can be explained by changes in the crop varieties cultivated during this same period. Similar trends have been observed in wheat grain [72,73] and potato tubers [74]. Marles [75] conceded that some modern varieties of vegetables and grains might have lower contents in some nutrients than older varieties due to a dilution effect of increased yield by the accumulation of carbohydrates without a proportional increase in certain other nutrients. Nevertheless, he argued that eating the WHO-recommended daily servings of fruits and vegetables would provide adequate nutrition [75]. Nonetheless, we know that most countries and the majority of the global population, especially in sub-Saharan Africa, are still well below the WHO-recommended daily intake levels of fruits and vegetables [17,18].

When aiming for high phytonutrient density and exciting flavors and tastes, it might well be worth exploring farmers' landraces, wild food plants, or populations found in a semi-wild or wild state, such as crops' wild relatives. Such species are often part of the conservation focus of national genebanks, e.g., the genebanks maintained by the USDA in the USA, or international ones, e.g., the Genebank maintained by the World Vegetable Center (WorldVeg). This idea of exploring landraces, wild food plants, or crops' wild relatives for microscale vegetable production has recently gained impetus [22,26,33,76,77].

The microgreens of wild plants and culinary herbs could constitute a source of functional food with attractive aromas, textures, and visual appeal, which could provide health benefits due to their elevated nutraceutical value and could be exploited in new gastronomic trends [25,38,39]. Studies of 13 wild edible plants from 11 families undertaken by Romojaro et al. [78] revealed that their outstanding nutritional value would merit promotion to provide health benefits. Fennel, which is commonly used for sprout and microgreen production, has higher radical scavenging activity, total phenolic, and total flavonoid contents in its wild form compared to medicinal and edible fennel [79]. Variations in the phytochemical content of wild fennel obtained from different geographical areas was also reported. For broccoli, kale, and pak choi, there is a variation of the concentrations of secondary plant metabolites among cultivars with ranges up to 10-fold (Table 2) [49].

Studies involving three wild leafy species, *Sanguisorba minor* (salad burnet), *Sinapis arvensis* (wild mustard), and *Taraxacum officinale* (common dandelion), at the microgreen and baby green stages were conducted by Lenzi et al. (Table 2) [33]. The authors recognized the potential of those wild edible plants in achieving competitive yields and contributing to the dietary intake of nutritionally essential macro- and microelements, as well as bioactive compounds.

Sprouted seeds of chia (*Salvia hispanica*), golden flax, evening primrose, phacelia and fenugreek are an excellent source of health-promoting phytochemicals, especially antioxidants and minerals [80]. Germination significantly increased the total phenolic content (e.g., from 1.40 to 4.54 mg GAE g^{-1} in fenugreek and from 0.33 to 5.88 mg GAE g^{-1} in phacelia), antioxidant activity (e.g., 1.5 to 52.5-fold in fenugreek and phacelia, respectively), and the content of phenolic acids and flavonoids in sprouts compared to the ungerminated seed of the mentioned species.

A rather exotic medicinal vegetable with a mild, bitter flavor is Korean ginseng (*Panax ginseng*). Sprouts of this crop can be grown to whole plants in 20 to 25 days in a soilless cultivation system [81,82]. Their main bioactive compounds are ginsenosides which have anti-cancer, anti-diabetic, immunomodulatory, neuroprotective, radioprotective, anti-amnestic, and anti-stress properties (see references in [81]). Korean ginseng sprouts can be included in salads, milkshakes, sushi, soups, and tea. It is also used in health food supplements and cosmetics.

In summary, underutilized plants, such as farmers' landraces, wild food plants, or crops' wild relatives, often conserved in genebanks, might offer valuable opportunities to produce sprouts and microgreens with high nutritional value and exciting flavors and tastes, thus meeting the demands of health-conscious consumers. However, additional research efforts are required to determine whether the germination performance of these novel plant materials is satisfactory for commercial microscale vegetable production.

Table 2. Underutilized plant material for sprouting and microgreen production, consumer acceptance, and highlights of secondary metabolites.

Family	Species	Type of Plant Material	Secondary Metabolites	References
Amaranthaceae	*Amaranthus caudatus* (foxtail amaranth)	old varieties	High total phenolics, total betalain, and total flavonoid content	[83]
	Amaranthus cruentus (red amaranth)	old varieties	Amaranth sprouts are a good source of anthocyanins and total phenolics with high antioxidant activity	[83,84]
	Amaranthus hypochondriacus (Prince's feather)	ornamental	Good source of antioxidants, especially the leaves	[83]
	Amaranthus tricolor (edible amaranth)	landrace	A genebank accession (VI044470) consistently received the highest ratings for appearance, texture, taste, and general acceptability at the sprout, microgreen, and fully grown stage compared to commercial cultivars	[45]
	Atriplex hortensis (red orach)	under-utilized	Ascorbic acid content	[69]
	Chenopodium album (pigweed)	under-utilized	Antioxidant activity and total phenolic content are enhanced in germinated *C. album* seeds	[85]
	Chenopodium quinoa (quinoa)	old variety	Quinoa sprouts are a good source of anthocyanins and total phenolics with high antioxidant activity	[84]
Apiaceae	*Anethum graveolens* (dill)	under-utilized	Total phenolic and total flavonoid content; antioxidant activity	[86]
	Coriandrum sativum (coriander)	under-utilized	A strong influence of the substrate on the content of carotenoids and total phenolics	[87]
Araliaceae	*Panax ginseng* (Korean ginseng)	under-utilized medicinal plant	Ginsenosides (triterpene glycoside saponin)	[81,82]
Asteraceae	*Artemisia dracunculus* (tarragon)	aromatic herb	N/A; red and blue LED exposure enhances germination and growth of tarragon sprouts	[88]
	Cichorium intybus (chicory)	medicinal herb	Total phenolics, tocopherols, anthocyanins, high levels of carotenoids	[89]
	Taraxacum officinale (common dandelion)	wild plants	Anthocyanins and carotenoids; high Fe content	[33]
Basellaceae	*Basella alba* (Malabar spinach)	underutilized vegetable	High ascorbic acid and total phenolic content	[90]
Boraginaceae	*Borago officinalis* (borage)	medicinal herb	Total phenolic and carotenoid content, antioxidant capacity	[91]
	Phacelia tanacetifolia (phacelia)	wildflower	Total phenolics, flavonoids, and antioxidant activity	[80]
Brassicaceae	*Brassica oleracea* var. *italica* (broccoli)	landrace	(1) High polyphenol content in broccoli landrace; (2) highest vitamin C content found in microgreens of broccoli landrace	[76]
	Brassica oleracea var. *acephala* (kale)	landrace	(1) Higher content of flavonoids (quercetin and kaempferol derivatives) in traditional cultivars than in modern cultivars (hybrids); (2) among 8 cultivars, higher concentrations of lutein and β-carotene were found in old cultivars	[49]
	Sinapis arvensis (field mustard)	under-utilized	Carotenoids and anthocyanins	[33]
	Wasabi japonica (wasabi)	under-utilized	Ascorbic acid, β-carotene, lutein/zeaxanthin content	[69]
Convolvulaceae	*Ipomoea aquatica* (water spinach)	under-utilized	High total phenolics and total flavonoid content; high antioxidant activity	[89,90]

Table 2. *Cont.*

Family	Species	Type of Plant Material	Secondary Metabolites	References
Cucurbitaceae	*Cucumis sativus* (cucumber)	under-utilized	High ascorbic acid content	[90]
	Cucurbita moschata pumpkin)	under-utilized	High total phenolics and total flavonoids content	[90]
	Lagenaria siceraria (bottle gourd)	under-utilized	High total phenolics content; high antioxidant activity; high Cu and Fe levels	[90]
Fabaceae	*Glycine max* (soybean)	landrace	Nutrient and antioxidant contents of soybean sprouts were superior to mungbean sprouts	[92]
	Medicago intertexta (hedgehog medick)	wild species	Total phenolic and flavonoid contents, antioxidant, and antidiabetic activities	[93]
	Medicago polymorpha (bur clover)	wild, invasive species	Total phenolic and flavonoid contents, antioxidant, and antidiabetic activities	[93]
	Melilotus indicus (annual yellow sweet clover)	wild species	Total phenolic and flavonoid contents, antioxidant, and antidiabetic activities	[93]
	Vigna radiata (mungbean)	landrace	(1) Old mungbean accessions were superior in protein, calcium (Ca), iron (Fe), zinc (Zn), carotenoid, and vitamin C content compared to improved mungbean lines at the fully mature stage; (2) compared to commercial mungbean varieties, a landrace from Taiwan (VI000323) showed the highest levels of caffeic acid and kaempferol at the sprouting and fully mature stage	[92]
Lamiaceae	*Ocimum basilicum* (Sweet basil)	culinary herb	High phylloquinone and total phenolics concentration	[39]
	Ocimum x africanum (lemon basil)	culinary herb	Total phenolic and total flavonoid content; antioxidant activity	[86]
	Ocimum sanctum (sacred basil)	medicinal herb	Total phenolic and total flavonoid content; antioxidant activity	[86]
	Salvia hispanica (chia)	under-utilized	Total phenolics, flavonoids, antioxidant activity.	[80]
Linaceae	*Linum flavum* (golden flax)	under-utilized	Total phenolics, flavonoids, antioxidant activity.	[80]
Malvaceae	*Corchorus olitorius* (jute mallow)	under-utilized	High ascorbic acid and total phenolics content; high antioxidant activity	[90]
	Hibiscus subdariffa (red roselle)	under-utilized culinary herb	Anthocyanins, flavonoids, and phenolic acids contribute to the antioxidative activity	[65]
Onagraceae	*Oenothera biennis* (evening primrose)	under-utilized	Total phenolics, flavonoids, antioxidant activity	[80]
Plantaginaceae	*Plantago coronopus* (buck's-horn plantain)	wild herb	Total phenolics, flavonoids, and antioxidant activity	[94]
Polygonaceae	*Rumex acetosa* (sorrel)	wild herb	Total phenolics, flavonoids, and antioxidant activity	[94]
Portulacaceae	*Portulaca oleracea* (purslane)	wild herb	Total phenolics, flavonoids, and antioxidant activity	[94]
Rosaceae	*Sanguisorba minor* (salad burnet)	under-utilized	Carotenoids and anthocyanins; high amounts of Mg, P, Zn, Mn, and Mo	[33]

4. Variation of Nutritional Value and Content of Phytochemicals According to Plant Growth Stages

Numerous studies have shown that the nutritional value and content of phytochemicals of vegetables and other crops may vary with plant growth and development. The concentration of essential minerals, vitamins, bioactive compounds, and antioxidant activity often increases in this sequence: raw seeds—sprouted seeds—microgreens (Figure 4) [41]. In many cases, sprouts and microgreens even exceed the nutritional value of fully grown plants (Figure 5) [95]. Examples of variations of the content of essential nutrients, vitamins, and phytochemicals according to plant growth stages (seeds, sprouts, microgreens, baby leaves, and fully grown) are listed in Table 3 and discussed below.

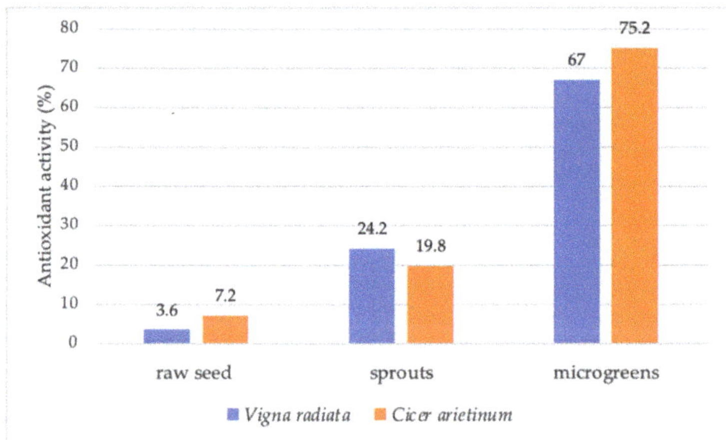

Figure 4. The antioxidant activity (%) in methanol extract (100 mg/mL) of raw seed, sprouts, and microgreens of *Vigna radiata* and *Cicer arietinum*; a graphical representation of data published by Kurian and Megha [42].

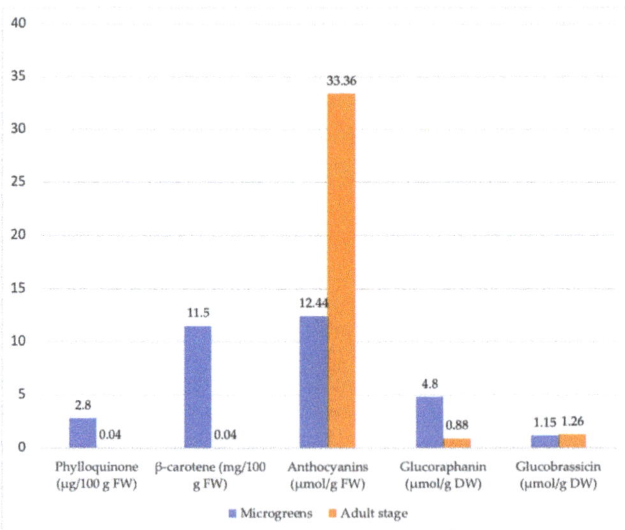

Figure 5. A comparison of selected phytochemical concentrations of red cabbage (*Brassica oleracea* var. *capitata*) at the microgreen and adult growth stage. FW = fresh weight; DW = dry weight; a graphical representation of data published by Choe et al. [95].

Research undertaken by Pająk et al. [96] and Khang et al. [97] has shown that seed germination can increase total phenolic content (TPC) levels and antioxidant activity in mungbean (*Vigna radiata*), adzuki bean (*V. angularis*), black bean (*V. cylindrica*), soybean (*Glycine max*), peanut (*Arachis hypogaea*), radish (*Raphanus sativus*), broccoli (*Brassica oleracea* var. *italica*), and sunflower (*Helianthus annuus*). Among the 13 phenolic compounds detected in high concentrations in the studies carried out by Khang et al. [97], sinapic acid, ellagic acid, ferulic acid, and cinnamic acid showed high correlations with antioxidant activities. High TPC levels have also been confirmed in sprouted seeds of several underutilized species, such as chia (*Salvia hispanica*), golden flax (*Linum flavum*), phacelia (*Phacelia tanacetifolifa*), fenugreek (*Trigonella foenum-graecum*), and evening primrose (*Oenothera biennis*) [80]. Evening primrose showed the highest TPC values and antioxidant activity among those underutilized species, both for sprouts and seeds. Compared to dry grains, seed sprouting enhanced TPC levels of chia from 0.92 to 4.40 mg gallic acid equivalent (GAE) g^{-1}, golden flax from 0.93 to 4.50 mg GAE g^{-1}, phacelia from 0.33 to 5.88 mg GAE g^{-1}, and fenugreek from 1.40 to 4.64 mg GAE g^{-1} [80].

Compared to ungerminated seeds, amaranth and quinoa sprouts showed higher contents of total flavonoids, phenolics, and antioxidant activity (Table 3) [83,84,98,99]. A substantial increase in vitamin C (ascorbic acid) content was observed from amaranth sprouts to microgreens (2.7-fold) and from amaranth microgreens to fully grown leafy amaranth (2.9-fold) (Table 3) [45]. Higher ascorbic acid and α-tocopherol levels were detected in spinach microgreens compared to the mature vegetable stage [67]. The often-higher ascorbic acid (vitamin C) content of microgreens compared to sprouts [25,92] can be explained by the presence of photosynthetic activity, which is absent in sprouts. Ascorbate is synthesized from photosynthetic hexose products and plays a significant role in photosynthesis as an enzyme cofactor (including in the synthesis of plant hormones and anthocyanins) and cell growth regulation [100]. Red cabbage microgreens had a 6-fold higher concentration of total ascorbic acids than mature cabbage [69]. With 131.6 mg/100 g FW, garnet amaranth also had a much higher total ascorbic acid (TAA) concentration than its mature counterpart [69]. With some exceptions (e.g., golden pea tendrils and sorrel), most of the 25 microgreens studied by Xiao et al. [69] showed higher TAA concentration than their mature counterparts.

The ascorbic acid content of fenugreek, spinach, and roselle microgreens reached 120%, 127%, and 119%, respectively, of their fully grown, mature stage (Table 3) [67]. The ascorbic acid levels of the studied microgreens ranged from 29.9–123.2 mg/100 g, and, therefore, were comparable to those of citrus fruits, which are generally known to be rich sources of vitamin C [101].

Tocopherols and tocotrienols belong to the vitamin E family. The α-tocopherol levels of fenugreek, spinach, and roselle microgreens were significantly higher than those of their respective mature leaves (Table 3) [67]. Among the 25 microgreens evaluated, green daikon radish microgreens exhibited the highest tocopherol concentrations in the α- and γ-forms, followed by coriander, opal radish, and peppercress [69]. Although golden pea tendrils had the lowest tocopherol concentrations of α (4.9 mg/100 g FW) and γ (3.0 mg/100 g FW), these values were still higher than those determined in mature spinach leaves.

In general, the microgreens' phylloquinone (vitamin K1) content is relatively high compared to the corresponding values of mature vegetables (Table 3) [95]. A total of 18 out of 25 commercially grown microgreens contain similar or greater phylloquinone concentrations than the commonly consumed vegetable broccoli. Green (pea tendrils) or bright red (garnet amaranth) microgreens often exhibit higher phylloquinone concentrations than yellow microgreens (popcorn and golden pea) [69].

Among 25 microgreens, Choe et al. [95] detected wide ranges of β-carotene concentrations. Red sorrel exhibited the highest β-carotene concentration (12.1 mg/100 g FW), while golden pea tendrils and popcorn microgreens had the lowest β-carotene concentrations (0.6 mg/100 g FW). With 11.7 mg/100 g FW, coriander microgreens had the second highest β-carotene concentration, a 3-fold higher concentration than found in mature coriander leaves. With 11.5 mg/100 g FW, red cabbage microgreens had a 260-fold higher β-carotene content than that found in mature red cabbage (0.044 mg/100 g FW) [95]. Except for golden pea tendrils and popcorn shoots, most microgreens were rich in β-carotene. Coriander (10.1 mg/100 g FW) and red cabbage (8.6 mg/100 g FW) microgreens had 11.2- and 28.6-fold higher lutein/zeaxanthin concentrations, respectively, than their mature crops [69]. Coriander microgreens also exhibited the highest violaxanthin concentration.

Microscale *Brassica* vegetables (sprouts, microgreens, and baby leaves) of broccoli, kale, and radish are good sources of health-promoting phytochemicals with high antioxidant capacities. These are, in general, found in higher concentrations at the sprout and microgreen stage than in the respective adult edible plant organs (Table 3) [76,102]. In the studies undertaken by Di Bella et al. [76], polyphenol profiles differed among the three novel food types (sprouts, microgreens, and baby leaves) and cultivars within the same food type. Sprouts showed the highest total polyphenol content of the broccoli cultivars and the highest antioxidant capacity of all three cultivars studied (Table 3) [76]. Ascorbic acid levels varied significantly among the studied cultivars and the three plant growth stages. Microgreens of the landrace 'Broccolo Nero' presented the highest ascorbic acid values [76]. Chicory, lettuce, and broccoli microgreens showed higher amounts of α-tocopherol and carotenoids than mature vegetables (Table 3) [89]. Health-promoting phytochemicals are more concentrated in cruciferous sprouts (e.g., broccoli and red radish) than in their respective adult plant edible organs.

In the Fabaceae vegetables chickpea and mungbean, the content of total phenolics and vitamins and the antioxidant activity increased in the sequence of raw seeds, sprouts, and microgreens [42]. The sprouting of mungbean seeds increased total phenolic and flavonoid (TF) levels and the antioxidant activity (AA) when compared to ungerminated seeds (Table 3) [96]. Compared to sprouts, flaxseed (*Linum usitatissimum*) microgreens exhibited higher chlorophyll, carotenoid, and phenol contents, as well as higher antioxidant capacity [103].

Yadav et al. [90] studied nine leafy summer vegetables' mineral content and antioxidant activity at the microgreen and mature stages. While microgreens had a higher content of K and Zn, no specific trend was observed for Cu, Fe, and Mn [90]. Microgreens of jute (*Corchorus olitorius*) and cucumber (*Cucumis sativus*) presented higher ascorbic acid levels (34.9 mg/100 g fresh weight (FW) and 25.0 mg/100 g FW, respectively) as compared to their mature stages (10.0 mg/100 g FW and 17.45 mg/100 g FW, respectively) [90]. The ascorbic acid content of water spinach (*Ipomoea aquatica*) was comparable at the microgreen and mature stages. For other vegetable species, including bottle gourd (*Lagenaria siceraria*), pumpkin (*Cucurbita moschata*), amaranth (*Amaranthus tricolor*), Malabar spinach (*Basella alba*), radish (*Raphanus raphanistrum*), and beet (*Beta vulgaris* var. *bengalensis*), the mature plants showed higher ascorbic acid contents in comparison with the microgreen stage.

Although the ascorbic acid content is often higher at the adult stage than the microgreen stage, the human body cannot appropriately benefit from this rich ascorbic acid source. Leafy vegetables at the mature stage are generally consumed after cooking, and ascorbic acid is known to be thermolabile [56]. In contrast, microgreens are usually consumed fresh; hence, the human body can fully benefit from this ascorbic acid source in microgreens. Jute (*Corchorus olitorius*) and water spinach (*Ipomoea aquatica*) are richer sources of phenolics and flavonoids compared to commonly consumed vegetable crops such as broccoli, lettuce, and carrot at the mature stage [90].

Weber [104] studied the mineral content of lettuce (*Lactuca sativa*) and cabbage (*Brassica oleracea* var. *capitata*) microgreens and compared them to their mature vegetable stage. The average ratios across ten nutrients (P, K, Ca, Mg, S, Mn, Cu, Zn, Na, and Fe) indicated that hydroponically grown lettuce and cabbage microgreens were 2.7 and 2.9 times, respectively, more nutrient-rich than their corresponding mature vegetables (Table 3). When microgreens were cultivated on vermicompost, their nutritional superiority over the adult stage was even more pronounced. In a similar experiment with broccoli microgreens, eight minerals were analyzed that are commonly reported in nutrition information facts for foods (P, K, Ca, Mg, Mn, Fe, Zn, and Na) [105]. In these studies, the average nutrient ratio of vermicompost-grown broccoli microgreens to fully grown broccoli was 1.73 [105]. Based on this experimentally verified ratio, Weber [105] argued that one would need to eat ca. 42% less mass of microgreens (ca. 53 g FW) to obtain the same amount of minerals present in a serving of raw broccoli florets (91 g). Furthermore, broccoli microgreens would require 158–236 times less water to grow than a nutritionally equivalent amount of broccoli vegetable in fields in California's Central Valley [105].

Pinto et al. [106] reported a high nutritional content of 2-week-old butterhead lettuce (*Lactuca sativa* var. *capitata*) microgreens. The content of essential minerals such as Ca, Mg, Fe, Mn, Zn, Se, and Mo was higher in lettuce microgreens than in mature lettuces. High nitrate levels (NO_3^-) may accumulate in leafy vegetable crops (e.g., cabbage, spinach, and lettuce), and breeders aim to breed leafy vegetables with low nitrate contents. Nitrate that remains unassimilated in vegetable plant tissues can be enzymatically converted to the more toxic nitrite (NO_2^-) during storage and food processing. NO_3^- ingested by humans can also be reduced to NO_2^- through the activity of gut microorganisms [107]. Nitrite is a potent carcinogen and may cause the accumulation of methemoglobin, a compound with potentially toxic effects on human health [108], particularly in infants and children. Therefore, it is essential to note that lettuce microgreens have a much lower NO_3^- content than mature lettuces and are thus safe for consumption by infants and children [106]. A lower concentration of nitrates in Swiss chard (*Beta vulgaris* subsp. *vulgaris*) and rocket (*Eruca sativa*) microgreens than typically found in the corresponding baby leaf or adult vegetables was reported by Bulgari et al. [109]. Withholding nutrient supplementation in the growing media of microgreens is another option almost suppressing nitrate accumulation completely. The production of brussels sprouts (*Brassica oleracea* var. *gemmifera*) and cabbage (*B. oleracea* var. *capitata*) microgreens without nutrient supplementation led to a 99% decrease in the nitrate content while maintaining steady calorimetric qualities and total phenolic acid contents with only minor yield reduction [110]. Under nutrient deprivation, cabbage microgreens even showed a 30% increase in total ascorbic acid and a 12% increase in total anthocyanins.

Protection against carcinogenesis, mutagenesis, and other forms of toxicity can be achieved by the induction of phase 2 detoxication enzymes [61]. Large quantities of inducers of enzymes that protect against carcinogens can be delivered through dietary means by small amounts of young crucifer sprouts. For example, three-day-old broccoli sprouts contained as much inducer activity as 10–100 times larger quantities of the corresponding mature vegetable [61]. This is a tremendous health benefit of the *Brassica* microscale vegetables which are easily accessible to consumers.

In addition to *Brassica* microscale vegetables, okra (*Abelmoschus esculentus*) and water spinach (*Ipomoea aquatica*) sprout extracts also exhibited anti-proliferative effects on gastric cancer, hepatoma, and melanoma cell lines [111]. However, alfalfa and pea sprout extracts showed negligible anti-cancer activity. Matsuo et al. [111] hypothesized that the water-soluble bioactive compounds in okra and water spinach sprouts are responsible for the observed anti-cancer activities.

Besides macro- and microelements, vitamins, polyphenols, and other bioactive compounds, dietary fiber (DF) is another essential component of the human diet. The macro-molecules of DF mainly consist of plant cell wall components, polysaccharides, and lignin. They resist digestion by endogenous enzymes in the human gut and promote the satiety sensation [112]. The health benefits of DF include weight loss, prevention and treatment of constipation, control of serum cholesterol levels, and improved glucose tolerance, among others [67,112]. In addition, the ability of DF to bind toxic compounds has been recognized as a protective mechanism against cancer.

The studies conducted by Paradiso et al. [89] indicated relatively low DF contents ranging from 0.3 to 0.7 g per 100 g FW in broccoli and chicory microgreens, respectively. In contrast, Ghoora et al. [67] reported total dietary fiber (TDF) contents ranging from 1.41 g/100 g FW in French basil and 3.63 g/100 g FW in sunflower to 4.28 g/100 g FW in fennel. Radish and roselle microgreens showed relatively high values of soluble dietary fiber of 0.28 and 0.29 g/100 g FW, respectively. Thus, the TDF values of sunflower and fennel microgreens are in the same range as mature leafy vegetables, known for their high TDF contents ranging from 3.0 g/100 g FW in cabbage to 4.9 g/100 g FW in fenugreek leaves [112]. It is obvious that with the increasing age of microgreens, their TDF content is expected to increase as well.

The examples in this section indicate that microscale vegetables are, in general, nutrient-dense and rich in phytochemicals, often with a reduced level of antinutrients as compared to the adult growth stage, hence constituting an attractive component as a functional food in the diet of health-conscious consumers.

Table 3. The variations in contents of nutrients and phytochemicals according to plant growth stages (seeds, sprouts, microgreens, baby leaves, and fully grown).

Family	Species	Secondary Metabolites	Reference
Amaranthaceae	*Amaranthus caudatus* (foxtail amaranth)	Amaranth sprouts showed significantly higher contents of total flavonoids, rutin, amaranthine, and iso-amaranthine than ungerminated seeds.	[83]
	Amaranthus cruentus (red amaranth)	Amaranth sprouts have a significantly higher antioxidant activity than seeds, which may be a result of the difference in the content of polyphenols, anthocyanins, and other compounds.	[84]
	Amaranthus tricolor (edible amaranth)	(1) Mean protein, Fe and Zn content were considerably higher in amaranth sprouts compared with amaranth microgreens; (2) a substantial increase in vitamin C content from amaranth sprouts to microgreens (2.7-fold) and from amaranth microgreens to fully grown leafy amaranth (2.9-fold); (3) α-carotene and β-carotene were detected in all three growth stages and content increased considerably from sprouts to microgreens.	[45]
	Chenopodium quinoa (quinoa)	Quinoa sprouts have a significantly higher antioxidant activity than seeds.	[84]
	Chenopodium quinoa	Total phenol content and antioxidant activity increase with the sprouting of seeds.	[98]
	Chenopodium quinoa	Sprouts have significantly higher antioxidant capacity values after four days of germination than raw seeds; (2) phenolic content values of 4-day-old sprouts are about 2.6 times higher than seeds.	[99]
	Spinacia oleracea (spinach)	Higher ascorbic acid and α-tocopherol levels in microgreens compared to the mature stage.	[67]
Asteraceae	*Helianthus annuus* (sunflower)	Sprouting increased total phenolic and flavonoid levels, as well as the antioxidant activity compared to ungerminated seeds.	[96]
	Lactuca sativa (lettuce)	Sprouts showed higher amounts of α-tocopherol and carotenoids compared to mature lettuce.	[89]
	Lactuca sativa	The average ratio of ten nutrients (P, K, Ca, Mg, S, Mn, Cu, Zn, Na, and Fe) indicated that hydroponically grown lettuce microgreens were 2.7 times more nutrient-rich than mature lettuce.	[104]
	Lactuca sativa var. *capitata* (butterhead lettuce)	The content of essential minerals such as Ca, Mg, Fe, Mn, Zn, Se, and Mo was higher and nitrate content was lower in lettuce microgreens than in mature lettuces.	[106]
Boraginaceae	*Phacelia tanacetifolia* (phacelia)	TPC and antioxidant activity were higher in sprouts than in ungerminated seeds.	[80]
Brassicaceae	*Brassica oleracea* var. *capitata* (cabbage)	The average ratio of ten nutrients (P, K, Ca, Mg, S, Mn, Cu, Zn, Na, and Fe) indicated that hydroponically grown cabbage microgreens were 2.9 times more nutrient-rich than mature cabbage.	[104]
	Brassica oleracea var. *capitata*	Higher total ascorbic acid, phylloquinone, β-carotene, and glucoraphanin in cabbage microgreens than in mature cabbage.	[95]

Table 3. *Cont.*

Family	Species	Secondary Metabolites	Reference
	Brassica oleracea var. *italica* (broccoli)	(1) Sprouts showed significantly higher polyphenol values than microgreens and baby leaves; (2) high increments of kaempferol and apigenin in broccoli landrace from the seed to the baby leaves growth stage; (3) antioxidant levels were highest in sprouts and tended to decrease with further growth.	[76]
	Brassica oleracea var. *italica*	Sprouting increased total phenolic and flavonoid levels, as well as the antioxidant activity compared to ungerminated seeds.	[96]
	Brassica oleracea var. *italica*	Health-promoting phytochemicals are more concentrated in cruciferous sprouts (e.g., broccoli and red radish) than in the adult plant edible organs.	[102]
	Brassica oleracea var. *italica*	3-day-old broccoli sprouts contained a much higher inducer activity of detoxication enzymes than the corresponding mature vegetable.	[61]
	Brassica oleracea var. *italica*	Broccoli sprouts showed higher amounts of α-tocopherol and carotenoids compared to mature broccoli.	[89]
	Brassica oleracea var. *italica*	10-fold higher content of glucobrassicin in broccoli microgreens compared to the mature stage.	[95]
	Brassica oleracea var. *acephala* (kale)	Sprouts showed significantly higher polyphenol values than microgreens and baby leaves.	[76]
	Brassica rapa subsp. *chinensis* (pak choi)	Decreasing content of 3-butenyl glucosinolates from sprouts to adult leaves.	[49]
	Cichorium intybus (chicory)	Sprouts showed higher amounts of α-tocopherol and carotenoids compared to mature chicory.	[89]
	Eruca sativa (arugula)	Higher content of total ascorbic acid, phylloquinone, and β-carotene in arugula sprouts compared to the mature stage.	[95]
	Raphanus sativus (radish)	Health-promoting phytochemicals are more concentrated in cruciferous sprouts (e.g., broccoli and red radish) than in the respective adult plant edible organs.	[102]
	Raphanus sativus	Sprouting increased total phenolic and flavonoid levels and the antioxidant activity compared to ungerminated seeds; radish (and sunflower) sprouts were the richest in phenolic compounds.	[96]
Fabaceae	*Cicer arietinum* (chickpea)	Chickpea microgreens contained higher vitamins and higher antioxidant activity than raw seeds and sprouts.	[42]
	Trigonella foenum-graecum (fenugreek)	Higher ascorbic acid and α-tocopherol levels in microgreens compared to the mature stage.	[67]
	Vigna radiata (mungbean)	Sprouting mungbean seeds enhanced vitamin C content 2.7-fold compared to mature mungbean grain.	[92]
	Vigna radiata	Mungbean sprouts showed increased total phenolic (TPC) and total flavonoid (TF) contents and higher antioxidant activity (AA) than ungerminated seeds; radish and sunflower sprouts were superior to mungbean sprouts regarding TPC, TF, and AA levels.	[96]
	Vigna radiata	The total phenolics and vitamins content increased in the sequence of raw seeds, sprouts, and microgreens.	[42]
	Glycine max (soybean)	(1) Isoflavones were found at high concentrations in soybean sprouts and could easily provide the recommended anticarcinogenic dose range from 1.5 to 2.0 mg/kg of body weight per day; (2) The vegetable soybean stage was nutritionally superior to soybean sprouts in terms of the content of protein (14% increase), Zn (45%), Ca (72%), and Fe (151%).	[92]
Linaceae	*Linum usitatissimum* (flaxseed)	Microgreens exhibited a higher chlorophyll (+62.6%), carotenoid (+24.4%), and phenol content (+37.8%), as well as higher antioxidant capacity (+25.1%) than sprouts.	[103]
Malvaceae	*Hibiscus sabdariffa* (roselle)	Higher ascorbic acid and α-tocopherol levels in microgreens compared to the mature stage.	[67]

5. Environmental and Priming Factors That Have an Impact on the Nutrient and Phytochemical Content of Sprouts and Microgreens

As shown in previous sections of this paper, many factors determine the contents of nutrients and phytochemicals in microscale vegetables, such as the selected crop and cultivar, the chosen genotype's breeding status, and the growth stage. Other factors that may impact the nutritional quality of microscale vegetables are the environment in which they are grown, the selected illumination, substrates used, nutrient biofortification, and salinity stress. On the other hand, packaging methods and storage temperature help retain nutrients and phytochemicals [37]. All these factors may influence microscale vegetables' photosynthetic and metabolic activities and may improve nutritional quality, depending on the crop/species and genotype used.

5.1. The Effect of Growth Environment and Growing Substrates

Microgreens can be easily self-produced by consumers at home or commercially grown using controlled environment agriculture (CEA). However, recent research has shown that the cultivation environment might influence the composition of secondary metabolites, such as polyphenols and glucosinolates. This has been the case with kale and broccoli microgreens grown under commercial (growth chamber) and home-grown (windowsill) environments [52]. Windowsill-grown microgreens showed higher concentrations of hydroxycinnamic acid esters of flavanols than those produced in a growth chamber. On the other hand, the contents of 4-methoxyglucobrassicin and neoglucobrassicin were higher in microgreens grown under a controlled environment.

The substrates used for microgreen production significantly impact the nutrient content per gram of fresh weight of plant material. Cabbage microgreens grown on vermicompost had considerably higher concentrations of K, S, Ca, Mg, Mn, Cu, Zn, Fe, and Na than hydroponically grown cabbage [104]. Exceptionally high nutrient ratios for Fe were detected in cabbage microgreens grown on vermicompost (54.6-fold content of mature cabbage), while cabbage microgreens grown hydroponically still exceeded mature cabbage by a factor of 5.4. Similarly, lettuce microgreens grown on vermicompost showed significantly larger quantities of K, S, Ca, Mn, Zn, Fe, and Na than hydroponically grown lettuce microgreens [104]. Regarding Zn, cabbage microgreens had a 5 to 7.5 times higher nutrient ratio than mature cabbage. Microgreens are apparently able to absorb significant amounts of essential micronutrients from nutrient-rich food wastes that accumulate in households (mainly fruit and vegetable wastes) and become bioavailable in vermicompost.

5.2. Response to Environmental Stresses

Polyphenols play a fundamental role in the defense system of plants against heavy metals, salinity, drought, extreme temperatures, pesticides, and ultraviolet (UV) radiations [113]. In response to environmental stresses, plants produce diverse metabolites, which also contribute to the functional quality of edible plant parts, such as mineral nutrients, amino acids, peptides, proteins, vitamins, pigments, and other primary and secondary metabolites [114]. The application of eustress, i.e., mild to moderate salinity or nutritional stress, can elicit targeted plant responses by activating physiological and biochemical mechanisms. These, in turn, may lead to the accumulation of desired bioactive compounds in the harvested produce (see the literature review by Rouphael and Kyriacou [115]). Salinity eustress may enhance health-promoting phytochemicals such as lycopene, β-carotene, vitamin C, and polyphenols in vegetables [115]. For example, exposing Se-biofortified maize grains to mild NaCl stress (i.e., 25 mM NaCl) during germination resulted in good sprout yields, increased the content of selenocysteine, and boosted the synthesis of pro-nutritional semipolar metabolites with antioxidant properties [116].

Nutrient deprivation in wild rocket (*Diplotaxis tenuifolia*) microgreen production elicited a substantial increase in secondary metabolites, such as lutein (110%), β-carotene (30%), total ascorbic acid (58%), and total anthocyanins (20%); however, with a concomitant significant yield reduction of 47% [110]. On the other hand, moderate nutrient stress (half-strength nutrient solution-NS) applied to red Salanova butterhead lettuce (*Lactuca sativa* var. *capitata*) enhanced the concentrations of total ascorbic acid, total phenolic acids, and anthocyanins by 266%, 162%, and 380%, respectively, compared to the control, grown under full-strength NS [117]. For the above reasons, mild salinity, unbalanced mineral nutrition, or complete nutrient deprivation in the growth solution of soil-less culture systems for microscale vegetable production may prove helpful to naturally modulate the levels of functional compounds, such as ascorbate, carotenoids, and phenols. Moreover, it may also curtail anti-nutrients such as nitrate [110].

Sulfur is essential in the biosynthesis of secondary metabolites, such as glucosinolates in *Brassica* crops. Levels of sulfur and/or nitrogen nutrition during plant growth may result in significant changes in the phenolic content of edible plant parts, especially flavonoids and hydroxycinnamic acid derivatives [52]. Sulfur fertilization significantly improved the antioxidant activity of two ecotypes of spring broccoli, also known as Italian turnip (*Brassica rapa* subsp. *sylvestris* var. *esculenta*). It was associated with a genotype-dependent significant reduction in leaf nitrate content [118].

Plant stress caused by withholding irrigation during the head formation of cabbage (*Brassica oleracea* var. *capitata*) led to an increase in the concentration of bioactive glucosinolates [119]. However, this gain in nutritional value must be balanced with an eventual yield loss.

Environmental shocks such as high light (exposure to a light intensity of 700 μmol m^{-2} s^{-1} for 1 day) and chilling (exposure to 4 °C at a light intensity of 120 μmol m^{-2} s^{-1} for 1 day) enhanced the total phenolic content in sprouts of alfalfa (*Medicago sativa*), broccoli (*Brassica oleracea* var. *italica*), and radish (*Raphanus sativus*) [120]. The enhanced phenolic content was correlated with higher antioxidant activity, and dry biomass accumulation was unaffected. High light produced a more robust response than chilling in enhancing the content of individual phenolic compounds. Similarly, kale sprouts (*B. oleracea* var. *acephala*) exposed to low-temperature stress (growth temperature of 8 °C with intermittent freezing for one hour at −8 °C) increased the total content of phenolic acids and glucosinolates. However, such a treatment should be used with caution, as it also led to a significant decrease in the content of carotenoids and total flavonoids [121].

Radiation with short wavelengths, such as ultraviolet (UV) lights (200–400 nm), stimulates the production of pigments that absorb light and enhance leaf coloring, such as chlorophylls and carotenoids [122]. UV radiation may also induce physiological and metabolic stress responses in plants, such as the production of antioxidant systems, the activation of reparative enzymes, the expression of genes involved in UV protection and repair, and the accumulation of UV-absorbing compounds (e.g., phenolics and carotenoids) and defense-related (e.g., glucosinolates) phytochemicals [123]. This effect of UV light has been applied to broccoli sprouts to induce the biosynthesis and accumulation of flavonoids and glucosinolates [124]. Within 24 h after application of low UV-B (280–320 nm) doses, the flavonoids kaempferol and quercetin and glucosinolates accumulated in broccoli sprouts. A single exposure of broccoli sprouts to UV-B and UV-A (320–400 nm) for 120 min before the harvest was shown to enhance the phenolic and glucosinolate contents [125]. A synergistic effect in the accumulation of neoglucobrassicin was observed by exposing broccoli sprouts to a combination of UV irradiation and sprays of the phytohormone methyl jasmonate (25 μM). A single application of UV-B triggered the production of aliphatic or specific indole glucosinolates [125].

Exposing kale (*Brassica oleracea* var. *sabellica*) sprouts to periodical low UV-B treatments on days 3, 5, 7, and 10 of sprouting, with the four treatments reaching a total dose of either 10 or 15 kJ m^{-2}, is a helpful tool to stimulate the biosynthesis of phytochemicals without compromising sprout growth [126]. During sprouting, repeated UV-B treatments increased the total phenolic content of kale sprouts by 30%, stimulating the synthesis of glucosinolates (glucoraphanin and glucobrassicin) by 30% and enhancing the antioxidant activity by 20%. Therefore, the periodic application of low UV-B doses during sprout growth can optimize the content of phytochemicals in microscale vegetables.

Postharvest exposure of broccoli and radish sprouts to abiotic stress treatment in the form of UV-B radiation enhanced total phenolic content (TPC) and total antioxidant capacity (TAC) after a shelf life of 10 days at 4 °C [127]. UV-B treatment also enhanced the glucosinolates content of both crops by about 30%, while the content of sulforaphane increased by 38% in broccoli sprouts and 72% in radish sprouts.

Zlotek et al. [128] compared the effect of thermal (2-day-old sprouts exposed for 2 h to 40 °C), osmotic (NaCl exposure), and oxidative (H_2O_2 exposure) stresses on adzuki bean (*Vigna angularis*) sprouts. Their research revealed that only thermal stress enhanced the antioxidant activity of extracts obtained from the adzuki bean seed coat [128]. Similarly, Świeca et al. [129] were able to demonstrate that both low (4 °C) and high (40 °C) temperature stress may cause an increase in the content of polyphenols and enhance the antioxidant properties of lentil (*Lens culinaris*) sprouts.

Randhir et al. [130] have shown that natural elicitors such as fish protein hydrolysates, lactoferrin, and oregano extract may significantly improve the phenolic, antioxidant, and antimicrobial properties of 1-day-old mungbean sprouts. Elicitation of broccoli sprouts with autoclaved cultures of *Saccharomyces cerevisiae* (brewer's yeast) and water extracts of *Salix daphnoides* (Daphne willow) bark was also effective [131]. The most effective elicitor concentrations for the increase in the content of phenolics and enhancement of antioxidant activity of broccoli sprouts were 1% bark extract of *S. daphnoides* and 0.5% of *S. cerevisiae* cultures.

The preharvest treatment of broccoli microgreens with 10 mM calcium chloride ($CaCl_2$) to extend their shelf life led to a significant increase in aliphatic and indolic glucosinolates [132]. The raised glucosinolate (GLS) levels may have been responsible for the strengthened stress tolerance and defense mechanisms of broccoli microgreens, which resulted in delayed postharvest decay of the microgreens. This positive effect of a 10 mM calcium chloride ($CaCl_2$) preharvest treatment was confirmed in experiments conducted by Lu et al. [133], which showed a significant increase in aliphatic glucosinolates levels and an overall improvement in visual quality and a longer storage life of broccoli microgreens.

The above examples have shown that applying environmental stresses might be a viable approach to enhance the health-promoting qualities of microscale vegetables. However, the most promising type of environmental stress and its intensity require crop-specific research.

5.3. Seed Priming and Biostimulants

Seed priming enhances seed germination, seedling growth, plant establishment, and crop performance. Priming techniques include hydro-priming (soaking seeds in water); osmo-priming (soaking seeds in osmotic solutions, such as polyethylene glycol); halo-priming (soaking seeds in sodium and potassium salts); solid matrix priming (mixing seeds with solid or semi-solid material and a specified amount of water); biopriming (coating seeds with beneficial fungi or bacteria); and treatment with plant growth regulators that are incorporated into the priming medium [134–137].

Seed priming with potassium nitrate (KNO_3) improves seedling establishment and plant vigor [138]. Sprouts of three *Medicago* species treated with KNO_3 showed increased total phenolic and flavonoid contents and enhanced antioxidant and antidiabetic activities [93]. The response of KNO_3 priming was species-specific, with *Medicago intertexta* showing the highest antioxidant and antidiabetic activities, followed by *M. polymorpha* and *M. indicus*.

The phytohormones jasmonic acid and methyl jasmonate (MeJA) (25–250 μM) and the amino acid DL-methionine (1–10 mM) were used as elicitors to enhance the total glucosinolate content of broccoli and radish sprouts [139]. The most effective treatments consisted of 24 h imbibition of seeds in priming solution, followed by exogenous sprays of elicitors on the cotyledons from days 4 to 7 of sprouting. MeJA priming in combination with exogenous sprays of elicitors led to the most significant increases of total glucosinolate content, from 34% to 100% in broccoli sprouts and from 45% to 118% in radish sprouts.

Commercial biostimulants containing beneficial fungi or bacteria promoting plant growth are often recommended as a sustainable strategy to increase plant performance productivity and produce quality under environmental stresses aggravated by climate change [140]. Plant biostimulants are commonly defined as *"Substance(s) and/or microorganisms whose function is to stimulate natural processes that enhance nutrient uptake, nutrient use efficiency, tolerance to abiotic stress, and crop quality"* [141]. Bioactive molecules in commercial biostimulants enhance the capability of plants to overcome adverse environmental conditions through their action on primary or secondary plant metabolism [142]. In addition, the presence of phytohormones and other secondary metabolites, vitamins, antioxidants, and inorganic nutrients in the extract of biostimulants may affect plant growth and production directly by enhancing plant tolerance against abiotic stresses [143].

The use of exogenous fungal polysaccharide elicitors obtained from the endophytic fungus *Bionectra pityrodes* (race Fat6) enhanced the sprout growth and flavonoid (rutin, quercetin) production of tartary buckwheat (*Fagopyrum tataricum*) [144]. Seed inoculation of common buckwheat (*Fagopyrum esculentum*) with the endophytic bacterium *Herbaspirillum* sp. (isolate ST-B2), isolated from common buckwheat seedling stems, enhanced the growth of sprouts and microgreens, promoted root elongation, and increased sprout and microgreen yields [145]. Soaking common buckwheat seeds in a solution containing *Ecklonia maxima* algae extract, which is known to enhance plant tolerance to abiotic stressors and plant growth, promoted the accumulation of dry matter in sprouts [48]. Buckwheat sprouts grown from seeds soaked in a solution containing nitrophenols, which occur naturally in plants (Biostimulant Asahi SL), and *Pythium oligandrum* oospores, showed a significantly higher level of crude protein [48]. *P. oligandrum* is a common oomycete found in soils worldwide and has a beneficial effect on pathogen control and induces resistance in the host plant [146]. Therefore, using fungal, bacterial, or other elicitors could be an efficient strategy for improving the nutritional and functional quality of sprouts and microgreens. During recent years, the study and use of plant biostimulants has been steadily growing. They may be applied singly or in combination, and there may be synergistic and additive effects of microbial and non-microbial plant biostimulators. Meanwhile, the design and development of the second generation of plant biostimulants are underway with specific modes of action to render agriculture more sustainable and resilient [141].

5.4. Biofortification

The biofortification of vegetables with micronutrients that are essential or beneficial to human health, including iodine, iron, zinc, and selenium, can be achieved in soil-less culture systems [115]. Accurate control of microelement concentrations and constant exposure of roots to the fortified nutrient solution without soil interaction can maximize their uptake, translocation, and accumulation in the edible plant parts.

Selenium (Se) is an essential microelement for living organisms and plays a significant role in antioxidant defense. Many studies have been undertaken on the biofortification of plants to produce Se-enriched foods and elicit the production of secondary metabolites that are beneficial to human health [147]. Biofortification with Se is also used to improve the nutritional quality of sprouts and microgreens via an increase in the overall content of bioactive compounds [148,149]. Bioactive compounds, including phenolics, flavonoids, vitamin C, anthocyanin, and antioxidant activity, significantly increased in wheat microgreens biofortified with moderate levels (0.25–0.50 mg/L) of Se [149]. The biofortification of buckwheat microgreens with a combination of Se and iodine increased microgreen yield by 50–70% compared to single applications of Se or iodine [150]. Moreover, the combination treatment of Se and iodine led to synergistic effects regarding Se accumulation (an increase of 50% over Se application alone). On the other hand, this combination treatment reduced iodine accumulation by 50% over iodine application alone.

Puccinelli et al. [148] noted a higher germination index, higher Se content, and higher antioxidant capacity in microgreens of sweet basil (*Ocimum basilicum*) grown hydroponically and supplemented with 4 or 8 mg Se L^{-1} as sodium selenate. Pannico et al. [151] studied the

Se biofortification of four microgreen genotypes (coriander—*Coriandrum sativum*; tatsoi—*Brassica rapa* subsp. *narinosa*; and green and purple basil—*Ocimum basilicum*) to produce Se-enriched foods with a high nutraceutical profile in a simple soil-less cultivation system (SCS). They concluded that the optimal Se dose that guarantees effective Se biofortification and improves the content of bioactive compounds was 16 μM in coriander and tatsoi and 8 μM in green and purple basil. A fresh portion (10 g) of coriander and tatsoi microgreens supplemented with 16 μM Se in SCS satisfied 61% and 90% of Se's recommended dietary allowance (RDA), respectively. A lower Se supplementation of green and purple basil microgreens with 8 μM in SCS was sufficient to supply 133% and 83% of the RDA of Se, respectively, when consuming a 10 g portion of fresh microgreens [151].

The biofortification of broccoli sprouts with selenium nanoparticles (NSePs) did not affect chlorophyll content, total carotenoid, and total phenols content. Still, it enhanced the antioxidant capacity of the treated sprouts [152]. Puccinelli et al. [94] reported that the wild herb species *Rumex acetosa* (garden sorrel), *Plantago coronopus* (buck's-horn plantain), and *Portulaca oleracea* (common purslane) are of interest to produce Se-biofortified microgreens. Especially *P. coronopus* showed a strong correlation between the Se concentration in the growth medium and the Se accumulation detected in the microgreens. Furthermore, Se-biofortified *P. coronopus* microgreens also showed the highest chlorophyll and flavonoid contents [94]. Therefore, consuming microgreens of all three wild herb species would benefit human health.

5.5. Effect of Light in Controlled Environments

Light is an essential environmental factor that affects the growth and development of plants. The light environment is critical for seed germination, seedling development, photosynthetic productivity, plant metabolism, and the production of secondary metabolites. For the successful production of plants in controlled environments, it is necessary to deliver the required photons to the plant through artificial means, either by redirecting sunlight indoors, employing artificial lights, or adopting a hybrid lighting system [153]. This also applies to the production of sprouts and microgreens, either in greenhouses or in plant factories (indoor vertical farming). Compared to other artificial light sources, such as fluorescent lamps and high-pressure sodium (HPS) lamps, light-emitting diodes (LEDs) offer several advantages, such as high photoelectric conversion efficiency, low radiant heat output, low energy requirement, and long lifespan [113,122,154]. Moreover, the use of LEDs allows the modification of the light spectra and adjusting light intensities, thus facilitating the regulation of plant growth and quality of the produce. A recent, comprehensive review looked at the effects of ultraviolet radiation alone (see Section 5.2) or in combination with visible spectrum LED lighting on the stimulation of bioactive compounds during the growth and shelf life of sprouts, microgreens, and baby leaves and sheds light on the possible modes of action of these elicitors [122].

5.5.1. Effects of Light Intensity, Exposure Time, and Light Sources

In a study involving three *Brassica* microgreens (purple kohlrabi—*Brassica oleracea* var. *gongylodes*; mizuna—*Brassica rapa* var. *japonica*; and mustard—*Brassica juncea*) grown in hydroponic tray systems on multilayer shelves, percent dry weight increased for kohlrabi, mizuna, and mustard microgreens with an increase in light intensity from 105 to 315 $\mu mol \cdot m^{-2} \cdot s^{-1}$ [154]. This effect was independent of the light quality used.

Irradiance levels can also impact the production of secondary metabolites such as carotenoids and glucosinolates. Studies conducted by Kopsell et al. [155] demonstrated that exposure of mustard (*Brassica juncea*) microgreens to high light intensity (463 μmol photons $m^{-2} s^{-1}$) before harvest increased the content of zeaxanthin (2.3-fold) and antheraxanthin (1.5-fold), but decreased β-carotene and neoxanthin levels [155]. Similarly, increasing light intensities led to a significant decrease in the content of total carotenoids of mizuna and mustard microgreens [156]. For kohlrabi, increasing light intensities enhanced the total concentration of anthocyanins compared with those grown under lower light intensities.

According to Kopsell et al. [155], plants adjust to different irradiance levels via the xanthophyll cycle. With exposure to high light intensity, violaxanthin is converted to zeaxanthin via antheraxanthin. This process is reversible under low light intensity.

Samuolienė et al. [157] studied the effect of the irradiance intensity of a combination of blue, red, and far-red LED lighting on the growth, nutritional quality, and antioxidant properties of kohlrabi (*Brassica oleracea* var. *gongylodes*, mustard (*Brassica juncea*), red pak choi (*Brassica rapa* var. *chinensis*), and tatsoi (*Brassica rapa* var. *rosularis*) microgreens. The light intensities ranged from 110 μmol m^{-2} s^{-1} to 545 μmol m^{-2} s^{-1}. The best growth and antioxidant activity results were reached with light intensities of 330 and 440 μmol m^{-2} s^{-1}. These irradiance levels led to larger leaf surface area, lower content of nitrates, and higher content of anthocyanins, phenols, and antioxidant capacity [157]. When evaluating the effects of LED photon flux density levels of 545, 440, 330, 220, and 110 μmol m^{-2} s^{-1}, Brazaitytė et al. [158] noted a crop-/species-specific reaction. The concentrations of various carotenoids in red pak choi (*Brassica rapa* var. *chinensis*) and tatsoi (*Brassica rapa* subsp. *narinosa*) were higher under the illumination of 330–440 μmol m^{-2} s^{-1}. In contrast, mustard (*Brassica juncea*) responded best at 110–220 μmol m^{-2} s^{-1}.

In line with the observations made by Samuolienė et al. [157] and Brazaitytė et al. [158] regarding the most effective irradiance levels, Harakotr et al. [86] reported similar findings in experiments with five traditional vegetable microgreens, namely water convolvulus (*Ipomoea aquatica*), red holy basil (*Ocimum sanctum*), lemon basil (*O. africanum*), dill (*Anethum graveolens*), and rat-tailed radish (*Raphanus sativus* var. *caudatus*). These authors concluded that an irradiance level of 330 μmol m^{-2} s^{-1} was optimal for microgreen growth and the accumulation of bioactive compounds of water convolvulus, red holy basil, dill, and lemon basil. In addition, this light intensity led to the highest dry weight, total phenolic and flavonoid contents, and free radical scavenging activity [86]. However, rat-tailed radish microgreens did not respond positively to the irradiance level.

Continuous (24 h per day) lighting, either with LED or fluorescent lighting, enhanced the growth and nutritional quality of microgreens of arugula (*Eruca sativa*), broccoli (*Brassica oleracea* var. *italica*), mizuna (*Brassica rapa* var. *nipposinica*), and radish (*Raphanus sativus* var. *radicula*) in growth chambers [159]. The mild oxidative stress induced by continuous light treatment led to an increase in non-enzymatic antioxidants (anthocyanin, flavonoid, and proline) and enhanced the activities of antioxidant enzymes. As the effects were more pronounced under LED lighting, Shibaeva et al. [159] concluded that one could produce microgreens of arugula, broccoli, mizuna, and radish under LED continuous light with economic and nutritional benefits.

In common buckwheat (*Fagopyrum esculentum*), LED lighting in phytotrons stimulated the production of health-promoting phenolic compounds. Still, it led to more compact plants and a decrease in above-ground biomass compared with solar light supplemented with high-pressure sodium (HPS) lamps [160].

When comparing the effect of different light sources (incandescent, fluorescent, and LED RGB (red, green, blue)) on kale sprouts, LED lights resulted in the highest yield and the highest amounts of chlorophylls, β-carotene, lutein, neoxanthin, and ascorbic acid content [161]. Furthermore, compared to incandescent/fluorescent lighting, broccoli microgreens under high proportions of blue (20%) and red (80%) LED lighting accumulated a higher proportion of chlorophylls, carotenoids, glucoraphanin, and minerals [162].

Traditional high-pressure sodium lamps (400 W; 600 nm) supplemented with blue LED light (450 nm) and blue-violet LED light (420 and 440 nm, respectively) under a 24-h photoperiod and a photon flux density of 300 $\pm$ 10 μmol m^{-2} s^{-1} increased the production of phenolic acids in basil (*Ocimum basilicum*) plants. In addition, flavonoids were enhanced in arugula (*Eruca sativa*) plants compared to the control (high-pressure sodium lamps only) [163].

5.5.2. Effects of Light Spectra

Blue (400–500 nm) wavelengths promote the photosynthetic process by inducing stomatal opening and chloroplast relocation and enhancing the accumulation of antioxidant compounds and pigments in vegetables and fruits [164]. Blue light also affects vegetative and leaf growth and is particularly important for young plants such as sprouts, microgreens, or baby leaves [122]. Red light (600–700 nm) is another crucial factor that enhances the growth rates of plants and promotes the synthesis of pigments and phytochemicals in different plant species, thus improving the nutritional quality of the product [164]. Red light interacts with blue light in regulating plant responses. When the appropriate light intensity is applied, the optimal, crop-specific R:B ratio enhances photosynthetic capacity and improves growth and yield. Far-red light (700–800 nm) contributes to plant elongation and triggers flowering and fruit production and biomass accumulation in plants [165]. Green light (550 nm) also contributes to photosynthesis and biomass accumulation and may influence secondary metabolism [164].

Recently, Appolloni et al. [166] undertook a review on the effects of LED lighting on the content of phytochemicals in medicinal and aromatic plants, microgreens, and edible flowers. Among the 40 papers reviewed, most studies used a combination of red and blue light (22%) or monochromatic blue light (23%), with a 16 h day^{-1} photoperiod (78%) and a light intensity greater than 200 μmol m^{-2} s^{-1} (77%). The application of red and blue light together, sometimes also in combination with other spectra (far-red and green), often showed beneficial effects regarding the accumulation of phytochemicals, particularly in the case of microgreens [166]. However, the impact of red and blue light on the synthesis of specialized metabolites may vary with the plant species studied.

Sprouts are usually grown in darkness. Compared to darkness, fluorescent light treatments with white, blue, and red at a moderate photon flux density (110 μmoL m^{-2}.s^{-1}) increased the contents of vitamin C and pigments such as carotenoids, chlorophylls, and anthocyanins in 5-day-old sprouts of radish, soybean, mungbean, and pumpkin [167]. In addition, increased contents of soluble proteins and sugars were observed in soybean and pumpkin sprouts, respectively, exposed to light treatments. However, polyphenols were increased only with soybean sprouts and only with red light applications. Similarly, fluorescent lighting and combined LED lighting treatments of blue (B) + red (R) and B + R + FR (far-red) improved total antioxidant activity. They increased the content of bioactive compounds in carrot (*Daucus carota*) sprouts compared to darkness [127]. Both combined LED treatments increased the phenolic content (phenolic acids and rutin) by 45% and 65%, respectively, compared to darkness, and 32% compared to fluorescent light. Moreover, the combined B + R LED treatment also enhanced the content of carotenoids, but not when far-red was added.

Comparing three light sources, namely fluorescent light lamps, blue (460 mm) LED lights, and red (625 nm) LED lights, at a light intensity of 35 μmol/m^2 s^{-1}, common buckwheat (*Fagopyrum esculentum*) sprouts exhibited the highest total phenolics and total flavonoids contents. Furthermore, they showed the most increased antioxidant activities when grown under blue light [168]. Experiments conducted with rapeseed (*Brassica napus*) sprouts exposed to white (380 nm), blue (470 nm), red (660 nm), and blue + red LED lighting under a 16-h photoperiod and a photon flux density of 50 μmol/s·m^2, revealed that red LED light enhances sprout growth. In contrast, blue light effectively increased the accumulation of glucosinolates and phenolics [169]. Similarly, phenolics, flavonoids, and glucoraphanin accumulated in broccoli sprouts under a photon flux density of 50 μmol m^{-2} s^{-1} of blue LED lighting on their own and/or combined with red LED lighting at an equal ratio (50:50) [170].

Basal LED spectra composed of a combination of blue, red, and far-red light and supplemented with green (510 nm) led to a significant increase in the content of total phenols, α-tocopherol, and vitamin C, and enhanced the antioxidant capacity in sprouted seeds of lentil (*Lens esculenta*) and wheat (*Triticum aestivum*) [171]. In addition, basal

LED spectra supplemented with amber (595 nm) significantly enhanced the antioxidant properties of radish (*Raphanus sativus*) sprouts.

Like in sprouts, the contents of anthocyanins and phenolic acids are primarily influenced by the proportion of red and blue light in microgreens. Blue and red light LED treatments have been shown to enhance the growth of basil (*Ocimum basilicum*) microgreens and increase phenolic compounds' contents [172]. However, the effect of light quality on the synthesis of phenolic substances and free radical scavenging activity was cultivar-specific. The green basil cultivar was strongly stimulated by a higher proportion of red light (2R:1B), while the red cultivar showed the best response when exposed to a higher ratio of blue light (1R:2B).

This cultivar-specific reaction of basil microgreens in response to different LED spectra was also reported by Chutimanukul et al. [173]. These authors used different LED light spectra ratios of red[®], green (G), and blue (B), and assessed their impact on photosynthesis, biomass, antioxidant capacity, and secondary metabolites of green and red holy basil (*Ocimum tenuiflorum*) plants grown hydroponically under a controlled environment. An R:B ratio of 1:3 (high percentage of the blue spectrum) significantly increased the total phenolics content (TPC) and free radical scavenging activity in leaves of both holy basil cultivars [173]. In contrast to TPC, the effect of LED light spectra on total flavonoid content was cultivar-dependent [173]. This example clearly illustrates that the red and blue LED lighting ratios need to be adjusted to the respective cultivar.

In experiments conducted by Brazaitytė et al. [174] with mustard (*B. juncea*) and kale (*B. napus*) microgreens, a high percentage (75 or 100%) of blue light was applied in the mix with red light (B75R25 and B100R0) at 220 μmol m^{-2} s^{-1} in an 18 h photoperiod for 5 days. This treatment positively affected the accumulation of macro- and micronutrients at the expense of a significant yield reduction [174]. Therefore, the authors recommended a lower proportion (25% or 50%) of blue light as a strategic tool for mustard and kale microgreen biofortification to increase chlorophyll, flavonol, anthocyanin, and carotenoid contents while maintaining high yields. Similar observations have been made by Samuolienė et al. [175] with mustard (*Brassica juncea*), beet (*Beta vulgaris* subsp. *vulgaris*), and parsley (*Petroselinum crispum*) microgreens.

The effect of three LED light sources emitting red/blue ratios of about 2, 5, and 9 units (RB2, RB5, and RB9, respectively) was tested on mustard, radish, green basil, red amaranth, garlic chives, borage, and pea microgreens grown in a vertical farming system [91]. The yield was enhanced in mustard, green basil, and pea microgreens when exposed to a high red portion (RB9), while garlic chives exhibited the highest fresh weight under RB2. In addition, RB9 increased the total phenolic content in all microgreens except for mustard, and increased the antioxidant capacity of pea, green basil, borage, red amaranth, and garlic chives.

Hytönen et al. [176] compared the effects of LED lighting with different spectral compositions on the growth, development, and nutritional quality of lettuce (*Lactuca sativa*). The authors were able to show that warm-white and warm-white supplemented with blue spectra provide equal growth and product quality compared to conventional HPS lighting, both in the absence and presence of daylight. Furthermore, their research demonstrated that a red + blue LED spectrum increased the concentration of several vitamins in lettuce but led to a decrease in biomass accumulation when daylight was excluded [176]. The far-red LED component proved to be more critical than green light or the red/blue ratio for biomass accumulation.

Purslane (*Portulaca oleracea*) microgreens grown for 21 days under saline conditions (80 mM of NaCl) had higher yields and lower contents of antinutrients when treated with combined B + R or combined B + R + FR LED lighting at a photon flux density of 150 μmol m^{-2} s^{-1} [177].

Even a short exposure of broccoli microgreens before harvest to blue LED light (41 μmol·m^{-2}·s^{-1}) significantly increased levels of shoot tissue β-carotene, violaxanthin, total xanthophyll cycle pigments, glucoraphanin, epiprogoitrin, aliphatic glucosinolates, essential micronutrients (Cu, Fe, B, Mn, Mo, Na, Zn), and essential macronutrients (Ca, P, K, Mg, and S) [178]. Hence, preharvest, short-duration blue light treatment effectively enhances broccoli microgreens' nutritional value and increases health-promoting phytochemical compounds such as carotenoids and glucosinolates.

Alrifai et al. [179] demonstrated that the incorporation of amber light (590 nm) into a combination of standard blue (455 nm) and red (655 nm) LED lights during the growth of four different *Brassica* microgreens improved the glucosinolate profile by modulating the biosynthesis of the precursors of isothiocyanates. The four species used were: mizunas (*Brassica rapa* var. *japonica*), pak choi (*Brassica rapa* var. *chinensis*), red radish (*Raphanus sativus*), and white mustard (*Brassica juncea*). In earlier experiments with identical light spectra and the same microgreen species and cultivars, these authors made similar observations regarding the biosynthesis of phenolic compounds [180]. While increasing amber and blue and concurrently decreasing the proportion of red light, overall positive correlations were observed for total phenolics, total flavonoid contents, and antioxidant activities. Alrifai et al. [180] concluded that the microgreens could be clustered into three groups based on phytochemical contents and sensitivity to the lighting as follows: (i) radish shows a high blue and amber dose-dependence regarding total phenolics and flavonoids content and antioxidant activity; (ii) mustards show a moderate to high sensitivity to overall lighting but no clear dose-dependence; and (iii) mizunas and pak choi show variable responses to lighting.

Experiments with *Brassica* microgreens grown under four different LED ratios (%) showed that supplemental lighting with 70% red, 10% green, and 20% blue (R_{70}:G_{10}:B_{20}) LEDs enhanced vegetative growth, while dominantly blue LEDs (R_{20}:B_{80}) increased the mineral and vitamin contents [181]. With the aim to balance microgreen yield with nutritional content, the best treatment proved to be the LED combination with green light (R_{70}:G_{10}:B_{20}), which resulted in the highest growth and nutritional content.

The exposure of soybean (*Glycine max*) microgreens to monochromatic blue LED light and ultraviolet-A (UV-A) radiation led to significant increases in total phenolic and total flavonoid content, as compared with the white LED light control [182]. Among four light spectra (100% red, 100% green, 100% blue, and an R:G:B ratio of 1:1:1) tested, flaxseed (*Linum usitatissimum*) microgreens treated with 100% blue light exhibited the highest content of flavonoids (2.48 mg catechin equivalents [CAE] g^{-1} FW), total phenols (3.76 mg GAE g^{-1} FW), chlorogenic acid (1.10 mg g^{-1} FW), and antioxidant capacity (8.06 μmol Trolox equivalent antioxidant capacity [TEAC] g^{-1} FW) [103].

In summary, significant progress has been made in recent years regarding the effects of light intensity, exposure time, light sources and light spectra on plant growth and the development of microscale vegetables in controlled environments. However, further studies are needed to elucidate the crop-specific interactions between light spectrum and light intensity and their relationship with other environmental factors [164].

6. Microgreen Market Trends and Outlook

The popularity of microscale vegetables is apparent, judging from their increasing use in upscale restaurants and the steadily growing availability of a wide range of microgreens in supermarkets. Once purchased, they continue actively growing on their media and are ready for cutting and use as perfectly fresh vegetables in homemade meals. Apart from human nutrition, cosmetics is another niche industry that drives the growth of microgreens [183]. Vitamin- and nutrient-dense microgreens are processed into oils and ingredients for skincare products, shampoos, and conditioners, making them attractive for health-conscious consumers.

Sprouts and microgreens are versatile, as they can quickly, easily, and cost-effectively be grown in urban and peri-urban settings where land is often a limiting factor. They can even be produced inside or around residential areas by the consumers themselves, independent of seasonal growth cycles [21,65,184], or through indoor farming (greenhouses and vertical farming) by specialized producers for sale in supermarkets [22,25,26,185].

Indoor farming is independent of arable land and external climate conditions. It can use different growing systems (hydroponics, aeroponics, aquaponics, or soil-based) and structures (glass or poly greenhouses, indoor vertical farms, or low-tech plastic hoop houses) and can be based in urban and peri-urban or rural areas, dependent on convenience for logistics, marketing, and costs [185,186]. Because of climate change discussions and the demand for more local food production to reduce the carbon footprint associated with transportation distances, there is now a clear trend to establish greenhouses and indoor vertical farms near urban and peri-urban areas to bring operations closer to the consumers [187].

The COVID-19 pandemic has accelerated this trend towards microscale vegetables as consumers are eager to strengthen their immune systems by consuming food rich in antioxidants and other health-promoting substances. Moreover, in the face of a global disruption of supply chains and significant changes in shopping habits due to the ongoing COVID-19 pandemic, do-it-yourself sprouts and microgreens offer an exciting and sustainable alternative. Growing sprouts and microgreens at home is easy to put into practice. It eliminates the need for long-distance transport, reduces fossil fuel consumption for product delivery, and provides consumer access to highly fresh, nutrient-dense produce that can be harvested as needed for meal preparation [105]. Local production and consumption of sprouts and microgreens would also reduce food waste partially caused by the food supply chain from farm via wholesale markets and consumer outlets.

A feasibility study on microgreens was conducted in India comprising ten microgreens: carrot, fenugreek, mustard, onion, radish, red roselle, spinach, sunflower, fennel, and French basil [65]. The market value of the microgreens was found to be five to eleven-fold greater than their production costs. Thus, microgreen production represents a viable enterprise that can support the economic stability of the rural and urban poor.

With the growing interest and demand from consumers, the global microgreens market is projected to register an estimated compound annual growth rate ranging from 7.5% during the period from 2021 to 2026 [183] to 13.1% during the period from 2020 to 2028 [188]. North America accounted for the most significant global market share in microgreens in 2020, and the Asia-Pacific region has the fastest-growing microgreens market [183].

Several varieties of the same crop or different crops can be grown together to enhance the diversity of tastes, colors, and textures [21]. They are then marketed as specialty mixes, such as "sweet," "mild," "colorful," or "spicy," fetching prices of US $66–110 per kg in the U.S. [34].

Based on a 2017 report of the intelligence platform Agrilyst [185], the five main crops grown under indoor farming in the U.S. were leafy greens, microgreens, herbs, flowers, and tomatoes. On average, leafy greens and microgreens had the highest profit margin at 40% across the various facility and system types, flowers reached a profit margin of 30%, and tomatoes only 10% [185]. In 2020, the percentage of the microgreen segment among total crop cultivation in greenhouses in the U.S. varied from 25% in the Midwest to 59% in the Northeast and 71% in the South [183].

Members of the Brassicaceae family (broccoli, cabbage, cauliflower, arugula, radish, and cress) dominated the global microgreen market in 2019 [189]. The top microgreen crops according to their market share in 2019 were: broccoli > arugula > cabbage > cauliflower > others > peas > basil > radish > cress. Hence, only peas (Fabaceae) and basil (Lamiaceae) were able to figure among the top microgreen crop types. This huge share of microgreen crop types of the Brassicaceae family in the global microgreen market can be attributed to the commonly known health benefits associated with Brassicaceae crops, which help fight diet-related NCDs [188]. Consumers usually buy fresh microgreens in retail stores (hypermarkets, supermarkets, or grocery stores), and this sector held a market share of 46.8% in 2019, followed by farmers markets [187]. The retail sector is forecast to grow annually by 11.4% from 2021 to 2028. The global microgreen market is expected to grow by 11.1% annually from 2021 to 2028 [189]. The global microgreens market is projected to reach \$3795.47 million by 2028, starting from \$1417.64 million in 2020 [188]. Prominent microgreen market trends include continued indoor, vertical, and greenhouse farming growth, a rise in the use of advanced production technologies, and a growing awareness for premium food products [188].

7. Conclusions

Sprouts and microgreens are novel functional food sources with great potential for sustainably diversifying global food systems, promoting human health, and facilitating the access of a steadily growing urban population to fresh microscale vegetables. These novel food sources have vivid colors, exciting textures, and diverse flavors and tastes, and they can be purchased in supermarkets or even home-grown for daily harvesting as needed. Furthermore, due to their short growth cycle, these nutrient-dense food sources can be produced with minimal input, without using pesticides; hence, they have low environmental impacts and a broad acceptance among health-conscious consumers. Furthermore, as sprouts and microgreens are usually consumed raw, there is hardly a loss or degradation of heat-sensitive micronutrients or vitamins through food processing.

Funding: This research received no external funding.

Institutional Review Board Statement: Not applicable.

Informed Consent Statement: Not applicable.

Conflicts of Interest: The author declares no conflict of interest.

References

1. Micha, R.; Mannar, V.; Afshin, A.; Allemandi, L.; Baker, P.; Battersby, J.; Bhutta, Z.; Chen, K.; Corvalan, C.; Di Cesare, M.; et al. *2020 Global Nutrition Report: Action on Equity to End Malnutrition*; Development Initiatives: Washington, DC, USA, 2020. Available online: http://eprints.mdx.ac.uk/30645/ (accessed on 12 November 2021).
2. Neufeld, L.M.; Hendriks, S.; Hugas, M. Healthy Diet: A Definition for the United Nations Food Systems Summit 2021. A Paper from the Scientific Group of the UN Food Systems Summit. 2021. Available online: https://www.un.org/sites/un2.un.org/files/healthy_diet_scientific_group_march-2021.pdf (accessed on 15 November 2021).
3. FAO; IFAD; UNICEF; WFP; WHO. The state of food security and nutrition in the world. In *Transforming Food Systems for Food Security, Improved Nutrition and Affordable Healthy Diets for All*; FAO: Rome, Italy, 2021; 240p. [CrossRef]
4. Herforth, A.; Bai, Y.; Venkat, A.; Mahrt, K.; Ebel, A.; Masters, W. Cost and affordability of healthy diets across and within countries. In *Background Paper for The State of Food Security and Nutrition in the World*; FAO: Rome, Italy, 2020. [CrossRef]
5. United Nations. Sustainable Development Goals. *Goal 2: Zero Hunger*. Available online: https://www.un.org/sustainabledevelopment/hunger/ (accessed on 18 November 2021).
6. World Health Organization (WHO). Noncommunicable Diseases. Available online: https://www.who.int/news-room/fact-sheets/detail/noncommunicable-diseases (accessed on 18 November 2021).
7. Bennett, J.E.; Stevens, G.A.; Mathers, C.D.; Bonita, R.; Rehm, J.; Kruk, M.E.; Riley, L.M.; Dain, K.; Kengne, A.P.; Chalkidou, K.; et al. NCD countdown 2030: Worldwide trends in non-communicable disease mortality and progress towards sustainable development goal target 3.4. *Lancet* **2018**, *392*, 1072–1088. [CrossRef]
8. Fischer, G.C.; Garnett, T. *Plates, Pyramids, and Planets: Developments in National Healthy and Sustainable Dietary Guidelines: A State of Play Assessment*; Food and Agriculture Organization of the United Nations: Rome, Italy, 2016.
9. Willett, W.; Rockström, J.; Loken, B.; Springmann, M.; Lang, T.; Vermeulen, S.; Garnett, T.; Tilman, D.; DeClerck, F.; Wood, A.; et al. Food in the anthropocene: The EAT-lancet commission on healthy diets from sustainable food systems. *Lancet* **2019**, *393*, 447–492. [CrossRef]
10. Vermeulen, S.J.; Park, T.; Khoury, C.K.; Béné, C. Changing diets and the transformation of the global food system. *Ann. N. Y. Acad. Sci.* **2020**, *1478*, 3–17. [CrossRef] [PubMed]
11. Padulosi, S.; Sthapit, B.; Lamers, H.; Kennedy, G.; Hunter, D. Horticultural biodiversity to attain sustainable food and nutrition security. *Acta Hortic.* **2018**, *1205*, 21–34. [CrossRef]
12. The Royal Society. *Reaping the Benefits: Science and the Sustainable Intensification of Global Agriculture*; The Royal Society: London, UK, 2009; 72p.
13. World Health Organization (WHO). Increasing Fruit and Vegetable Consumption to Reduce the Risk of Noncommunicable Diseases. Available online: https://www.who.int/elena/titles/fruit_vegetables_ncds/en/ (accessed on 19 November 2021).
14. Oyebode, O.; Gordon-Dseagu, V.; Walker, A.; Mindell, J.S. Fruit and vegetable consumption and all-cause, cancer and CVD mortality: Analysis of Health Survey for England data. *J. Epidemiol. Community Health* **2014**, *68*, 856–862. [CrossRef]

15. Siegel, K.R.; Ali, M.K.; Srinivasiah, A.; Nugent, R.A.; Narayan, K.V. Do we produce enough fruits and vegetables to meet global health need? *PLoS ONE* **2014**, *9*, e104059. [CrossRef]

16. Bahadur, K.K.C.; Dias, G.M.; Veeramani, A.; Swanton, C.J.; Fraser, D.; Steinke, D.; Lee, E.; Wittman, H.; Farber, J.M.; Dunfield, K.; et al. When too much isn't enough: Does current food production meet global nutritional needs? *PLoS ONE* **2018**, *13*, e0205683. [CrossRef]

17. Mason-D'Croz, D.; Sulser, T.B.; Wiebe, K.; Rosegrant, M.W.; Lowder, S.K.; Nin-Pratt, A.; Willenbockel, D.; Robinson, S.; Zhu, T.; Cenacchi, N.; et al. Agricultural investments and hunger in Africa modeling potential contributions to SDG2–zero hunger. *World Dev.* **2019**, *116*, 38–53. [CrossRef]

18. Mensah, D.O.; Nunes, A.R.; Bockarie, T.; Lillywhite, R.; Oyebode, O. Meat, fruit, and vegetable consumption in sub-Saharan Africa: A systematic review and meta-regression analysis. *Nutr. Rev.* **2021**, *79*, 651–692. [CrossRef]

19. Ebert, A.W. Sprouts, microgreens, and edible flowers: The potential for high value specialty produce in Asia. In Proceedings of the SEAVEG 2012 High Value Vegetables Southeast Asia Production, Supply Demand, Chiang Mai, Thailand, 16 September 2013; pp. 216–227.

20. Voinea, L.; Vrânceanu, D.M.; Filip, A.; Popescu, D.V.; Negrea, T.M.; Dina, R. Research on food behavior in Romania from the perspective of supporting healthy eating habits. *Sustainability* **2019**, *11*, 5255. [CrossRef]

21. Ebert, A.W. High value specialty vegetable produce. In *Handbook of Vegetables*, 1st ed.; Peter, K.V., Hazra, P., Eds.; Studium Press LLC.: Houston, TX, USA, 2015; Chapter 4; Volume 2, pp. 119–143.

22. Kyriacou, M.C.; Rouphael, Y.; Di Gioia, F.; Kyratzis, A.; Serio, F.; Renna, M.; De Pascale, S.; Santamaria, P. Micro-scale vegetable production and the rise of microgreens. *Trends Food Sci. Technol.* **2016**, *57*, 103–115. [CrossRef]

23. Di Gioia, F.; Renna, M.; Santamaria, P. Sprouts, microgreens and "baby leaf" vegetables. In *BT-Minimally Processed Refrigerated Fruits and Vegetables*; Food Engineering Series; Yildiz, F., Wiley, R.C., Eds.; Springer: Boston, MA, USA, 2017; pp. 403–432, ISBN 978-1-4939-7018-6. [CrossRef]

24. Benincasa, P.; Falcinelli, B.; Lutts, S.; Stagnari, F.; Galieni, A. Sprouted Grains: A Comprehensive Review. *Nutrients* **2019**, *11*, 421. [CrossRef] [PubMed]

25. Kyriacou, M.C.; El-Nakhel, C.; Graziani, G.; Pannico, A.; Soteriou, G.A.; Giordano, M.; Ritieni, A.; De Pascale, S.; Rouphael, Y. Functional quality in novel food sources: Genotypic variation in the nutritive and phytochemical composition of thirteen microgreens species. *Food Chem.* **2019**, *277*, 107–118. [CrossRef] [PubMed]

26. Galieni, A.; Falcinelli, B.; Stagnari, F.; Datti, A.; Benincasa, P. Sprouts and microgreens: Trends, opportunities, and horizons for novel research. *Agronomy* **2020**, *10*, 1424. [CrossRef]

27. Abdallah, M.M.F. Seed sprouts, a pharaoh's heritage to improve food quality. *Arab Univ. J. Agric. Sci.* **2008**, *16*, 469–478. [CrossRef]

28. Vidal-Valverde, C.; Frias, J.; Sierra, I.; Blazquez, I.; Lambein, F.; Kuo, Y.H. New functional legume foods by germination: Effect on the nutritive value of beans, lentils and peas. *Eur. Food Res. Technol.* **2002**, *215*, 472–477. [CrossRef]

29. Márton, M.; Mandoki, Z.S.; Csapo-Kiss, Z.S.; Csapo, J. The role of sprouts in human nutrition. A review. *Acta Univ. Sapientiae* **2010**, *3*, 81–117.

30. Wanasundara, P.; Shahidi, F.; Brosnan, M.E. Changes in flax (*Linum usitatissimum*) seed nitrogenous compounds during germination. *Food Chem.* **1999**, *65*, 289–295. [CrossRef]

31. Lee, J.S.; Pill, W.G.; Cobb, B.B.; Olszewski, M. Seed treatments to advance greenhouse establishment of beet and chard microgreens. *J. Hort. Sci. Biotechnol.* **2004**, *79*, 565–570. [CrossRef]

32. Di Gioia, F.; Mininni, C.; Santamaria, P. How to grow microgreens. In *Microgreens: Microgreens: Novel Fresh and Functional Food to Explore All the Value of Biodiversity*; Di Gioia, F., Santamaria, P., Eds.; ECO-Logica: Bari, Italy, 2015; pp. 51–79.

33. Lenzi, A.; Orlandini, A.; Bulgari, R.; Ferrante, A.; Bruschi, P. Antioxidant and mineral composition of three wild leafy species: A comparison between microgreens and baby greens. *Foods* **2019**, *8*, 487. [CrossRef]

34. Treadwell, D.; Hochmuth, R.; Landrum, L.; Laughlin, W. *Microgreens: A New Specialty Crop*; HS1164, rev. 9/2020, IFAS Extension; University of Florida: Gainesville, FL, USA, 2020.

35. Turner, E.R.; Luo, Y.; Buchanan, R.L. Microgreen nutrition, food safety, and shelf life: A review. *J. Food Sci.* **2020**, *85*, 870–882. [CrossRef] [PubMed]

36. Anonymous. Specialty Greens Pack a Nutritional Punch. Available online: https://agresearchmag.ars.usda.gov/AR/archive/20 14/Jan/greens0114.pdf (accessed on 3 December 2021).

37. Zhang, Y.; Xiao, Z.; Ager, E.; Kong, L.; Tan, L. Nutritional quality and health benefits of microgreens, a crop of modern agriculture. *J. Future Foods* **2021**, *1*, 58–66. [CrossRef]

38. Verlinden, S. Microgreens: Definitions, Product Types, and Production Practices. *Hortic. Rev.* **2020**, *47*, 85–124. [CrossRef]

39. Xiao, Z.; Lester, G.E.; Park, E.; Saftner, R.A.; Luo, Y.; Wang, Q. Evaluation and correlation of sensory attributes and chemical compositions of emerging fresh produce: Microgreens. *Postharvest Biol. Technol.* **2015**, *110*, 140–148. [CrossRef]

40. Ghani, M.; Kulkarni, K.P.; Song, J.T.; Shannon, J.G.; Lee, J.D. Soybean sprouts: A review of nutrient composition, health benefits and genetic variation. *Plant Breed. Biotechnol.* **2016**, *4*, 398–412. [CrossRef]

41. Butkutė, B.; Taujenis, L.; Norkevičienė, E. Small-seeded legumes as a novel food source. Variation of nutritional, mineral and phytochemical profiles in the chain: Raw seeds-sprouted seeds-microgreens. *Molecules* **2019**, *24*, 133. [CrossRef]

42. Kurian, M.S.; Megha, P.R. Assessment of variation in nutrient concentration and antioxidant activity of raw seeds, sprouts and microgreens of *Vigna radiata* (L.) Wilczek and *Cicer arietinum* L. In *AIP Conference Proceedings*; AIP Publishing LLC.: Melville, NY, USA, 2020; Volume 2263, 030005p. [CrossRef]

43. Lemmens, E.; Moroni, A.V.; Pagand, J.; Heirbaut, P.; Ritala, A.; Karlen, Y.; Lê, K.A.; Van den Broeck, H.C.; Brouns, F.J.; De Brier, N.; et al. Impact of cereal seed sprouting on its nutritional and technological properties: A critical review. *Compr. Rev. Food Sci. Food Saf.* **2019**, *18*, 305–328. [CrossRef]

44. Pirzadah, T.B.; Malik, B. Pseudocereals as super foods of 21st century: Recent technological interventions. *J. Agric. Food Res.* **2020**, *2*, 100052. [CrossRef]

45. Ebert, A.W.; Wu, T.H.; Yang, R.Y. Amaranth sprouts and microgreens—A homestead vegetable production option to enhance food and nutrition security in the rural-urban continuum. In Proceedings of the Regional Symposium on Sustaining Small-Scale Vegetable Production and Marketing Systems for Food and Nutrition Security (SEAVEG 2014), Bangkok, Thailand, 25–27 February 2014; pp. 25–27.

46. Janovská, D.; Stocková, L.; Stehno, Z. Evaluation of buckwheat sprouts as microgreens. *Acta Agric. Slov.* **2010**, *95*, 157. [CrossRef]

47. Le, L.; Gong, X.; An, Q.; Xiang, D.; Zou, L.; Peng, L.; Wu, X.; Tan, M.; Nie, Z.; Wu, Q.; et al. Quinoa sprouts as potential vegetable source: Nutrient composition and functional contents of different quinoa sprout varieties. *Food Chem.* **2021**, *357*, 129752. [CrossRef]

48. Witkowicz, R.; Biel, W.; Chłopicka, J.; Galanty, A.; Gleń-Karolczyk, K.; Skrzypek, E.; Krupa, M. Biostimulants and microorganisms boost the nutritional composition of buckwheat (*Fagopyrum esculentum* Moench) sprouts. *Agronomy* **2019**, *9*, 469. [CrossRef]

49. Neugart, S.; Baldermann, S.; Hanschen, F.S.; Klopsch, R.; Wiesner-Reinhold, M.; Schreiner, M. The intrinsic quality of brassicaceous vegetables: How secondary plant metabolites are affected by genetic, environmental, and agronomic factors. *Sci. Hortic.* **2018**, *233*, 460–478. [CrossRef]

50. Jahangir, M.; Kim, H.K.; Choi, Y.H.; Verpoorte, R. Health-Affecting Compounds in *Brassicaceae*. *Compr. Rev. Food Sci. Food Saf.* **2009**, *8*, 31–43. [CrossRef]

51. Ramirez, D.; Abellán-Victorio, A.; Beretta, V.; Camargo, A.; Moreno, D.A. Functional ingredients from Brassicaceae species: Overview and perspectives. *Int. J. Mol. Sci.* **2020**, *21*, 1998. [CrossRef]

52. Liu, Z.; Shi, J.; Wan, J.; Pham, Q.; Zhang, Z.; Sun, J.; Yu, L.; Luo, Y.; Wang, T.T.; Chen, P. Profiling of Polyphenols and Glucosinolates in Kale and Broccoli Microgreens Grown under Chamber and Windowsill Conditions by Ultrahigh-Performance Liquid Chromatography High-Resolution Mass Spectrometry. *ACS Food Sci. Technol.* **2021**, *2*, 101–113. [CrossRef]

53. Marchioni, I.; Martinelli, M.; Ascrizzi, R.; Gabbrielli, C.; Flamini, G.; Pistelli, L.; Pistelli, L. Small Functional Foods: Comparative Phytochemical and Nutritional Analyses of Five Microgreens of the Brassicaceae Family. *Foods* **2021**, *10*, 427. [CrossRef]

54. Ng, T.B.; Ng, C.C.W.; Wong, J.H. Health benefits of Brassica species. In *Brassicaceae: Characterization, Functional Genomics and Health Benefits*; Lang, M., Ed.; Nova Science Publishers: New York, NY, USA, 2013; pp. 1–18.

55. Sanlier, N.; Guler, S.M. The benefits of *Brassica* vegetables on human health. *J. Hum. Health Res.* **2018**, *1*, 1–13.

56. Francisco, M.; Velasco, P.; Moreno, D.A.; García-Viguera, C.; Cartea, M.E. Cooking methods of *Brassica rapa* affect the preservation of glucosinolates, phenolics and vitamin C. *Food Res. Int.* **2010**, *43*, 1455–1463. [CrossRef]

57. Rice-Evans, C.; Miller, N.; Paganga, G. Antioxidant properties of phenolic compounds. *Trends Plant Sci.* **1997**, *2*, 152–159. [CrossRef]

58. Verkerk, R.; Schreiner, M.; Krumbein, A.; Ciska, E.; Holst, B.; Rowland, I.; de Schrijver, R.; Hansen, M.; Gerhäuser, C.; Mithen, R.; et al. Glucosinolates in *Brassica* Vegetables: The Influence of the Food Supply Chain on Intake, Bioavailability and Human Health. *Mol. Nutr. Food Res.* **2009**, *53*, S219–S265. [CrossRef]

59. Odongo, G.A.; Schlotz, N.; Herz, C.; Hanschen, F.S.; Baldermann, S.; Neugart, S.; Trierweiler, B.; Frommherz, L.; Franz, C.M.; Ngwene, B.; et al. The role of plant processing for the cancer preventive potential of Ethiopian kale (*Brassica carinata*). *Food Nutr. Res.* **2017**, *61*, 1. [CrossRef] [PubMed]

60. Rose, P.; Huang, Q.; Ong, C.N.; Whiteman, M. Broccoli and watercress suppress matrix metalloproteinase-9 activity and invasiveness of human MDA-MB-231 breast cancer cells. *Toxicol. Appl. Pharmacol.* **2005**, *209*, 105–113. [CrossRef] [PubMed]

61. Fahey, J.W.; Zhang, Y.; Talalay, P. Broccoli sprouts: An exceptionally rich source of inducers of enzymes that protect against chemical carcinogens. *Proc. Natl. Acad. Sci. USA* **1997**, *94*, 10367–10372. [CrossRef] [PubMed]

62. Higdon, J.; Delage, B.; Williams, D.; Dashwood, R. Cruciferous Vegetables and Human Cancer Risk: Epidemiologic Evidence and Mechanistic Basis. *Pharmacol. Res.* **2007**, *55*, 224. [CrossRef]

63. De la Fuente, B.; López-García, G.; Máñez, V.; Alegría, A.; Barberá, R.; Cilla, A. Evaluation of the Bioaccessibility of Antioxidant Bioactive Compounds and Minerals of Four Genotypes of *Brassicaceae* Microgreens. *Foods* **2019**, *8*, 250. [CrossRef] [PubMed]

64. Michell, K.A.; Isweiri, H.; Newman, S.E.; Bunning, M.; Bellows, L.L.; Dinges, M.M.; Grabos, L.E.; Rao, S.; Foster, M.T.; Heuberger, A.L.; et al. Microgreens: Consumer sensory perception and acceptance of an emerging functional food crop. *J. Food Sci.* **2020**, *85*, 926–935. [CrossRef]

65. Ghoora, M.D.; Srividya, N. Micro-farming of greens: A viable enterprise for enhancing economic, food and nutritional security of farmers. *Int. J. Nutr. Agric. Res.* **2018**, *5*, 10–16.

66. Dimita, R.; Min Allah, S.; Luvisi, A.; Greco, D.; De Bellis, L.; Accogli, R.; Mininni, C.; Negro, C. Volatile Compounds and Total Phenolic Content of *Perilla frutescens* at Microgreens and Mature Stages. *Horticulturae* **2022**, *8*, 71. [CrossRef]

67. Ghoora, M.D.; Babu, D.R.; Srividya, N. Nutrient composition, oxalate content and nutritional ranking of ten culinary microgreens. *J. Food Compos. Anal.* **2020**, *91*, 103495. [CrossRef]
68. Caracciolo, F.; El-Nakhel, C.; Raimondo, M.; Kyriacou, M.C.; Cembalo, L.; De Pascale, S.; Rouphael, Y. Sensory Attributes and Consumer Acceptability of 12 Microgreens Species. *Agronomy* **2020**, *10*, 1043. [CrossRef]
69. Xiao, Z.; Lester, G.E.; Luo, Y.; Wang, Q. Assessment of Vitamin and Carotenoid Concentrations of Emerging Food Products: Edible Microgreens. *J. Agric. Food Chem.* **2012**, *60*, 7644–7651. [CrossRef] [PubMed]
70. Trifunovic, S.; Topalovic, A.; Knezevic, M.; Vajs, V. Free radicals and antioxidants: Antioxidative and other properties of Swiss chard (*Beta vulgaris* L. subsp. Cicla). *Poljopr. Sumar.* **2015**, *61*, 73. [CrossRef]
71. Davis, D.R.; Epp, M.D.; Riordan, H.D. Changes in USDA food composition data for 43 garden crops, 1950 to 1999. *J. Am. Coll. Nutr.* **2004**, *23*, 669–682. [CrossRef] [PubMed]
72. Garvin, D.F.; Welch, R.M.; Finely, J.W. Historical shifts in the seed mineral micronutrient concentration of US hard red winter wheat germplasm. *J. Sci. Food Agric.* **2006**, *86*, 2213–2220. [CrossRef]
73. Fan, M.-S.; Zhao, F.-J.; Fairweather-Tait, S.J.; Poulton, P.R.; Dunham, S.J.; McGrath, S.P. Evidence of decreasing mineral density in wheat grain over the last 160 years. *J. Trace Elem. Med. Biol.* **2008**, *22*, 315–324. [CrossRef] [PubMed]
74. White, P.J.; Bradshaw, J.E.; Finlay, M.; Dale, B.; Ramsay, G.; Hammond, J.P.; Broadley, M.R. Relationships between yield and mineral concentrations in potato tubers. *Hort Sci.* **2009**, *44*, 6–11. [CrossRef]
75. Marles, R.J. Mineral nutrient composition of vegetables, fruits and grains: The context of reports of apparent historical declines. *J. Food Compos. Anal.* **2017**, *56*, 93–103. [CrossRef]
76. Di Bella, M.C.; Niklas, A.; Toscano, S.; Picchi, V.; Romano, D.; Lo Scalzo, R.; Branca, F. Morphometric characteristics, polyphenols and ascorbic acid variation in Brassica oleracea L. novel foods: Sprouts, microgreens and baby leaves. *Agronomy* **2020**, *10*, 782. [CrossRef]
77. Chatzopoulou, E.; Carocho, M.; Di Gioia, F.; Petropoulos, S.A. The beneficial health effects of vegetables and wild edible greens: The case of the Mediterranean diet and its sustainability. *Appl. Sci.* **2020**, *10*, 9144. [CrossRef]
78. Romojaro, A.; Botella, M.Á.; Obón, C.; Pretel, M.T. Nutritional and antioxidant properties of wild edible plants and their use as potential ingredients in the modern diet. *Int. J. Food Sci. Nutr.* **2013**, *64*, 944–952. [CrossRef]
79. Faudale, M.; Viladomat, F.; Bastida, J.; Poli, F.; Codina, C. Antioxidant activity and phenolic composition of wild, edible, and medicinal fennel from different Mediterranean countries. *J. Agric. Food Chem.* **2008**, *56*, 1912–1920. [CrossRef] [PubMed]
80. Pająk, P.; Socha, R.; Broniek, J.; Królikowska, K.; Fortuna, T. Antioxidant properties, phenolic and mineral composition of germinated chia, golden flax, evening primrose, phacelia and fenugreek. *Food Chem.* **2019**, *275*, 69–76. [CrossRef] [PubMed]
81. Hong, J.; Gruda, N.S. The potential of introduction of Asian vegetables in Europe. *Horticulturae* **2020**, *6*, 38. [CrossRef]
82. Song, J.S.; Jung, S.; Jee, S.; Yoon, J.W.; Byeon, Y.S.; Park, S.; Kim, S.B. Growth and bioactive phytochemicals of *Panax ginseng* sprouts grown in an aeroponic system using plasma-treated water as the nitrogen source. *Sci. Rep.* **2021**, *11*, 2924. [CrossRef]
83. Li, H.; Deng, Z.; Liu, R.; Zhu, H.; Draves, J.; Marcone, M.; Sun, Y.; Tsao, R. Characterization of phenolics, betacyanins and antioxidant activities of the seed, leaf, sprout, flower and stalk extracts of three Amaranthus species. *J. Food Compos. Anal.* **2015**, *37*, 75–81. [CrossRef]
84. Paśko, P.; Bartoń, H.; Zagrodzki, P.; Gorinstein, S.; Fołta, M.; Zachwieja, Z. Anthocyanins, total polyphenols and antioxidant activity in amaranth and quinoa seeds and sprouts during their growth. *Food Chem.* **2009**, *115*, 994–998. [CrossRef]
85. Jan, R.; Saxena, D.C.; Singh, S. Physico-chemical, textural, sensory and antioxidant characteristics of gluten–Free cookies made from raw and germinated Chenopodium (Chenopodium album) flour. *LWT Food Sci. Technol.* **2016**, *71*, 281–287. [CrossRef]
86. Harakotr, B.; Srijunteuk, S.; Rithichai, P.; Tabunhan, S. Effects of Light-Emitting Diode Light Irradiance Levels on Yield, Antioxidants and Antioxidant Capacities of Indigenous Vegetable Microgreens. *Sci. Technol. Asia* **2019**, *24*, 59–66. [CrossRef]
87. Kyriacou, M.C.; El-Nakhel, C.; Pannico, A.; Graziani, G.; Soteriou, G.A.; Giordano, M.; Palladino, M.; Ritieni, A.; De Pascale, S.; Rouphael, Y. Phenolic Constitution, Phytochemical and Macronutrient Content in Three Species of Microgreens as Modulated by Natural Fiber and Synthetic Substrates. *Antioxidants* **2020**, *9*, 252. [CrossRef]
88. Enache, I.-M.; Livadariu, O. Preliminary results regarding the testing of treatments with light-emitting diode (LED) on the seed germination of *Artemisia dracunculus* L. *Sci. Bull. Ser. F Biotechnol.* **2016**, *20*, 51–56.
89. Paradiso, V.M.; Castellino, M.; Renna, M.; Gattullo, C.E.; Calasso, M.; Terzano, R.; Allegretta, I.; Leoni, B.; Caponio, F.; Santamaria, P. Nutritional characterization and shelf-life of packaged microgreens. *Food Funct.* **2018**, *9*, 5629–5640. [CrossRef] [PubMed]
90. Yadav, L.P.; Koley, T.K.; Tripathi, A.; Singh, S. Antioxidant potentiality and mineral content of summer season leafy greens: Comparison at mature and microgreen stages using chemometric. *Agric. Res.* **2019**, *8*, 165–175. [CrossRef]
91. Bantis, F. Light Spectrum Differentially Affects the Yield and Phytochemical Content of Microgreen Vegetables in a Plant Factory. *Plants* **2021**, *10*, 2182. [CrossRef] [PubMed]
92. Ebert, A.W.; Chang, C.-H.; Yan, M.-R.; Yang, R.-Y. Nutritional composition of mungbean and soybean sprouts compared to their adult growth stage. *Food Chem.* **2017**, *237*, 15–22. [CrossRef]
93. Zrig, A.; Saleh, A.; Hamouda, F.; Okla, M.K.; Al-Qahtani, W.H.; Alwasel, Y.A.; Al-Hashimi, A.; Hegab, M.Y.; Hassan, A.H.; AbdElgawad, H. Impact of Sprouting under Potassium Nitrate Priming on Nitrogen Assimilation and Bioactivity of Three *Medicago* Species. *Plants* **2022**, *11*, 71. [CrossRef]
94. Puccinelli, M.; Pezzarossa, B.; Pintimalli, L.; Malorgio, F. Selenium Biofortification of Three Wild Species, *Rumex acetosa* L., *Plantago coronopus* L., and *Portulaca oleracea* L., Grown as Microgreens. *Agronomy* **2021**, *11*, 1155. [CrossRef]

95. Choe, U.; Yu, L.L.; Wang, T.T. The science behind microgreens as an exciting new food for the 21st century. *J. Agric. Food Chem.* **2018**, *66*, 11519–11530. [CrossRef]
96. Pająk, P.; Socha, R.; Gałkowska, D.; Rożnowski, J.; Fortuna, T. Phenolic profile and antioxidant activity in selected seeds and sprouts. *Food Chem.* **2014**, *143*, 300–306. [CrossRef]
97. Khang, D.T.; Dung, T.N.; Elzaawely, A.A.; Xuan, T.D. Phenolic profiles and antioxidant activity of germinated legumes. *Foods* **2016**, *5*, 27. [CrossRef]
98. Alvarez-Jubete, L.; Wijngaard, H.; Arendt, E.K.; Gallagher, E. Polyphenol composition and in vitro antioxidant activity of amaranth, quinoa buckwheat and wheat as affected by sprouting and baking. *Food Chem.* **2010**, *119*, 770–778. [CrossRef]
99. Laus, M.; Cataldi, M.; Soccio, M.; Alfarano, M.; Amodio, M.; Colelli, G.; Flagella, Z.; Pastore, D. Effect of germination and sprout storage on antioxidant capacity and phenolic content in quinoa (Chenopodium quinoa Willd). In Proceedings of the SIBV-SIGA Joint Congress Sustainability of Agricultural Environment: Contributions of Plant Genetics and Physiology, Pisa, Italy, 19–22 September 2017.
100. Smirnoff, N.; Wheeler, G.L. Ascorbic acid in plants: Biosynthesis and function. *Crit. Rev. Plant Sci.* **2000**, *19*, 267–290. [CrossRef]
101. Khosravi, F.; Rastakhiz, N.; Iranmanesh, B.; Olia, S.S.S.J. Determination of Organic Acids in Fruit juices by UPLC. *Int. J. Life Sci.* **2015**, *9*, 41–44. [CrossRef]
102. Baenas, N.; Gómez-Jodar, I.; Moreno, D.A.; García-Viguera, C.; Periago, P.M. Broccoli and radish sprouts are safe and rich in bioactive phytochemicals. *Postharvest Biol. Technol.* **2017**, *127*, 60–67. [CrossRef]
103. Puccinelli, M.; Maggini, R.; Angelini, L.G.; Santin, M.; Landi, M.; Tavarini, S.; Castagna, A.; Incrocci, L. Can Light Spectrum Composition Increase Growth and Nutritional Quality of *Linum usitatissimum* L. Sprouts and Microgreens? *Horticulturae* **2022**, *8*, 98. [CrossRef]
104. Weber, C.F. Nutrient concentration of cabbage and lettuce microgreens grown on vermicompost and hydroponic growing pads. *J. Hortic.* **2016**, *3*, 190. [CrossRef]
105. Weber, C.F. Broccoli microgreens: A mineral-rich crop that can diversify food systems. *Front. Nutr.* **2017**, *4*, 7. [CrossRef]
106. Pinto, E.; Almeida, A.A.; Aguiar, A.A.; Ferreira, I.M.P.L.V.O. Comparison between the mineral profile and nitrate content of microgreens and mature lettuces. *J. Food Comp. Anal.* **2015**, *37*, 38–43. [CrossRef]
107. Liang, G.; Zhang, Z. Reducing the nitrate content in vegetables through joint regulation of short-distance distribution and long-distance transport. *Front. Plant Sci.* **2020**, *11*, 1079. [CrossRef]
108. Salehzadeh, H.; Maleki, A.; Rezaee, R.; Shahmoradi, B.; Ponnet, K. The nitrate content of fresh and cooked vegetables and their health-related risks. *PLoS ONE* **2020**, *15*, e0227551. [CrossRef]
109. Bulgari, R.; Baldi, A.; Ferrante, A.; Lenzi, A. Yield and quality of basil, Swiss chard, and rocket microgreens grown in a hydroponic system. *N. Z. J. Crop Hortic. Sci.* **2017**, *45*, 119–129. [CrossRef]
110. El-Nakhel, C.; Pannico, A.; Graziani, G.; Kyriacou, M.C.; Gaspari, A.; Ritieni, A.; De Pascale, S.; Rouphael, Y. Nutrient Supplementation Configures the Bioactive Profile and Production Characteristics of Three *Brassica L.* Microgreens Species Grown in Peat-Based Media. *Agronomy* **2021**, *11*, 346. [CrossRef]
111. Matsuo, T.; Asano, T.; Mizuno, Y.; Sato, S.; Fujino, I.; Sadzuka, Y. Water spinach and okra sprouts inhibit cancer cell proliferation. *Vitr. Cell. Dev. Biol.-Anim.* **2022**, 1–6. [CrossRef] [PubMed]
112. Khanum, F.; Siddalinga Swamy, M.; Sudarshana Krishna, K.R.; Santhanam, K.; Viswanathan, K.R. Dietary fiber content of commonly fresh and cooked vegetables consumed in India. *Plant Foods Hum. Nutr.* **2000**, *55*, 207–218. [CrossRef] [PubMed]
113. Loi, M.; Villani, A.; Paciolla, F.; Mulè, G.; Paciolla, C. Challenges and Opportunities of Light-Emitting Diode (LED) as Key to Modulate Antioxidant Compounds in Plants. A Review. *Antioxidants* **2021**, *10*, 42. [CrossRef]
114. Teklić, T.; Parađiković, N.; Špoljarević, M.; Zeljković, S.; Lončarić, Z.; Lisjak, M. Linking abiotic stress, plant metabolites, biostimulants and functional food. *Ann. Appl. Biol.* **2021**, *178*, 169–191. [CrossRef]
115. Rouphael, Y.; Kyriacou, M.C. Enhancing Quality of Fresh Vegetables Through Salinity Eustress and Biofortification Applications Facilitated by Soilless Cultivation. *Front. Plant Sci.* **2018**, *9*, 1254. [CrossRef]
116. Benincasa, P.; D'Amato, R.; Falcinelli, B.; Troni, E.; Fontanella, M.C.; Frusciante, S.; Guiducci, M.; Beone, G.M.; Businelli, D.; Diretto, G. Grain Endogenous Selenium and Moderate Salt Stress Work as Synergic Elicitors in the Enrichment of Bioactive Compounds in Maize Sprouts. *Agronomy* **2020**, *10*, 735. [CrossRef]
117. El-Nakhel, C.; Pannico, A.; Kyriacou, M.C.; Giordano, M.; De Pascale, S.; Rouphael, Y. Macronutrient deprivation eustress elicits differential secondary metabolites in red and green-pigmented butterhead lettuce grown in a closed soilless system. *J. Sci. Food Agric.* **2019**, *99*, 6962–6972. [CrossRef]
118. De Pascale, S.; Maggio, A.; Pernice, R.; Fogliano, V.; Barbieri, G. Sulphur fertilization may improve the nutritional value of *Brassica rapa* L. subsp. *sylvestris*. *Eur. J. Agron.* **2007**, *26*, 418–424. [CrossRef]
119. Radovich, T.J.; Kleinhenz, M.D.; Streeter, J.G. Irrigation timing relative to head development influences yield components, sugar levels, and glucosinolate concentrations in cabbage. *J. Am. Soc. Hortic. Sci.* **2005**, *130*, 943–949. [CrossRef]
120. Oh, M.M.; Rajashekar, C.B. Antioxidant content of edible sprouts: Effects of environmental shocks. *J. Sci. Food Agric.* **2009**, *89*, 2221–2227. [CrossRef]
121. Šamec, D.; Ljubej, V.; Redovnikovi'c, I.R.; Fistani'c, S.; Salopek-Sondi, B. Low Temperatures Affect the Physiological Status and Phytochemical Content of Flat Leaf Kale (*Brassica oleracea* var. acephala) Sprouts . *Foods* **2022**, *11*, 264. [CrossRef] [PubMed]

122. Artés-Hernández, F.; Castillejo, N.; Martínez-Zamora, L. UV and Visible Spectrum LED Lighting as Abiotic Elicitors of Bioactive Compounds in Sprouts, Microgreens, and Baby Leaves—A Comprehensive Review including Their Mode of Action. *Foods* **2022**, *11*, 265. [CrossRef] [PubMed]
123. Jenkins, G.I.; Brown, B.A. UV-B perception and signal transduction. In *Light and Plant Development*; Blackwell Publishing Ltd.: Oxford, UK, 2007; pp. 155–182.
124. Mewis, I.; Schreiner, M.; Nguyen, C.N.; Krumbein, A.; Ulrichs, C.; Lohse, M.; Zrenner, R. UV-B irradiation changes specifically the secondary metabolite profile in broccoli sprouts: Induced signaling overlaps with defense response to biotic stressors. *Plant Cell Physiol.* **2012**, *53*, 1546–1560. [CrossRef] [PubMed]
125. Moreira-Rodríguez, M.; Nair, V.; Benavides, J.; Cisneros-Zevallos, L.; Jacobo-Velázquez, D.A. UVA, UVB Light, and Methyl Jasmonate, Alone or Combined, Redirect the Biosynthesis of Glucosinolates, Phenolics, Carotenoids, and Chlorophylls in Broccoli Sprouts. *Int. J. Mol. Sci.* **2017**, *18*, 2330. [CrossRef] [PubMed]
126. Castillejo, N.; Martínez-Zamora, L.; Artés–Hernández, F. Periodical UV-B Radiation Hormesis in Biosynthesis of Kale Sprouts Nutraceuticals. *Plant Physiol. Biochem.* **2021**, *165*, 274–285. [CrossRef]
127. Martínez-Zamora, L.; Castillejo, N.; Gómez, P.A.; Artés-Hernández, F. Amelioration Effect of LED Lighting in the Bioactive Compounds Synthesis during Carrot Sprouting. *Agronomy* **2021**, *11*, 304. [CrossRef]
128. Złotek, U.; Szymanowska, U.; Baraniak, B.; Karaś, M. Antioxidant activity of polyphenols of adzuki bean (*Vigna angularis*) germinated in abiotic stress conditions. *Acta Sci. Pol. Technol. Aliment.* **2015**, *14*, 55–63. [CrossRef]
129. Świeca, M.; Surdyka, M.; Gawlik-Dziki, U.; Złotek, U.; Baraniak, B. Antioxidant potential of fresh and stored lentil sprouts affected by elicitation with temperature stresses. *Int. J. Food Sci. Technol.* **2014**, *49*, 1811–1817. [CrossRef]
130. Randhir, R.; Lin, Y.T.; Shetty, K. Stimulation of phenolics, antioxidant and antimicrobial activities in dark germinated mung bean sprouts in response to peptide and phytochemical elicitors. *Process Biochem.* **2004**, *39*, 637–646. [CrossRef]
131. Gawlik-Dziki, U.; Świeca, M.; Dziki, D.; Sugier, D. Improvement of nutraceutical value of broccoli sprouts by natural elicitors. *Acta Sci. Pol.-Hortorum Cultus* **2013**, *12*, 129–140.
132. Sun, J.; Kou, L.; Geng, P.; Huang, H.; Yang, T.; Luo, Y.; Chen, P. Metabolomic Assessment Reveals an Elevated Level of Glucosinolate Content in $CaCl_2$ Treated Broccoli Microgreens. *J. Agric. Food Chem.* **2015**, *63*, 1863–1868. [CrossRef] [PubMed]
133. Lu, Y.; Dong, W.; Alcazar, J.; Yang, T.; Luo, Y.; Wang, Q.; Chen, P. Effect of preharvest CaCl2 spray and postharvest UV-B radiation on storage quality of broccoli microgreens, a richer source of glucosinolates. *J. Food Compos. Anal.* **2018**, *67*, 55–62. [CrossRef]
134. Mercado, M.F.O.; Fernandez, P.G. Solid matrix priming of soybean seeds. *Philipp. J. Crop Sci.* **2002**, *27*, 27–35.
135. Pandita, V.K.; Anand, A.; Nagarajan, S.; Seth, R.; Sinha, S.N. Solid matrix priming improves seed emergence and crop performance in okra. *Seed Sci. Technol.* **2010**, *38*, 665–674. [CrossRef]
136. Tanha, A.; Golzardi, F.; Mostafavi, K. Seed Priming to Overcome Autotoxicity of Alfalfa (Medicago sativa). *World J. Environ. Biosci.* **2017**, *6*, 1–5.
137. Mondal, S.; Bose, B. An impact of seed priming on disease resistance: A review. In *Microbial Diversity and Biotechnology in Food Security*; Kharwar, R., Upadhyay, R., Dubey, N., Raghuwanshi, R., Eds.; Springer: New Delhi, India, 2014; pp. 193–203. [CrossRef]
138. Ali, M.M.; Javed, T.; Mauro, R.P.; Shabbir, R.; Afzal, I.; Yousef, A.F. Effect of seed priming with potassium nitrate on the performance of tomato. *Agriculture* **2020**, *10*, 498. [CrossRef]
139. Baenas, N.; Villaño, D.; García-Viguera, C.; Moreno, D.A. Optimizing elicitation and seed priming to enrich broccoli and radish sprouts in glucosinolates. *Food Chem.* **2016**, *204*, 314–319. [CrossRef]
140. Sangiorgio, D.; Cellini, A.; Donati, I.; Pastore, C.; Onofrietti, C.; Spinelli, F. Facing Climate Change: Application of Microbial Biostimulants to Mitigate Stress in Horticultural Crops. *Agronomy* **2020**, *10*, 794. [CrossRef]
141. Rouphael, Y.; Colla, G. Synergistic biostimulatory action: Designing the next generation of plant biostimulants for sustainable agriculture. *Front. Plant Sci.* **2018**, *9*, 1655. [CrossRef]
142. Bulgari, R.; Franzoni, G.; Ferrante, A. Biostimulants Application in Horticultural Crops under Abiotic Stress Conditions. *Agronomy* **2019**, *9*, 306. [CrossRef]
143. Ali, Q.; Shehzad, F.; Waseem, M.; Shahid, S.; Hussain, A.I.; Haider, M.Z.; Habib, N.; Hussain, S.M.; Javed, M.T.; Perveen, R. Plant-based biostimulants and plant stress responses. In *Plant Ecophysiology and Adaptation under Climate Change: Mechanisms and Perspectives I*; Springer: Singapore, 2020; pp. 625–661. [CrossRef]
144. Zhao, J.L.; Zou, L.; Zhong, L.Y.; Peng, L.X.; Ying, P.L.; Tan, M.L.; Zhao, G. Effects of polysaccharide elicitors from endophytic *Bionectria pityrodes* Fat6 on the growth and flavonoid production in tartary buckwheat sprout cultures. *Cereal Res. Commun.* **2015**, *43*, 661–671. [CrossRef]
145. Briatia, X.; Jomduang, S.; Park, C.H.; Lumyong, S.; Kanpiengjai, A.; Khanongnuch, C. Enhancing growth of buckwheat sprouts and microgreens by endophytic bacterium inoculation. *Int. J. Agric. Biol.* **2017**, *19*, 374–380. [CrossRef]
146. Gerbore, J.; Vallance, J.; Yacoub, A.; Delmotte, F.; Grizard, D.; Regnault-Roger, C.; Rey, P. Characterization of *Pythium oligandrum* populations that colonize the rhizosphere of vines from the Bordeaux region. *FEMS Microbiol. Ecol.* **2014**, *90*, 153–167. [CrossRef]
147. D'Amato, R.; Regni, L.; Falcinelli, B.; Mattioli, S.; Benincasa, P.; Dal Bosco, A.; Pacheco, P.; Proietti, P.; Troni, E.; Santi, C.; et al. Current Knowledge on Selenium Biofortification to Improve the Nutraceutical Profile of Food: A Comprehensive Review. *J. Agric. Food Chem.* **2020**, *68*, 4075–4097. [CrossRef]
148. Puccinelli, M.; Malorgio, F.; Rosellini, I.; Pezzarossa, B. Production of selenium-biofortified microgreens from selenium-enriched seeds of basil. *J. Sci. Food Agric.* **2019**, *99*, 5601–5605. [CrossRef]

149. Islam, M.Z.; Park, B.J.; Kang, H.M.; Lee, Y.T. Influence of selenium biofortification on the bioactive compounds and antioxidant activity of wheat microgreen extract. *Food Chem.* **2020**, *309*, 125763. [CrossRef]

150. Germ, M.; Stibilj, V.; Šircelj, H.; Jerše, A.; Kroflič, A.; Golob, A.; Maršić, N.K. Biofortification of common buckwheat microgreens and seeds with different forms of selenium and iodine. *J. Sci. Food Agric.* **2019**, *99*, 4353–4362. [CrossRef]

151. Pannico, A.; El-Nakhel, C.; Graziani, G.; Kyriacou, M.C.; Giordano, M.; Soteriou, G.A.; Zarrelli, A.; Ritieni, A.; De Pascale, S.; Rouphael, Y. Selenium biofortification impacts the nutritive value, polyphenolic content, and bioactive constitution of variable microgreens genotypes. *Antioxidants* **2020**, *9*, 272. [CrossRef]

152. Vicas, S.I.; Cavalu, S.; Laslo, V.; Tocai, M.; Costea, T.O.; Moldovan, L. Growth, Photosynthetic Pigments, Phenolic, Glucosinolates Content and Antioxidant Capacity of Broccoli Sprouts in Response to Nanoselenium Particles Supply. *Not. Bot. Horti Agrobot. Cluj-Napoca* **2019**, *47*, 821–828. [CrossRef]

153. Neo, D.C.J.; Ong, M.M.X.; Lee, Y.Y.; Teo, E.J.; Ong, Q.; Tanoto, H.; Xu, J.; Ong, K.S.; Suresh, V. Shaping and Tuning Lighting Conditions in Controlled Environment Agriculture: A Review. *ACS Agric. Sci. Technol.* **2022**, *2*, 3–16. [CrossRef]

154. Gerovac, J.R.; Craver, J.K.; Boldt, J.K.; Lopez, R.G. Light intensity and quality from sole-source light-emitting diodes impact growth, morphology, and nutrient content of *Brassica* microgreens. *HortScience* **2016**, *51*, 497–503. [CrossRef]

155. Kopsell, D.A.; Pantanizopoulos, N.I.; Sams, C.E.; Kopsell, D.E. Shoot tissue pigment levels increase in 'Florida Broadleaf' mustard (*Brassica juncea* L.) microgreens following high light treatment. *Sci. Hortic.* **2012**, *140*, 96–99. [CrossRef]

156. Craver, J.K.; Gerovac, J.R.; Lopez, R.G.; Kopsell, D.A. Light intensity and light quality from sole-source light-emitting diodes impact phytochemical concentrations within Brassica microgreens. *J. Am. Soc. Hortic. Sci.* **2017**, *142*, 3–12. [CrossRef]

157. Samuoliene, G.; Brazaityte, A.; Jankauskiene, J.; Viršile, A.; Sirtautas, R.; Novičkovas, A.; Sakalauskiene, S.; Sakalauskaite, J.; Duchovskis, P. LED irradiance level affects growth and nutritional quality of *Brassica* microgreens. *Open Life Sci.* **2013**, *8*, 1241–1249. [CrossRef]

158. Brazaitytė, A.; Sakalauskienė, S.; Samuolienė, G.; Jankauskienė, J.; Viršilė, A.; Novičkovas, A.; Sirtautas, R.; Miliauskienė, J.; Vaštakaitė, V.; Dabašinskas, L.; et al. The effects of LED illumination spectra and intensity on carotenoid content in Brassicaceae microgreens. *Food Chem.* **2015**, *173*, 600–606. [CrossRef]

159. Shibaeva, T.G.; Sherudilo, E.G.; Rubaeva, A.A.; Titov, A.F. Continuous LED Lighting Enhances Yield and Nutritional Value of Four Genotypes of Brassicaceae Microgreens. *Plants* **2022**, *11*, 176. [CrossRef] [PubMed]

160. Hornyák, M.; Dziurka, M.; Kula-Maximenko, M.; Pastuszak, J.; Szczerba, A.; Szklarczyk, M.; Płażek, A. Photosynthetic efficiency, growth and secondary metabolism of common buckwheat (*Fagopyrum esculentum* Moench) in different controlled-environment production systems. *Sci. Rep.* **2022**, *12*, 257. [CrossRef]

161. Fiutak, G.; Michalczyk, M. Effect of artificial light source on pigments, thiocyanates and ascorbic acid content in kale sprouts (*Brassica oleracea* L. var. *Sabellica* L.). *Food Chem.* **2020**, *330*, 127189. [CrossRef]

162. Kopsell, D.A.; Sams, C.E.; Barickman, T.C.; Morrow, R.C. Sprouting broccoli accumulate higher concentrations of nutritionally important metabolites under narrow-band light-emitting diode lighting. *J. Am. Soc. Hortic. Sci.* **2014**, *139*, 469. [CrossRef]

163. Taulavuori, K.; Pyysalo, A.; Taulavuori, E.; Julkunen-Tiitto, R. Responses of phenolic acid and flavonoid synthesis to blue and blue-violet light depends on plant species. *Environ. Exp. Bot.* **2018**, *150*, 183–187. [CrossRef]

164. Paradiso, R.; Proietti, S. Light-Quality Manipulation to Control Plant Growth and Photomorphogenesis in Greenhouse Horticulture: The State of the Art and the Opportunities of Modern LED Systems. *J. Plant Growth Regul.* **2021**, *41*, 742–780. [CrossRef]

165. Kalaitzoglou, P.; van Ieperen, W.; Harbinson, J.; van der Meer, M.; Martinakos, S.; Weerheim, K.; Nicole, C.C.S.; Marcelis, L.F.M. Effects of continuous or end-of-day far-red light on tomato plant growth, morphology, light absorption, and fruit production. *Front. Plant Sci.* **2019**, *10*, 322. [CrossRef] [PubMed]

166. Appolloni, E.; Pennisi, G.; Zauli, I.; Carotti, L.; Paucek, I.; Quaini, S.; Orsini, F.; Gianquinto, G. Beyond vegetables: Effects of indoor LED light on specialized metabolite biosynthesis in medicinal and aromatic plants, edible flowers, and microgreens. *J. Sci. Food Agric.* **2022**, *102*, 472–487. [CrossRef] [PubMed]

167. Mastropasqua, L.; Dipierro, N.; Paciolla, C. Effects of Darkness and Light Spectra on Nutrients and Pigments in Radish, Soybean, Mung Bean and Pumpkin Sprouts. *Antioxidants* **2020**, *9*, 558. [CrossRef]

168. Nam, T.G.; Kim, D.O.; Eom, S.H. Effects of light sources on major flavonoids and antioxidant activity in common buckwheat sprouts. *Food Sci. Biotechnol.* **2018**, *27*, 169–176. [CrossRef]

169. Park, C.H.; Kim, N.S.; Park, J.S.; Lee, S.Y.; Lee, J.W.; Park, S.U. Effects of light-emitting diodes on the accumulation of glucosinolates and phenolic compounds in sprouting canola (*Brassica napus* L.). *Foods* **2019**, *8*, 76. [CrossRef]

170. Yang, L.; Fanourakis, D.; Tsaniklidis, G.; Li, K.; Yang, Q.; Li, T. Contrary to Red, Blue Monochromatic Light Improves the Bioactive Compound Content in Broccoli Sprouts. *Agronomy* **2021**, *11*, 2139. [CrossRef]

171. Samuolienė, G.; Urbonavičiūtė, A.; Brazaitytė, A.; Šabajevienė, G.; Sakalauskaitė, J.; Duchovskis, P. The impact of LED illumination on antioxidant properties of sprouted seeds. *Cent. Eur. J. Biol.* **2011**, *6*, 68–74. [CrossRef]

172. Lobiuc, A.; Vasilache, V.; Oroian, M.; Stoleru, T.; Burducea, M.; Pintilie, O.; Zamfirache, M.-M. Blue and Red LED Illumination Improves Growth and Bioactive Compounds Contents in Acyanic and Cyanic *Ocimum basilicum* L. Microgreens. *Molecules* **2017**, *22*, 2111. [CrossRef]

173. Chutimanukul, P.; Wanichananan, P.; Janta, S.; Toojinda, T.; Darwell, C.T.; Mosaleeyanon, K. The influence of different light spectra on physiological responses, antioxidant capacity and chemical compositions in two holy basil cultivars. *Sci. Rep.* **2022**, *12*, 588. [CrossRef] [PubMed]

174. Brazaityte, A.; Miliauskiene, J.; Vaštakaite-Kairiene, V.; Sutuliene, R.; Laužike, K.; Duchovskis, P.; Małek, S. Effect of Different Ratios of Blue and Red LED Light on Brassicaceae Microgreens under a Controlled Environment. *Plants* **2021**, *10*, 801. [CrossRef] [PubMed]

175. Samuolienė, G.; Viršilė, A.; Brazaitytė, A.; Jankauskienė, J.; Sakalauskienė, S.; Vaštakaitė, V.; Novičkovas, A.; Viškelienė, A.; Sasnauskas, A.; Duchovskis, P. Blue light dosage affects carotenoids and tocopherols in microgreens. *Food Chem.* **2017**, *228*, 50–56. [CrossRef] [PubMed]

176. Hytönen, T.; Pinho, P.; Rantanen, M.; Kariluoto, S.; Lampi, A.; Edelmann, M.; Joensuu, K.; Kauste, K.; Mouhu, K.; Piironen, V.; et al. Effects of LED light spectra on lettuce growth and nutritional composition. *Lighting Res. Technol.* **2018**, *50*, 880–893. [CrossRef]

177. Giménez, A.; Martínez-Ballesta, M.C.; Egea-Gilabert, C.; Gómez, P.A.; Artés-Hernández, F.; Pennisi, G.; Orsini, F.; Crepaldi, A.; Fernández, J.A. Combined Effect of Salinity and LED Lights on the Yield and Quality of Purslane (*Portulaca oleracea* L.) Microgreens. *Horticulturae* **2021**, *7*, 180. [CrossRef]

178. Kopsell, D.A.; Sams, C.E. Increases in shoot tissue pigments, glucosinolates, and mineral elements in sprouting broccoli after exposure to short-duration blue light from light emitting diodes. *J. Am. Soc. Hortic. Sci.* **2013**, *138*, 31–37. [CrossRef]

179. Alrifai, O.; Mats, L.; Liu, R.; Hao, X.; Marcone, M.F.; Tsao, R. Effect of combined light-emitting diodes on the accumulation of glucosinolates in *Brassica* microgreens. *Food Prod. Process Nutr.* **2021**, *3*, 30. [CrossRef]

180. Alrifai, O.; Hao, X.; Liu, R.; Lu, Z.; Marcone, M.F.; Tsao, R. Amber, red and blue LEDs modulate phenolic contents and antioxidant activities in eight Cruciferous microgreens. *J. Food Bioact.* **2020**, *11*, 95–109. [CrossRef]

181. Kamal, K.Y.; Khodaeiaminjan, M.; El-Tantawy, A.A.; Moneim, D.A.; Salam, A.A.; Ash-shormillesy, S.M.; Attia, A.; Ali, M.A.; Herranz, R.; El-Esawi, M.A.; et al. Evaluation of growth and nutritional value of *Brassica* microgreens grown under red, blue and green LEDs combinations. *Physiol. Plant.* **2020**, *169*, 625–638. [CrossRef]

182. Zhang, X.; Bian, Z.; Li, S.; Chen, X.; Lu, C. Comparative analysis of phenolic compound profiles, antioxidant capacities, and expressions of phenolic biosynthesis-related genes in soybean microgreens grown under different light spectra. *J. Agric. Food Chem.* **2019**, *67*, 13577–13588. [CrossRef]

183. Mordor Intelligence. Microgreens Market—Growth, Trends, Covid-19 Impact, and Forecasts (2021–2026). Available online: https://www.mordorintelligence.com/industry-reports/microgreens-market (accessed on 10 December 2021).

184. Renna, M.; Di Gioia, F.; Leoni, B.; Mininni, C.; Santamaria, P. Culinary assessment of self-produced microgreens as basic ingredients in sweet and savory dishes. *J. Culin. Sci. Technol.* **2017**, *15*, 126–142. [CrossRef]

185. Agrilyst. State of Indoor Farming 2017. Available online: https://www.cropscience.bayer.com/sites/cropscience/files/inline-files/stateofindoorfarming-report-2017.pdf (accessed on 10 December 2021).

186. Rajan, P.; Lada, R.R.; MacDonald, M.T. Advancement in indoor vertical farming for microgreen production. *Am. J. Plant Sci.* **2019**, *10*, 1397. [CrossRef]

187. Stein, E.W. The Transformative Environmental Effects Large-Scale Indoor Farming May Have on Air, Water, and Soil. *Air Soil Water Res.* **2021**, *14*, 1178622121995819. [CrossRef]

188. ReportLinker. Global Microgreens Market Analysis & Trends—Industry Forecast to 2028. Available online: https://www.reportlinker.com/p06127645/Global-Microgreens-Market-Analysis-Trends-Industry-Forecast-to.html?utm_source=GNW (accessed on 10 December 2021).

189. Allied Market Research. Global Microgreens Market—Opportunities and Forecast, 2021–2028. Available online: https://www.alliedmarketresearch.com/microgreens-market-A08733 (accessed on 11 December 2021).

Article

Effect of Wheat Crop Nitrogen Fertilization Schedule on the Phenolic Content and Antioxidant Activity of Sprouts and Wheatgrass Obtained from Offspring Grains

Beatrice Falcinelli [1], Angelica Galieni [2,*], Giacomo Tosti [1], Fabio Stagnari [3], Flaviano Trasmundi [3], Eleonora Oliva [3], Annalisa Scroccarello [3], Manuel Sergi [3], Michele Del Carlo [3] and Paolo Benincasa [1]

[1]   Dipartimento di Scienze Agrarie, Alimentari ed Ambientali, Università di Perugia, Borgo XX Giugno 74, 06121 Perugia, Italy

[2]   Research Centre for Vegetable and Ornamental Crops, Council for Agricultural Research and Economics—CREA, Via Salaria 1, 63077 Monsampolo del Tronto, Italy

[3]   Faculty of Bioscience and Agro-Food and Environmental Technology, University of Teramo, Via Renato Balzarini 1, 64100 Teramo, Italy

*   Correspondence: angelica.galieni@crea.gov.it

Abstract: This work was aimed at investigating the effects of rate and timing of nitrogen fertilization applied to a maternal wheat crop on phytochemical content and antioxidant activity of edible sprouts and wheatgrass obtained from offspring grains. We hypothesized that imbalance in N nutrition experienced by the mother plants translates into transgenerational responses on seedlings obtained from the offspring seeds. To this purpose, we sprouted grains of two bread wheat cultivars (Bologna and Bora) grown in the field under four N fertilization schedules: constantly well N fed with a total of 300 kg N ha^{-1}; N fed only very early, i.e., one month after sowing, with 60 kg N ha^{-1}; N fed only late, i.e., at initial shoot elongation, with 120 kg N ha^{-1}; and unfertilized control. We measured percent germination, seedling growth, vegetation indices (by reflectance spectroscopy), the phytochemical content (total phenols, phenolic acids, carotenoids, chlorophylls), and the antioxidant activity (by gold nanoparticles photometric assay) of extracts in sprout and wheatgrass obtained from the harvested seeds. Our main finding is that grains obtained from crops subjected to late N deficiency produced wheatgrass with much higher phenolic content (as compared to the other N treatments), and this was observed in both cultivars. Thus, we conclude that late N deficiency is a stressing condition which elicits the production of phenols. This may help counterbalance the loss of income related to lower grain yield in crops subjected to such an imbalance in N nutrition.

Keywords: bioactive compounds; carotenoid; chlorophyll; gold nanoparticles photometric assay; phenolic acid; seedling; spectroscopy; vegetation index

Citation: Falcinelli, B.; Galieni, A.; Tosti, G.; Stagnari, F.; Trasmundi, F.; Oliva, E.; Scroccarello, A.; Sergi, M.; Del Carlo, M.; Benincasa, P. Effect of Wheat Crop Nitrogen Fertilization Schedule on the Phenolic Content and Antioxidant Activity of Sprouts and Wheatgrass Obtained from Offspring Grains. *Plants* **2022**, *11*, 2042. https://doi.org/10.3390/plants11152042

Academic Editor: Petko Denev

Received: 9 July 2022
Accepted: 2 August 2022
Published: 4 August 2022

Publisher's Note: MDPI stays neutral with regard to jurisdictional claims in published maps and institutional affiliations.

1. Introduction

Sprouted seeds represent micro-scale vegetables harvested at the initial and very earliest growth stages, particularly appreciated for their high content of bioactive molecules. Among species exploited for sprouting purposes, cereals are particularly prone to edible sprouts and wheatgrass production. Both (especially wheatgrass) are characterized by a high content of secondary metabolites and antioxidants [1], which contribute to numerous beneficial properties for human health [2]; wheat (*Triticum aestivum* L.) grains represent the main source of basic nutrients, especially carbohydrates, for human nutrition, as well as contain secondary metabolites [3].

Although there are many techniques available to increase phytochemicals accumulation in cereal sprouts and wheatgrass during sprouting [4,5], some recent studies demonstrated that most of the phytochemical content found after sprouting is principally related to the initial phytochemical concentration in grains (i.e., seed lot, cultivars, etc.) [1,4], which can be modulated through the growing conditions of the mother plants [6–9]. Nitrogen (N) fertilization is one of the main agronomic techniques affecting wheat growth, yield (i.e., photosynthetic activity and sink capacity such as grain number and size) [10], and grain quality, especially protein content [11], which in turn affects seed germination performances and seedling vigour [12]. In recent years, many studies have mainly focused on observing the effect of N fertilization on phenolic compounds and antioxidant activity in wheat grains [13–15]; results were not always consistent, due to the experimental design approach, data processing, and the limitation of phenolic/antioxidant assays [15].

During germination and sprouting, the mobilization of the major storage reserves occurs to provide nutrients to embryo and seedling growth [16]. Proteins, thanks to the increased amylase activity, are converted into free amino acid [16], i.e., phenylalanine, tyrosine, and tryptofane represent the substrate for phenylpropanoids synthesis (e.g., phenolic compounds) [17].

It might be thus hypothesized that different amounts and timing of N fertilization application to the mother plant may affect the protein content in the offspring grains and, consequently, the phenolic content in the obtained sprouts and wheatgrass. Only Engert et al. [18] have investigated the effect of N fertilization on the phenolic content and antioxidant activity on wheat sprouts, restricting the study to only two N doses and one seedling stage (i.e., two-day-old sprouts). The N fertilization schedule is also expected to affect seedling growth and the accumulation of photosynthetic pigments, such as chlorophylls and carotenoids, which are important for both determining the consumer acceptability by affecting the colour of sprouts and for the nutritional properties of sprouts, given the health benefits associated to these compounds [19]. However, no literature is available regarding the effect of a more diversified N fertilization schedule (i.e., including rates and timing of N application) on the comprehensive profiles of health-related compounds of sprouts and wheatgrass obtainable from offspring seeds. Such compounds can be quantified by classical laboratory analysis, even though it is possible to alternatively estimate some quality parameters by quick non-destructive methods. Spectroscopy of vegetation represents a high-throughput technology utilized in precision agriculture and for plant phenotyping purposes, and it is based on optical properties of plant tissues—such as leaf and/or canopy reflectance—which allow estimating several vegetable traits [20]. The information conveyed by spectral reflectance can be used, considering the whole spectrum as a plant "fingerprint", and/or for the construction of vegetation indices (VIs). The latter have been widely used for the estimation of the pigment concentration [21] as well as of quality attributes of vegetables [22]. Despite its potential, reflectance spectroscopy applications on quality assessment remain underexplored and, as far as our knowledge, no studies are available on the application of this technique on young seedlings yielded for consumption.

Based on these assumptions, our work was aimed at studying the effects, determined by the application of different N fertilization schedules—supplied at optimal and/or sub-optimal levels—to the mother plants of two wheat genotypes, on the germination performances and accumulation of nutritional and functional molecules of the harvested grain seeds and their sprouts and wheatgrass. In particular, the main objective was to provide a first attempt to delve into the contribution of N fertilization, applied to the seed's plants, on quality attributes of the obtained seedlings intended for consumption. Our hypothesis is that N imbalance acts as an elicitor, leading to transgenerational effects in terms of the higher nutraceutical profile of sprouts and wheatgrass obtained from the offspring seeds. The suitability of spectroscopy of vegetation to further assess quality performances of sprouts and wheatgrass was also evaluated.

2. Results

2.1. Germination and Seedling Growth Parameters

Germination performances and growth parameters of sprouts and wheatgrass are reported in Table 1.

Table 1. Total germination % (G), mean germination time (MGT, expressed as days after sowing—DAS), and individual sprout and wheatgrass parameters: lengths (L, mm) and dry weights (DW, mg). Sprouts and wheatgrass were obtained from seeds of two *Triticum aestivum* cultivars (Bologna, BL; Bora, BR) subjected to four different N fertilization schedules (N0: unfertilized control; N300: constantly well N fertilized throughout the growth cycle; N60-0: N fertilized only one month after sowing; N0-120: N fertilized only late at initial shoot elongation). For wheatgrass, L and DW refer to shoot biomass.

Treatment	Germination Performances		Sprouts Growth Parameters		Wheatgrass Growth Parameters	
	G %	MGT	L	DW	L	DW
BL						
N0	96 (0.5)	1.4 (0.01)	41.3	25.7	87	6.1
N300	99 (0.5)	1.5 (0.04)	42.6	20.5	108	7.8
N60-0	97 (1.0)	1.3 (0.04)	41.3	24.2	91	7.0
N0-120	95 (0.5)	1.4 (0.00)	39.2	23.9	90	7.2
BR						
N0	96 (3.0)	1.3 (0.05)	40.5	32.1	102	8.3
N300	96 (1.5)	1.3 (0.00)	43.7	33.9	116	9.4
N60-0	95 (2.0)	1.3 (0.00)	42.0	31.3	95	8.4
N0-120	95 (0.5)	1.3 (0.04)	38.5	32.3	110	9.8
F-test						
CV	-	**	*ns*	**	**	**
N	-	*	**	*ns*	**	**
CV × N	-	*ns*	*ns*	*ns*	*ns*	*ns*
LSD						
CV	-	0.05	0.90	4.84	6.96	0.39
N	-	0.07	1.27	6.85	9.85	0.56
CV × N	-	0.10	1.80	9.68	13.93	0.79

Two-factor analysis of variance (ANOVA) at 5% level of probability: cultivar (CV); N fertilization schedules (N). * $p < 0.05$; ** $p < 0.01$; *ns* = not significant. LSD = Least Significant Difference ($p < 0.05$). G% values were submitted to ANOVA after arcsin transformation, but no significant differences were recorded, thus, only the G% values without arcsin transformation were shown here.

Germination percentage (G) was not significantly affected by cultivar (CV) and N treatments (N) and ranged between 95% and 99%. The mean germination time (MGT) was significantly affected by either the cultivar or the N treatment. The lowest values were recorded in all BR treatments and in BL_N60-0, while the highest one was recorded in BL_N300.

As far as growth parameters are concerned, the effect of CV and N was different between sprouts and wheatgrass. In sprouts, the length was significantly affected only by N treatments, and in both BL and BR, the lowest values were observed in N0-120. On the other hand, sprouts' dry weight was significantly affected only by CV and, averaging among treatments, BL always showed the lowest values. In wheatgrass, both the lengths and dry weights were affected by CV and N: averaging among treatments, BR showed the highest values of both shoot length and dry weight, while, concerning the effect of N, N300 showed the highest values of lengths in both cultivars and of dry weight in BL.

2.2. Pigment Concentration

In sprouts, no appreciable differences were recorded in terms of pigment concentrations, while in wheatgrass, both the CV and N treatments significantly affected chlorophyll A (Chl A), chlorophyll B (Chl B), and carotenoids (Car) (Figure 1). BL showed the highest values (Chl A + B: 2125 vs. 1837 μg g^{-1} fresh weight, FW; Car: 269 vs. 207 μg g^{-1} FW) and

N300 gave greater Chl concentration in wheatgrass (+15% on average); regardless of CV, N60-0 induced the highest Car values (269 µg g^{-1} FW) (Figure 1).

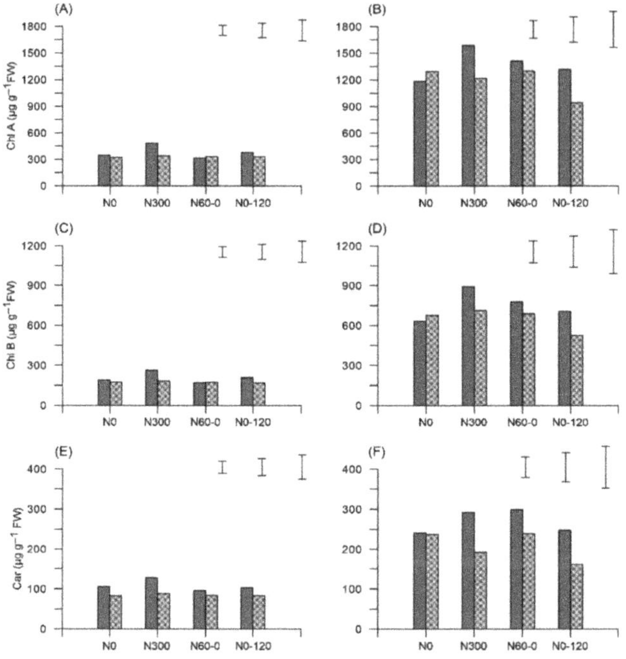

Effects	Sprouts Chl A	Sprouts Chl B	Sprouts Car	Wheatgrass Chl A	Wheatgrass Chl B	Wheatgrass Car
CV	*ns*	*ns*	**	**	*	**
N	*ns*	*ns*	*ns*	**	*	*
CV × N	*ns*	*ns*	*ns*	*	*ns*	*ns*

BL
BR

Figure 1. Chlorophyll A (Chl A), Chlorophyll B (Chl B), and carotenoid (Car) concentration (µg g^{-1} fresh weight, FW) as determined in sprouts (**A**,**C**,**E** for Chl A, Chl B, and Car, respectively) and wheatgrass (**B**,**D**,**F** for Chl A, Chl B, and Car, respectively) obtained from seeds of two *Triticum aestivum* cultivars (Bologna, BL; Bora, BR) subjected to four different N fertilization schedules (N0: unfertilized control; N300: constantly well N fertilized throughout the growth cycle; N60-0: N fertilized only one month after sowing; N0-120: N fertilized only late at initial shoot elongation). In the box, the F-test from the analysis of variance (ANOVA): two-factor ANOVA at 5% level of probability; cultivar (CV); N fertilization schedules (N). * $p < 0.05$; ** $p < 0.01$; *ns* = not significant. Bars represent the Least Significant Difference (LSD; $p < 0.05$) for CV, N, and CV × N (from left to right).

2.3. Changes in the Levels of Bioactive Compounds and Antioxidant Activities

As expected, the total phenolic content (TPC)—as determined by photometric assay (FC)—sharply increased from seeds to sprouts (+82% averaged over CV and N) to wheatgrass (+329% averaged over CV and N) (Figure 2). No clear trends were observed in response to N treatments except for the higher values recorded in N60-0. TCP was pretty stable through the growth phases for BL, whereas it increased for BR passing from seeds (333–421 µg g^{-1}) to sprouts (423–1167 µg g^{-1}) to wheatgrass (1247–2194 µg g^{-1}).

Focusing on wheatgrass—where TPC refers only to shoots—significantly higher values were recorded for N60-0 in both BL and BR. On average over the two cultivars, TPC was 2823 µg g^{-1} DW in N60-0, against 1451 µg g^{-1} DW in N0 1220 in N300 and 1797 µg g^{-1} DW in N0-120 (Figure 2C). However, a significant CV × N interaction was observed, with BR showing the greatest increases in TPC in case of imbalances in N nutrition (N60-0 and N0-120).

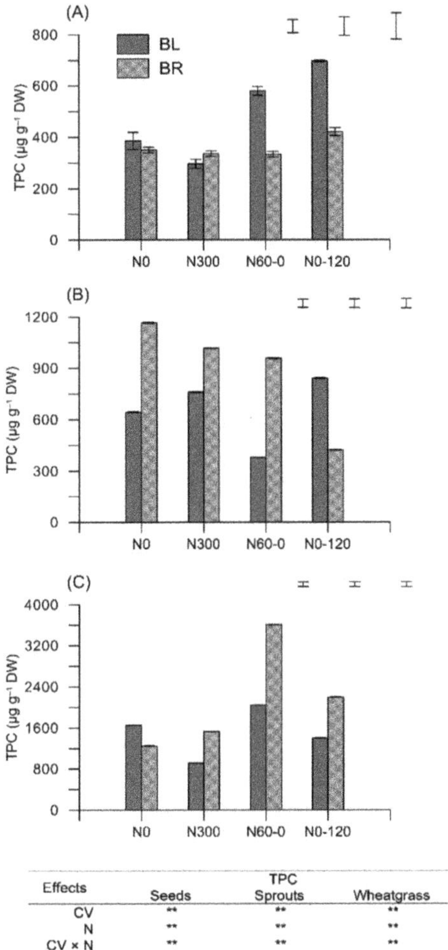

Effects	TPC		
	Seeds	Sprouts	Wheatgrass
CV	**	**	**
N	**	**	**
CV × N	**	**	**

Figure 2. Total phenolic content (TPC) (Folin–Ciocalteu, FC; µg gallic acid equivalents g^{-1} dry weight, DW) as determined in seeds (**A**), sprouts (**B**), and wheatgrass (**C**) obtained from two *Triticum aestivum* cultivars (Bologna, BL; Bora, BR) subjected to four different N fertilization schedules (N0: unfertilized control; N300: constantly well N fertilized throughout the growth cycle; N60-0: N fertilized only one month after sowing; N0-120: N fertilized only late at initial shoot elongation). In the box, F-test from the analysis of variance (ANOVA): two-factor ANOVA at 5% level of probability; cultivar (CV); N fertilization schedules (N). ** $p < 0.01$; Bars represent the Least Significant Difference (at $p < 0.05$) for CV, N, and CV × N (from left to right).

Additionally, antioxidant capacity (AOC) showed an increasing trend as germination went on, ranging from a 1.2- to 11.0-fold increase in ABTS-based assay in sprouts and wheatgrass, respectively (compared to seeds) (Table 2).

This trend was confirmed by the gold nanoparticles (AuNPs)-based assay, with values ranging from a 2.8- to 3.2-fold increase as compared to seeds (Table 2).

For the reported trend, neglecting the effects of CV and for all the investigated growth stages (i.e., seeds, sprouts, and wheat grass), the highest values were observed in treatments that involved strong N stress on the mother plant, either for the entire crop cycle (i.e., N0) or in late growth phases (i.e., N60-0) (Table 2).

Table 2. Antioxidant capacity (AOC) and TPC: ABTS-based assay (ABTS, µg gallic acid equivalents (GAE) g^{-1} dry weight (DW)) and gold nanoparticles photometric assay (AuNPs, µg GAE g^{-1} DW) as determined in seeds, sprouts, and wheatgrass of two *Triticum aestivum* cultivars (Bologna, BL; Bora, BR) subjected to four different N fertilization schedules (N0: unfertilized control; N300: constantly well N fertilized throughout the growth cycle; N60-0: N fertilized only one month after sowing; N0-120: N fertilized only late at initial shoot elongation).

Treatment	Seeds		Sprouts		Wheatgrass	
	ABTS	AuNPs	ABTS	AuNPs	ABTS	AuNPs
BL						
N0	21.2	374	32.6	1277	238	1614
N300	14.1	439	30.3	1131	232	1437
N60-0	41.8	436	13.8	1120	219	1412
N0-120	35.1	418	31.0	1425	386	1255
BR						
N0	8.2	526	59.8	1610	247	1499
N300	30.6	366	20.6	881	312	1396
N60-0	25.7	603	32.0	1504	280	1676
N0-120	25.2	505	28.6	1326	315	1436
F-test						
CV	**	**	**	**	**	**
N	**	**	**	**	**	**
CV × N	**	**	**	**	**	**
LSD						
CV	1.13	1.83	1.38	1.37	2.44	25.86
N	1.61	2.58	1.95	1.94	3.45	36.57
CV × N	2.27	3.65	2.75	2.74	4.88	51.72

Two-factor analysis of variance (ANOVA) at 5% level of probability: cultivar (CV); N fertilization schedules (N). ** $p < 0.01$; LSD = Least Significant Difference ($p < 0.05$).

The targeted analysis by LC-MS/MS allowed for characterizing the phenolic content in the different matrices (Table 3). A set of 34 standard analytes was selected for the identification and quantification of phenolic compounds (PhCs), belonging to the class of phenolic acids or flavonoids, on the basis of the retention time (tR) and molecular weight (MW) of each analyte. On the basis of their phenolic content, three sets of samples (seeds, sprouts, and wheatgrass) were discriminated. In total, 19 compounds were identified belonging to the classes of phenylethanoids (hydroxytyrosol—OH-Tyr), phenolic acids (hydrobenzoic acids: protocatechuic acid—PrCA, chlorogeninc acid—CGA, 4-hydroxybenzoic acid—4-OH-BA, vanillic acid—VA, syringic acid—SRA, ellagic acid— EA; hydrocinnamic acids: caffeic acid—CFA, *p*-coumaric acid—*p*-CUA, ferulic acid—FRA, sinapic acid—SNA), flavonols (aglycon: quercetin—Quer; glycoside: quercetin–hexoside— Quer-hexo, rutin—Rut), flavones (aglycon: luteolin—Lut, apigenin—Api, diosmetin—Dios; glycoside: orientin, Ori), and flavanols (aglycon: epigallocatechin—Epigall). For each detected phenolic compound, recoveries ranging from 65 to 89% in the three matrices (seeds, sprouts, and wheatgrass) were observed, further proving the extraction effectiveness. Indeed, the optimized extraction, coupled with a clean-up step (see Section 4.6.1 of Materials and Methods), allows reduction of the matrix effect (<14%) and remotion of interference compounds and then achieving a limit of quantification (LOQs) between 9.0×10^{-4} and 7.0×10^{-2} ng mg^{-1}, further proving the robustness of the method with precision and accuracy values included between ±10% near LOQs.

Table 3. Phenolic compounds (hydroxytyrosol—OH-Tyr, protocatechuic acid—PrCA, chlorogenic acid—CGA, 4-hydroxybenzoic acid—4-OH-BA, vanillic acid—VA, syringic acid—SRA, ellagic acid—EA, caffeic acid—CFA, *p*-coumaric acid—*p*-CUA, ferulic acid—FRA, sinapic acid—SNA, quercetin—Quer, quercetin-hexoside—Quer-hexo, rutin—Rut, luteolin—Lut, apigenin—Api, diosmetin—Dios, orientin—Ori, epigallocatechin—Epigall) determined by LC-MS/MS in seeds, sprouts, and wheatgrass extracts of two *Triticum aestivum* cultivars (Bologna, BL; Bora, BR) subjected to four different N fertilization schedules (N0: unfertilized control; N300: constantly well N fertilized throughout the growth cycle; N60-0: N fertilized only one month after sowing; N0-120: N fertilized only late at initial shoot elongation).

| Treatments | Phenylethanoid | Phenolic Acids | | | | | | | | | | | | Flavonoids | | | | | | |
	OH-Tyr	PrCA	CGA	4-OH-BA	VA	SRA	EA	CFA	*p*-CUA	FRA	SNA	Ori	Quer-hexo	Rut	Api	Lut	Quer	Epigall	Dios	Total
Seeds																				
BL																				
N0	-	45	11	20	2.3	0.4	1.4	1.7	235	4.6	4.7	8.1	0.92	38	0.007	0.112	0.05	7.0	9.3	390
N300	-	58	16	14	1.4	1.2	1.4	2.3	302	4.5	5.5	9.1	0.06	15	0.001	0.019	0.02	4.4	4.9	439
N60+0	-	39	12	12	1.6	0.9	2.4	2.0	590	4.6	3.6	14.4	0.03	15	0.000	0.003	0.27	4.7	8.8	712
N0-120	0.4	62	19	22	2.7	0.9	1.6	2.9	265	3.8	5.2	8.5	0.01	52	0.038	0.570	0.93	32.9	6.7	486
BR																				
N0	-	43	17	26	2.5	1.0	1.0	1.2	356	4.3	4.4	8.7	0.03	49	0.002	0.064	0.19	2.0	6.2	522
N300	-	34	12	60	4.5	2.4	2.0	2.1	328	5.1	8.3	15.0	0.58	58	0.007	0.158	0.40	35.3	4.9	574
N60+0	-	51	14	51	3.2	1.2	2.0	1.3	411	5.7	4.7	11.0	0.15	71	0.030	0.104	0.16	13.2	6.9	648
N0-120	-	15	8	16	1.3	0.5	0.2	1.3	186	4.6	2.6	10.3	0.11	51	0.023	0.574	1.86	39.1	3.7	343
F-test																				
CV	-	**	**	**	**	**	**	**	**	**	**	**	**	**	**	**	**	**	**	**
N	-	**	**	**	**	**	**	**	**	**	**	**	**	**	**	**	**	**	**	**
CV × N	-	**	**	**	**	**	**	**	**	**	**	**	**	**	**	**	**	**	**	**
LSD																				
CV	-	0.847	0.165	0.721	0.070	0.021	0.078	0.02	7.23	0.158	0.125	0.314	0.010	1.209	0.001	0.013	0.013	0.796	0.182	7.74
N	-	1.198	0.233	1.019	0.099	0.030	0.110	0.03	10.22	0.223	0.177	0.444	0.014	1.709	0.001	0.019	0.019	1.125	0.257	10.94
CV × N	-	1.694	0.330	1.441	0.139	0.042	0.155	0.05	14.46	0.316	0.251	0.628	0.020	2.417	0.002	0.027	0.027	1.591	0.364	15.47
Sprouts																				
BL																				
N0	79	137	35	1806	9.1	25	26	3.5	143	10	30	1748	0.05	24	0.041	0.11	15.2	9.9	3.0	4103
N300	120	88	25	2942	12.4	33	38	3.7	68	13	39	2454	1.47	18	0.095	0.18	2.5	39.7	3.7	5902
N60+0	25	90	28	3336	18.0	17	16	3.2	109	13	33	1577	1.03	24	0.089	0.27	3.6	14.2	7.0	5315
N0-120	82	68	11	5361	18.0	21	17	1.7	82	11	27	930	0.55	11	0.036	0.48	1.0	3.3	4.9	6651
BR																				
N0	222	121	20	2549	13.7	29	41	3.8	84	15	46	2364	0.38	24	0.051	0.10	6.4	18.5	3.2	5560
N300	97	64	8	3334	19.5	39	31	3.1	84	27	54	742	4.13	33	0.097	0.18	5.2	2.1	7.0	4555

Table 3. *Cont.*

Treatments	Phenylethanoid				Phenolic Acids									Flavonoids						
	OH-Tyr	PrCA	CGA	4-OH-BA	VA	SRA	EA	CFA	*p*-CUA	FRA	SNA	Ori	Quer-hexo	Rut	Api	Lut	Quer	Epigall	Dios	Total
N60+0	144	95	22	2820	15.1	28	43	4.7	186	16	46	1968	0.10	24	0.046	0.14	3.8	2.2	3.2	5422
N0-120	60	55	11	1754	9.2	21	23	3.2	73	11	30	809	0.12	23	0.073	0.78	0.5	5.5	5.0	2894
F-test																				
CV	**	**	**	**	*ns*	**	**	**	**	**	**	**	**	**	*ns*	**	**	**	*	**
N	**	**	**	**	**	**	**	**	**	**	**	**	**	**	**	**	**	**	**	**
CV × N	**	**	**	**	**	**	**	**	**	**	**	**	**	**	**	**	**	**	**	**
LSD																				
CV	5.56	5.33	0.398	41.81	0.088	0.320	0.416	0.033	2.192	0.189	0.227	4.93	0.264	0.650	0.002	0.012	0.735	2.93	0.072	44.6
N	7.86	7.54	0.564	59.12	0.124	0.453	0.589	0.046	3.101	0.268	0.320	6.98	0.373	0.920	0.003	0.017	1.039	4.14	0.101	63.0
CV × N	11.12	10.67	0.797	83.61	0.176	0.641	0.832	0.066	4.385	0.379	0.453	9.87	0.528	1.301	0.004	0.025	1.470	5.86	0.143	89.1
Wheatgrass																				
BL																				
N0	1.4	33	313	561	608	73	47	8.7	42	92	55	6035	0.7	643	0.10	22.9	3.4	-	10	8548
N300	0.8	33	204	643	527	103	53	8.1	121	69	56	3725	0.2	32	0.24	12.0	1.1	-	15	5602
N60+0	10.6	284	380	397	330	69	34	6.6	92	42	65	9160	10.0	118	0.24	2.6	1.9	-	10	11,011
N0-120	6.3	103	244	371	385	53	23	9.8	286	54	43	4655	12.2	547	0.14	19.5	9.2	-	13	6835
BR																				
N0	10.9	206	191	250	165	85	47	6.0	195	55	83	4242	3.6	246	0.14	2.3	75.4	-	11	5876
N300	10.0	71	91	543	162	174	79	8.7	63	79	90	4180	11.9	435	0.11	4.1	0.1	-	20	6022
N60+0	14.5	543	227	797	768	195	99	11.8	120	81	155	7774	9.2	1316	1.26	36.4	2.1	-	20	12,170
N0-120	9.1	95	61	857	356	178	81	3.6	65	41	101	5187	2.3	33	0.36	1.9	0.0	-	22	7094
F-test																				
CV	**	**	**	**	**	**	**	**	**	*ns*	**	**	*ns*	**	**	**	**	-	**	**
N	**	**	**	**	**	**	**	**	**	**	**	**	**	**	**	**	**	-	**	**
CV × N	**	**	**	**	**	**	**	**	**	**	**	**	**	**	**	**	**	-	**	**
LSD																				
CV	0.552	3.18	13.0	17.6	17.4	2.73	1.82	0.275	2.58	1.89	3.36	6.88	1.55	15.8	0.094	0.383	3.96	-	1.29	37.5
N	0.781	4.50	18.4	24.9	24.7	3.86	2.57	0.389	3.65	2.67	4.76	9.72	2.19	22.3	0.132	0.542	5.60	-	1.83	53.0
CV × N	1.104	6.36	26.0	35.2	34.9	5.45	3.64	0.550	5.16	3.77	6.73	13.75	3.09	31.5	0.187	0.766	7.92	-	2.58	75.0

Two-factor analysis of variance (ANOVA) at 5% level of probability: cultivar (CV); N fertilization schedules (N). * $p < 0.05$; ** $p < 0.01$; *ns* = not significant. LSD = Least Significant Difference ($p < 0.05$).

Of note, good correlations among methods employed for TPC analysis were observed. In detail, Pearson coefficients of 0.853 for HPLC-MS/MS vs. Folin–Ciocalteu (FC) and 0.872 for HPLC-MS/MS vs. AuNPs were reported. While a lower correlation was reported among HPLC-MS/MS and ABTS (R = 0.677). Hence—despite the different principle methods—the AuNPs assay, as well as the FC, returns a quantitative estimation of the TPC. Table 3 shows how the germination process sharply affected the amount of total polyphenols (as the sum of all the investigated molecules), inducing an increase by 9.8-fold and 15.4-fold for sprouts and wheatgrass, respectively (averaged over CV and N; seeds: 514 μg g^{-1} DW; sprouts: 5050 μg g^{-1} DW; wheatgrass: 7895 μg g^{-1} DW).

The trend of each PhC varied coherently—in terms of amount—according to the wheat growth phases (Table 3). In any case, representative phenolic compounds were identified for each seedling stage with significantly different AOC and belonging to different phenolic classes. Then, the phenolic patterns were affected during wheat development: phenolic acids, in detail, *p*-CUA, resulted in the most representative compound in seeds (65% of the total) and 4-OH-BA and Ori in sprouts (59% and 31%, respectively). In addition, for wheatgrass, Ori was the most representative polyphenol (71% of the total). Moreover, OH-Tyr was greatly abundant in sprouts and then detectable only after the germination process, while Epigall decreased to not-detectable amounts in wheatgrass (Table 3).

Differences among CV, for both seeds and wheatgrass, although significant, can be considered negligible; indeed, considering the sum of the single compounds, a more pronounced difference was observed only at the sprouts stage, with BL showing the highest value (5493 vs. 4607 μg g^{-1} DW, averaged over N treatments).

Among the N-treatments, N60-0 induced higher concentrations in seeds, which resulted in sprouts and, especially, wheatgrass richer in bioactive compounds (over CV: 680, 5368, and 11,591 μg g^{-1} DW for seeds, sprouts, and wheatgrass, respectively) (Table 3). In wheatgrass, N60-0 favoured a 60.7%, 66.4%, and 99.4% increase compared with N0, N0-120, and N300, respectively; the observed differences were principally attributable to the most representative compounds: *p*-CUA, Ori, and 4-OH-BA. The highest values for most of the investigated compounds were obtained in the wheatgrass of BR_N60-0 (Table 3).

2.4. Use of the Reflectance Spectroscopy as a Non-Destructive Alternative to Estimate Wheat-Seedlings Traits

From reflectance spectroscopy data of sprouts and wheatgrass, we selected 10 suitable indicators (VIs, listed in Table S1) related to pigment content and chemical composition of the crop.

Treatments under comparison did not result in significant differences for the selected VIs (data not shown). On the other hand, some VIs highlighted clear trends, showing correlations with the analytical data (Pearson's correlation coefficients at $p < 0.05$). To this purpose, principal component analysis (PCA) was independently performed on sprouts and wheatgrass in order to explore the response patterns and summarize the correlations among variables (i.e., chemical and physiological) as well as to evaluate the association between treatments and variables. The results are graphically displayed in a correlation biplot (Figures 3 and 4). In sprouts, the first and second principal components (PCs) explained 52.4% of the total data variability, while in wheatgrass, PC1 and PC2 captured 56.6% of the total data variability (Figures 3 and 4, respectively). As expected, the strongest correlations among VIs and chemical data were obtained in combination with pigments' concentrations and for the indices' red-edge NDVI (mNDVI) and modified red-edge ratio (mSR); VIs performed better at the sprout growth stage (Pearson's correlation coefficient of 0.72 and 0.74 for mNDVI and mSR, respectively; data not shown). Interestingly, in wheatgrass, the modified anthocyanin reflectance index (mARI) resulted in a clear tendency to be associated with some single polyphenols identified in LC-MS/MS: it showed a negative linear relationship with PrCA (−0.68) and Api (−0.67) (data not shown).

Both in sprouts and wheatgrass, BR_N60-0 was the treatment mostly related to the content of bioactive compounds (Figures 3 and 4).

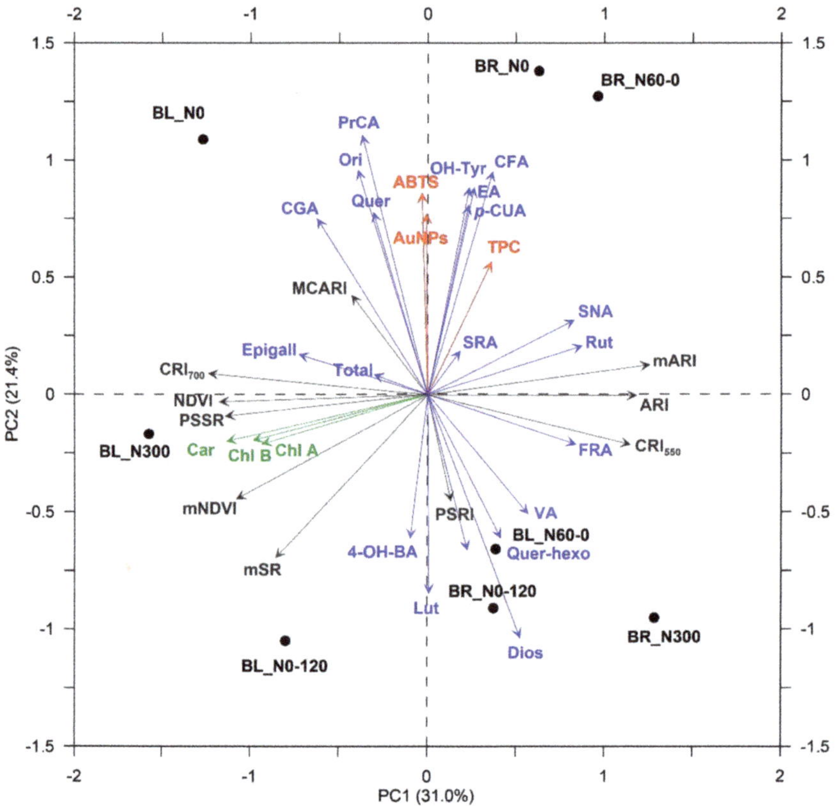

Figure 3. Two-dimensional correlation bi-plot from principal component analysis (PCA) performed on data observed in sprouts obtained from two *Triticum aestivum* cultivars (Bologna, BL; Bora, BR) subjected to four different N fertilization schedules (N0: unfertilized control; N300: constantly well N fertilized throughout the growth cycle; N60-0: N fertilized only one month after sowing; N0-120: N fertilized only late at initial shoot elongation). Symbols show the standardised scores on PC1 (x-axis) and PC2 (y-axis) for the eight treatments (BL_N0; BL_N300; BL_N60-0; BL_N0-120; BR_N0; BR_N300; BR_N60-0; BR_N0-120); vectors' coordinates represent the correlations between standardised variables (green group: pigments; red group: results from photometric analysis; blue group: phenolic profile from LC-MS/MS; grey group: reflectance-derivate vegetation indices—see text for labels) and PCs.

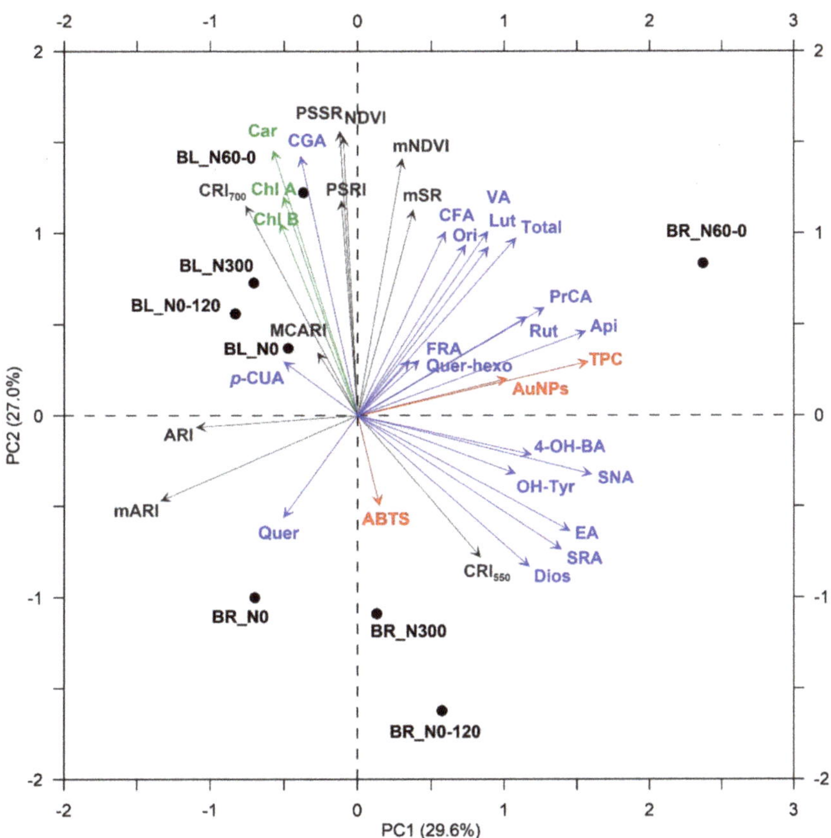

Figure 4. Two-dimensional correlation bi-plot from principal component analysis (PCA) performed on data observed in wheatgrass obtained from two *Triticum aestivum* cultivars (Bologna, BL; Bora, BR) subjected to four different N fertilization schedules (N0: unfertilized control; N300: constantly well N fertilized throughout the growth cycle; N60-0: N fertilized only one month after sowing; N0-120: N fertilized only late at initial shoot elongation). Symbols show the standardised scores on PC1 (x-axis) and PC2 (y-axis) for the eight treatments (BL_N0; BL_N300; BL_N60-0; BL_N0-120; BR_N0; BR_N300; BR_N60-0; BR_N0-120); vectors' coordinates represent the correlations between standardised variables (green group: pigments; red group: results from photometric analysis; blue group: phenolic profile from LC-MS/MS; grey group: reflectance-derivate vegetation indices—see text for labels) and PCs.

3. Discussion

As well known, germination leads to a deep modification of biochemical, nutritional, and sensorial traits of sprouted seeds [16,23]. However, the potential chemical composition of seedlings results from the combination of factors acting before, during, and after the actual germination process. In this study, we experienced that, beyond the genotype, the growing conditions endured by the mother plants affected the fitness of the next generation in terms of seedling vigour and, more interestingly, the nutritional characteristics of the offspring sprouts and wheatgrass obtained by the harvested seeds. We observed two clear and opposite transgenerational responses linked to specific traits—i.e., (i) morphological/growth traits and (ii) secondary metabolism compounds—towards full nitrogen availability or stress induced by N deficiency during the crop cycle.

Firstly, N rates applied to parent plants satisfying crop needs have positive effects on seedling vigour. Not only does N fertilization accelerate the germination process [24,25], but it also results in more effective growth of the germinated seeds, as was distinctly observable at the wheatgrass stage. Indeed, seedling vigour is related to seed size and protein concentration and, more strictly, to protein content per seed [12,26,27]. Moreover, N affects the wheat ovary size and, therefore, the size of the grain itself [10]. Seedlings of the treatments subjected to late N availability (i.e., N0-120) showed significantly greater growth than treatments with early N applications or no N applications (i.e., N60-0 and N0, respectively). As expected, the two wheat cultivars differed in growth parameters, given their different seed size (small grain for BL; large grain for BR) [10]. As previously observed [28], chlorophyll content in wheatgrass tissues (i.e., young seedlings) decreased under low parental N levels, probably due to the accumulation and/or reduction of specific proteins, following exposure to sub-optimal growth conditions [29,30]. Conversely, late N deficiency appeared to induce Car accumulation, as revealed by the N60-0 treatment. This is likely because carotenoids play a key antioxidant role against the stress caused by nutrient deficiency [31].

Secondly, as observed for Car, N deficiency acted as an environmental stressor (elicitor) whose effects on parentals were transmitted to the next generation, generally inducing the accumulation of secondary metabolites, such as the phenolic compounds, in the seeds [32]. However, in this study, we highlighted the performances only in the first stages of growth, up to 10 days from germination of the harvested seeds, and those transgenerational consequences translated into an enrichment of the nutritional quality of edible sprouts and wheatgrass. To our knowledge, this is the first study investigating the phytochemical modifications in sprouts and wheatgrass due to the maternal N environment manipulation. Benincasa et al. [8] underlined the effect of salt stress experienced by mother plants on rapeseed edible sprouts, highlighting a higher content of single phenolic acids and antioxidant activity in salted treatments. Secondary metabolites are widely recognized as key factors in stress tolerance. However, tracing a global phenolic pattern profile along the whole plant's development is still a challenge because the production of the phenolic compounds is strongly related to the growing environment and harvesting conditions as well as to the plant genotypes [33]. In our study, regardless of wheat cultivar, grains obtained from crops fertilized at early stages and then subjected to late N deficiency (i.e., N60-0) produced wheatgrass with a much higher phenolic content. The effect of N deficiency on seeds was already appreciable, characterized by higher total polyphenol content. Indeed, Stumpf et al. [32] proved that under N-deficiency, the production of carbon-rich secondary metabolites increased, according to the carbon–nitrogen balance theory. However, recent studies on wheat seeds showing the effect of N fertilization on some single phenolic compounds appear to be not univocal: Ma et al. [13] observed that *p*-CUA had a decreasing trend with increasing N levels; moreover they also found that FRA, *p*-CUA, and VA were improved when N fertilizer ranged between 180 kg N ha^{-1} and 300 kg N ha^{-1}; Stumpf et al. [32] reported that the content of soluble FRA decreased with high amount of N fertilizer. These differences might be a consequence of different cultivation systems and, especially, climatic conditions [3].

It is worth highlighting that the differences among N treatments registered in the various growth stages depended on the different contributions of specific individual PhCs. Indeed, during germination, besides an overall increase in secondary metabolites, a different phenolic pattern was observed. According to literature, the most dominant compounds of wheat—from seeds to wheatgrass—are the hydroxybenzoic acids (i.e., PrCA, 4-OH-BA, VA, and SRA) and hydroxycinammic acids (i.e., *p*-CA, FRA, and SNA) [33–35]. Furthermore, the glycosylate flavones (i.e., Ori and Rut) increased with germination. *p*-CA typically occurs in cereal seeds (i.e., oat, barley, wheat, and corn), mainly located in the pericarp [35,36], whereas, in sprouts, the most abundant compounds were 4-OH-BA, which are coupled to an emerging complex structure (i.e., Ori) typical of wheat germination and the growth step [37]. Later, in wheatgrass, a proportional increase in whole phenolic species with an interesting increase in glycosylate flavones was reported. It is known that a greater amount and variety of phytochemicals are observable after germination because biochemical and structural changes lead to a greater PhC bioaccessibility [1,37].

Photometric assays confirmed the main evolution trend of phytochemicals during seedling growth: the FC for the TPC evaluation, the ABTS for the antioxidant capacity (AOC) assessment, and the AuNPs-based assay. The latter relies on the direct formation of AuNPs—starting from a metal cation precursor (Au(III))—mediated by phenolic compounds then providing information on both the amount and intrinsic reducing ability/reactivity of polyphenols [38,39]. The AuNPs-based assay results were more coherent with FC results, with respect to ABTS, than providing information about TPC. However, the results obtained in terms of TPC and AOC on the progeny seedlings are apparently contradictory. It is important to point out that the reported results are strictly related to the intrinsic reactivity of PhCs. Indeed, a marked increase in TPC was reported compared to the AOC (at least one order of magnitude greater). The significantly lower values of AOC compared to TPC values could be ascribed to the observed phenolic pattern, which is characterized by low radical scavenging activity [38,39].

Vis-NIR reflectance spectroscopy, to assess in a rapid and non-destructive way the quality attributes of sprouts and wheatgrass shoots [40], gave pretty inconclusive results, with the exclusion of those related to photosynthetic pigments estimation [41]. This could be attributed to (i) the use of VIs rather than the full reflectance [42–44] or (ii) the effect of the anatomical and biochemical tissue traits (i.e., surface texture or thickness of cuticle, shape, and thickness of the palisade and spongy mesophyll) [45]. However, such investigations deserve to be deepened in future works, involving the preparation of dedicated experiments with a larger number of samples as well as various pre-processing and processing methods of the raw spectra.

PCA allowed for summarizing the differences among treatments, confirming individual results obtained on single traits. Transgenerational responses to N-stressed environments experienced by the mother plants can be observed even at the sprout growth stage, with N0 and N60-0 inducing the higher secondary metabolite accumulation. The effects were appreciable for single PhCs: Ori, Quer, and PrCA resulted in the most represented molecules in the offspring sprouts obtained from plants grown under N-stressful conditions prolonged throughout the crop cycle (e.g., N0). Interestingly, PCA separated CV and, especially, at the wheatgrass stage highlighted the greatest content in bioactive molecules to be associated with seedlings obtained from germinated larger seeds, i.e., seeds of BR plants exposed to N-deficiency in the later stages of crop cycles (e.g., N60-0). Seeds' dimensions (BR vs. BL) seemed to be also associated with the pigment content of both sprouts and wheatgrass.

4. Materials and Methods

4.1. The Field Trial Source of the Grains Used for Sprouts and Wheatgrass Production

The wheat (*Triticum aestivum* L.) grains were harvested in June 2018 from a field experiment carried out at the experimental station of the Department of Agricultural, Food and Environmental Sciences of the University of Perugia. The field was characterized by a clay–loam soil (clay: 33%; silt: 37%; sand: 30%) and 1.05 g kg^{-1} total N; grain sorghum was selected as the previous crop to deplete the soil from mineral N. In total, 640 mm of rainfall was recorded during crop cycle. Two bread wheat cultivars (CV) were used, having very different grain size: Bologna (BL, seed weight of 28–32 mg) and Bora (BR, seed weight of 45–50 mg), which had been already studied by Benincasa et al. [10] for other purposes. Both cultivars were subjected to four different N fertilization schedules (N), according to a split-plot design with four replicates, with the cultivar in the main plot and the N treatment in the sub-plot: (1) constantly well N fed (N300), i.e., fertilized with 300 kg N ha^{-1}, split into five applications of 60 kg N ha^{-1} each on 16 December (early tillering), 10 January and 12 February (tillering), 15 March (early shoot elongation), and 8 April (late shoot elongation); (2) N fed only very early (N60-0), i.e., fertilized with 60 kg N ha^{-1} on 16 December;

and (3) N fed only late (N0-120), i.e., fertilized with 120 kg N ha^{-1} on 15 March; (4) never N fed (N0). It is worth underlining that N300 represented a non-limiting N availability, N0 represented a constantly N deficient crop, and N0-120 was intended to cause an imbalance in N nutrition with early N deficiency and very high late N availability, whereas N60-0 was intended to cause an opposite imbalance in N nutrition, with a good early N availability and late N deficiency. It is also worth pinpointing that, in field grown rain-fed wheat, late soil N depletion after early N supply can be obtained only if the early N supply is moderate and the fall–winter rainfall is very high. Luckily, the fall of 2017 and winter of 2018 were very rainy, thus, the four fertilization treatments were effective and differences among them were very marked for both cultivars in terms of crop growth indices, leaf greenness, biomass accumulation, and grain yield and quality. Offspring seeds obtained from each treatment were mixed, confounding the four field replicates, and were then used for germination tests and sprouting.

4.2. Germination Trials

The germination test was performed in Petri dishes with 2 replicates of 100 seeds per treatment by laying seeds over Whatman paper wetted with 9 mL of distilled water. Seeds were incubated in a controlled temperature chamber at 20 °C in the dark. A seed was considered germinated when the roots measured at least 2 mm. The number of germinated seeds was recorded daily and used to calculate the total germination percentage (G) and the mean germination time (MGT). MGT was calculated as follows:

$$\text{MGT} = \frac{\sum(ni \times Di)}{N}$$

where *ni* represents the number of seeds newly germinated at time *i*, *Di* the corresponding day from the beginning of the germination test at time *i*, and *N* the number of total germinated seeds.

4.3. Sprouts and Wheatgrass Production

Grains were sprouted following the methodology of Benincasa et al. [1]. Seeds from the eight treatments (2 CV × 4 N) were incubated on plastic trays with distilled water. Treatments were laid down according to a completely randomized block design with four replicates (trays). Each tray contained 15 g of grains. In each tray, grains were positioned on filter paper laid over cotton wetted with distilled water, to guarantee constant water availability while preventing anoxia. Distilled water was periodically added to trays to restore initial tray weight, assuming that weight change was mainly due to water evaporation. The trays were placed at a light/dark regime of 16/8 h with a light intensity of 200 μmol photons m^{-2}·s^{-1}.

Sprouts were harvested 4 days after sowing (DAS), collecting the whole seedlings (shoot and roots). Wheatgrass was harvested 10 DAS, collecting only the shoots. Fresh (FW) and oven-dry weights (DW) and the lengths (L) of shoots were measured on a subsample of 10 individuals per replicate.

4.4. Chemicals and Stock Solutions

All the chemicals were of analytical reagent grade. Apigenin, caffeic acid, chlorogenic acid (3-O-caffeoylquinic acid), catechin, diosmetin (luteolin-4-methyl ether), epicatechin, (−)-epigallocatechin, (−)-epigallocatechin gallate, ferulic acid, gallic acid, hesperidin, hydroxytyrosol, 3-hydroxybenzoic acid, hyperoside (quercetin-3-d-galactoside), isoquercetin (quercetin-3-b-d-glucoside), isoxhanthohumol, kaempferol, luteolin, myricetin, naringenin, o-coumaric acid, orientin (luteolin-8-glucoside), oleuropein, p-coumaric acid, protocatechuic acid, quercetin, rosmarinic acid, rutin, sinapic acid, siringic acid, trans-cinnamic acid, tyrosol, vanillic acid, and xanthohumol were purchased from Sigma-Aldrich (St Louis, MO, USA). Methanol (MeOH), cetyltrimethylammonium chloride (CTAC; 25% in water), hydrogen tetracholoroaurate (HAuCl$_4$·3H$_2$O, 99.9%), 2,2-azino-bis(3-ethylbenzothiazoline-

6-sulphonic acid) (ABTS), sodium carbonate, Folin and Ciocalteu's reagent (FC), sodium phosphate monobasic monohydrate anhydrous, and sodium phosphate dibasic anhydrous were purchased from Sigma-Aldrich (St. Louis, MO, USA). Ultrapure water (H_2O), acetic acid, ethanol (EtOH), MeOH, and acetonitrile (ACN) were UPLC-MS grade and were purchased from VWR (Radnor, PA, USA).

The phenolic compounds' stock solutions were prepared in methanol at a concentration of 1.0×10^{-2} mol L^{-1} and stored at $-20\ ^\circ$C in the dark. Milli-Q water (18.2 MΩ) was used for all the experiments' and reagents' stock solutions preparation.

4.5. Pigments Evaluation

Shoots, from both sprouts and wheatgrass, were analysed for their leaf chlorophyll and carotenoid contents (Chl A, Chl B, and Car, respectively), following the method described by Lichtenthaler and Buschmann [46]. The results were expressed in μg g^{-1} of fresh weight (FW).

4.6. Phenolic Compounds Evaluation

4.6.1. Seeds, Sprouts, and Wheatgrass Extraction

To 100 mg of each sample was added 1 mL of methanol and acidified water (0.1% acetic acid) mixture (90:10 v/v). Then, the samples were extracted by means of a Precellys® Evolution homogenizer (Bertin Technologies SAS, Montigny-le-Bretonneux, France) at 6500 rpm with a 10 sec pause (3 times) for 60 sec and centrifuged at 10,000 rpm for 10 min at 4 $^\circ$C. The supernatant was collected, and the pellet was extracted again under the same conditions.

The supernatants were combined and subjected to a clean-up procedure by using a SPE Strata XL cartridge (330 mg, 1 mL) from Phenomenex (Torrance, CA, USA), according to Oliva et al. [47].

In brief, the SPE cartridge was activated with 1 mL of MeOH and then conditioned with 1 mL of a phosphate buffer mixture (50 mM) at pH 3:MeOH (90:10 v/v). Each extract was diluted in 1 mL of the conditioning solution and loaded onto the cartridge. The cartridge was then washed with 1 mL of acidified H_2O (pH 3) to remove the interferents. Finally, the analytes were eluted with 1 mL of MeOH, and the samples were injected into the HPLC-MS/MS system.

4.6.2. Phenolic Compounds Evaluation via HPLC–MS/MS Targeted Analysis

A Nexera XR HPLC system (Shimadzu, Tokyo, Japan) coupled to a 4500 Qtrap mass spectrometer (Sciex, Toronto, ON, Canada) equipped with a heated ESI source (V source) was used for the analysis, following Oliva et al. [47]. The different PCs were separated using an Excel 2 C18-PFP (10 cm $\times$ 2.1 mm ID) column from Advanced Chromatography Technologies (Aberdeen, UK) with 2 μm particles and safety protection. H_2O with 1% acetic acid was used as mobile phase (A) and ACN as phase (B). The injection volume was set to 6 μL and the flow rate was set to 0.300 mL min^{-1}. All analytes were detected in negative ionization with a capillary voltage of -4500, nebulizer gas (air) at 40 psi, and turbogas (nitrogen) at 40 psi and 500 $^\circ$C. Data collection and processing were performed with Analyst 1.7.2 software and quantification with Multiquant 3.0 software (Sciex).

4.6.3. Phenolic Compounds Evaluation via Photometric Assay

For the photometric assays, all the absorbance measurements were performed by JENWAY 6400 spectrophotometer from Barloworld Scientific (Staffordshire, UK).

Total Phenolic Content Evaluation

The total phenolic content (TPC) determination was carried out through the Folin–Ciocalteu (FC) photometric assay. In brief, 20 μL of properly diluted sample extract was mixed with 20 μL of FC reagent and stirred for 3 min with an orbital shaker (VDRL 711/CT orbital shaker from Asal, Florence, Italy). Then, 400 μL of 7.5% sodium carbonate followed by 550 μL of deionized water were added and stirred for 60 min, at room temperature, in the dark. Finally, the absorbance at 760 nm was recorded and evaluated against the blank (reaction mix without sample). The gallic acid was employed as a reference standard to calibrate the method.

Gold Nanoparticles Formation-Based Assay

Gold nanoparticles (AuNPs) based assay was used for the TPC evaluation, according to Della Pelle et al. [38]. The assay was performed in a final volume of 1 mL. In brief, to 910 μL of 100 mM phosphate buffer solution (pH 8.0), 20 μL of 25% CTAC, 50 μL of 20 mM HAuCl$_4$ solution, and 20 μL of properly diluted sample extract were subsequently added. Then, the reaction mix was stirred for 3 min with an orbital shaker (VDRL 711/CT orbital shaker from Asal, Florence, Italy) and incubated in a water bath at 45 °C for 10 min (720 D thermostat digital group water bath from Asal, Florence, Italy). Finally, the reaction was blocked at −20 °C for 10 min to allow measurements in series. The absorbance value at 540 nm was recorded and evaluated against the blank (reaction mix without sample). The gallic acid was employed as a reference standard to calibrate the method.

Phenolic's Antioxidant Capacity Evaluation

The antioxidant capacity (AOC) was evaluated with the ABTS method. The radical ABTS•+ reagent stock solution was prepared according to Re et al. [48] and stored at −20 °C. Before analysis, the ABTS•+ radical solution was promptly diluted in MeOH up to an absorbance value of 0.7 ± 0.05 (λ = 734 nm) and directly used for the assay. In brief, 20 μL of properly diluted sample extract was mixed with 980 μL of ABTS•+ and incubated 5 min at room temperature in the dark. Final absorbance at 734 nm was recorded against the control, prepared by adding MeOH instead of sample extract. The gallic acid was employed as a reference standard to calibrate the method.

4.7. Reflectance Measurements

Reflectance was recorded by contact with a portable non-imaging spectroradiometer (FieldSpec® 4 Hi-Res, ASD Inc., USA) in the range of 350–2500 nm, using an optical fibre contact probe (ASD Plant Probe; ASD Inc., Baltimore, MD, USA) with a 10 mm field of view and an integrated halogen reflector lamp, equipped with a leaf clip (ASD Leaf Clip; ASD Inc., Boulder, CO, USA). The Spectralon® panel was used as a white reference for calibration purposes; each sample scan represented an average of 20 reflectance spectra. Starting from reflectance data, 10 common spectral vegetation indices (VIs), based on two or more wavelength combinations, were calculated (listed in Table S1) [49–59]. We selected literature indices related to both pigment and phenolics contents [41,60], such as the normalized difference vegetation index (NDVI), the red-edge NDVI (mNDVI), the modified chlorophyll absorption ratio index (MCARI), the plant senescence reflectance index (PSRI), the modified red-edge ratio (mSR), the pigment specific simple ratio (PSSR), the carotenoid reflectance index-1 (CRI$_{550}$), the carotenoid reflectance index-2 (CRI$_{700}$), the anthocyanin reflectance index (ARI), and the modified anthocyanin reflectance index (mARI).

The measurements were carried out on sub-samples of three randomly-selected individuals (i.e., sprouts or wheatgrass) for each tray by placing them under the sensor, using the support structure (rotating head) of the clip; three measurements were performed for each replicate. The sub-samples were then stored at −20 °C until an analysis was performed on fresh basis, i.e., pigment content, or dry (frozen-dry) basis, i.e., phenolics.

4.8. Statistical Analysis

A two-way analysis of variance (ANOVA) was applied to test (F-test) the effects of the CV and N fertilization schedules, according to a complete randomized design. ANOVA assumptions were tested through graphical methods. Separation of the means was set at 5% ($p < 0.05$) level of significance by LSD test.

The PCA was performed on standardized data: variables, NDVI, mNDVI, MCARI, PSRI, mSR, PSSR, CRI_{550}, CRI_{700}, ARI, mARI, Chl A, Chl B, Car, TPC, ABTS, Au_NPs, OH-Tyr, PrCA, CGA, 4-OH-BA, VA, SRA, EA, CFA, *p*-CUA, FRA, SNA, Ori, Quer-Hexo, Rut, Api, Lut, Quer, Epigall, Dios, and Total; treatments were derived from the combination of two CV (BL and BR) and four fertilization schedules (N0, N300, N60-0, N0-120), BL_N0, BL_N300, BL_N60-0, BL_N120-0, BR_N0, BR_N300, BR_N60-0, and BR_N120-0.

All the statistical analyses were performed using the R software (version 4.0.2) [61].

5. Conclusions

Some general conclusions can be drawn from this work. One is that strong imbalances in N nutrition (i.e., early or late N deficiency) have little effect on seedling vigour, as compared to crops constantly well N fertilized throughout the growth cycle. Another piece of evidence is that vegetation indexes, based on spectrometric measurements, were not efficient in detecting the different seedlings' physiological and quality attributes. However, this outcome can be considered as preliminary and needs to be further ascertained. As far as edible sprouts and wheatgrass are concerned, our results confirm that their phenolic content and antioxidant activity are much higher than in ungerminated seeds, especially for wheatgrass. What is new here is that the N fertilization schedule applied to the mother plants can greatly affect the phenolic content of the offspring seedlings. In particular, grains obtained from crops fertilized at early stages and then subjected to late N deficiency produced wheatgrass with much higher phenolic content, and this was observed in both cultivars. Thus, we can conclude that late N deficiency represents a stressing condition which elicits the production of phenolic compounds. Of course, such an imbalance in crop nutrition is not desired but cannot be excluded a priori, in light of the need to reduce the economic and environmental costs of N fertilization and due to climate changes, which can result in unusual but possible heavy rainfall in spring.

Supplementary Materials: The following supporting information can be downloaded at: https://www.mdpi.com/article/10.3390/plants11152042/s1, Table S1: List of the reflectance vegetation indices used in this study. Indices are listed in order of their appearance in the main text.

Author Contributions: Conceptualization, P.B. and B.F.; methodology, P.B., B.F., G.T., A.G., E.O. and A.S.; software, A.G., B.F. and E.O.; validation, A.G., E.O. and A.S.; formal analysis, F.S., A.G., B.F., F.T., E.O. and A.S.; investigation, B.F., A.G., F.S., E.O. and A.S.; resources, P.B., F.S., A.G., M.S. and M.D.C.; data curation, B.F., A.G., E.O. and A.S.; writing—original draft preparation, P.B., B.F., A.G., F.S. and A.S.; writing—review and editing, P.B., B.F., A.G. and A.S.; supervision, P.B., M.S. and M.D.C.; project administration, P.B.; funding acquisition, P.B., M.D.C. and M.S. All authors have read and agreed to the published version of the manuscript.

Funding: The experiment was supported by the Project Ricerca di Base 2020 of the Department of Agricultural, Food and Environmental Sciences of the University of Perugia; Coordinator: Paolo Benincasa.

Institutional Review Board Statement: Not applicable.

Informed Consent Statement: Not applicable.

Data Availability Statement: Not applicable.

Acknowledgments: We gratefully acknowledge Silvano Locchi for his help in running the sprouting experiment.

Conflicts of Interest: The authors declare no conflict of interest.

References

1. Benincasa, P.; Galieni, A.; Manetta, A.C.; Pace, R.; Guiducci, M.; Pisante, M.; Stagnari, F. Phenolic compounds in grains, sprouts and wheatgrass of hulled and non-Hulled wheat species. *J. Sci. Food Agric.* **2015**, *95*, 1795–1803. [CrossRef] [PubMed]
2. Tufail, T.; Saeed, F.; Afzaal, M.; Suleria, H.A.R. Wheatgrass Juice: A nutritional weapon against various maladies. In *Human Health Benefits of Plant Bioactive Compounds: Potentials and Prospects*; Apple Academic Press: Palm Bay, FL, USA, 2019; pp. 245–253.

3.   Ma, D.; Wang, C.; Feng, J.; Xu, B. Wheat grain phenolics: A review on composition, bioactivity, and influencing factors. *J. Sci. Food Agric.* **2021**, *101*, 6167–6185. [CrossRef]
4.   Galieni, A.; Falcinelli, B.; Stagnari, F.; Datti, A.; Benincasa, P. Sprouts and microgreens: Trends, opportunities, and horizons for novel research. *Agronomy* **2020**, *10*, 1424. [CrossRef]
5.   Liu, H.; Kang, Y.; Zhao, X.; Liu, Y.; Zhang, X.; Zhang, S. Effects of elicitation on bioactive compounds and biological activities of sprouts. *J. Funct. Foods* **2019**, *53*, 136–145. [CrossRef]
6.   Puccinelli, M.; Malorgio, F.; Rosellini, I.; Pezzarossa, B. Production of selenium-Biofortified microgreens from selenium-Enriched seeds of basil. *J. Sci. Food Agric.* **2019**, *99*, 5601–5605. [CrossRef]
7.   Benincasa, P.; D'Amato, R.; Falcinelli, B.; Troni, E.; Fontanella, M.C.; Frusciante, S.; Guiducci, M.; Beone, G.M.; Businelli, D.; Diretto, G. Grain endogenous selenium and moderate salt stress work as synergic elicitors in the enrichment of bioactive compounds in maize sprouts. *Agronomy* **2020**, *10*, 735. [CrossRef]
8.   Benincasa, P.; Bravi, E.; Marconi, O.; Lutts, S.; Tosti, G.; Falcinelli, B. Transgenerational effects of salt stress imposed to rapeseed (*Brassica napus* var. *oleifera* Del.) plants involve greater phenolic content and antioxidant activity in the edible sprouts obtained from offspring seeds. *Plants* **2021**, *10*, 932. [CrossRef]
9.   Wijewardana, C.; Reddy, K.R.; Krutz, L.J.; Gao, W.; Bellaloui, N. Drought stress has transgenerational effects on soybean seed germination and seedling vigor. *PLoS ONE* **2019**, *14*, e0214977. [CrossRef] [PubMed]
10.   Benincasa, P.; Reale, L.; Tedeschini, E.; Ferri, V.; Cerri, M.; Ghitarrini, S.; Falcinelli, B.; Frenguelli, G.; Ferranti, F.; Ayano, B.E.; et al. The relationship between grain and ovary size in wheat: An analysis of contrasting grain weight cultivars under different growing conditions. *Field Crops Res.* **2017**, *210*, 175–182. [CrossRef]
11.   Zörb, C.; Ludewig, U.; Hawkesford, M.J. Perspective on wheat yield and quality with reduced nitrogen supply. *Trends Plant Sci.* **2018**, *23*, 1029–1037. [CrossRef] [PubMed]
12.   Wen, D.; Hou, H.; Meng, A.; Meng, J.; Xie, L.; Zhang, C. Rapid evaluation of seed vigor by the absolute content of protein in seed within the same crop. *Sci. Rep.* **2018**, *8*, 5569. [CrossRef]
13.   Ma, D.; Sun, D.; Li, Y.; Wang, C.; Xie, Y.; Guo, T. Effect of nitrogen fertilisation and irrigation on phenolic content, phenolic acid composition, and antioxidant activity of winter wheat grain. *J. Sci. Food Agric.* **2015**, *95*, 1039–1046. [CrossRef] [PubMed]
14.   Stumpf, B.; Yan, F.; Honermeier, B. Influence of nitrogen fertilization on yield and phenolic compounds in wheat grains (*Triticum aestivum* L. ssp. *aestivum*). *J. Plant Nutr. Soil Sci.* **2019**, *182*, 111–118. [CrossRef]
15.   Tian, W.; Zheng, Y.; Wang, W.; Wang, D.; Tilley, M.; Zhang, G.; Zhonghu, H.; Li, Y. A comprehensive review of wheat phytochemicals: From farm to fork and beyond. *Compr. Rev. Food Sci. Food Saf.* **2022**, *21*, 2274–2308. [CrossRef] [PubMed]
16.   Benincasa, P.; Falcinelli, B.; Lutts, S.; Stagnari, F.; Galieni, A. Sprouted grains: A comprehensive review. *Nutrients* **2019**, *11*, 421. [CrossRef]
17.   Saltveit, M.E. Synthesis and metabolism of phenolic compounds. *Fruit Veget. Phytochem. Chem. Hum. Health* **2017**, *2*, 115.
18.   Engert, N.; John, A.; Henning, W.; Honermeier, B. Effect of sprouting on the concentration of phenolic acids and antioxidative capacity in wheat cultivars (*Triticum aestivum* ssp. *aestivum* L.) in dependency of nitrogen fertilization. *J. Appl. Bot. Food Qual.* **2012**, *84*, 111–118.
19.   Wojdyło, A.; Nowicka, P.; Tkacz, K.; Turkiewicz, I.P. Sprouts vs. microgreens as novel functional foods: Variation of nutritional and phytochemical profiles and their in vitro bioactive properties. *Molecules* **2020**, *25*, 4648. [CrossRef] [PubMed]
20.   Calzone, A.; Cotrozzi, L.; Lorenzini, G.; Nali, C.; Pellegrini, E. Hyperspectral detection and monitoring of salt stress in pomegranate cultivars. *Agronomy* **2021**, *11*, 1038. [CrossRef]
21.   Verma, B.; Prasad, R.; Srivastava, P.K.; Yadav, S.A.; Singh, P.; Singh, R.K. Investigation of optimal vegetation indices for retrieval of leaf chlorophyll and leaf area index using enhanced learning algorithms. *Comput. Electron. Agric.* **2022**, *192*, 106581. [CrossRef]
22.   Sytar, O.; Zivcak, M.; Neugart, S.; Brestic, M. Assessment of hyperspectral indicators related to the content of phenolic compounds and multispectral fluorescence records in chicory leaves exposed to various light environments. *Plant Physiol. Biochem.* **2020**, *154*, 429–438. [CrossRef] [PubMed]
23.   Bianchi, G.; Falcinelli, B.; Tosti, G.; Bocci, L.; Benincasa, P. Taste quality traits and volatile profiles of sprouts and wheatgrass from hulled and non-Hulled *Triticum* species. *J. Food Biochem.* **2019**, *43*, e12869. [CrossRef] [PubMed]
24.   Farhadi, E.; Daneshyan, J.; Hamidi, A.; Rad, A.S.; Valadabadi, H. Effects of parent plant nutrition with different amounts of nitrogen and irrigation on seed vigor and some characteristics associated with hybrid 704 in kermanshah region. *J. Novel Appl. Sci.* **2014**, *3*, 551–556.
25.   Paneru, P.; Bhattachan, B.K.; Amgain, L.P.; Dhakal, S.; Yadav, B.P.; Gyawaly, P. Effect of mother plant nutrition on seed quality of wheat (*Triticum aestivum* L.) in central Terai region of Nepal. *Int. J. Appl. Sci. Biotechnol.* **2017**, *5*, 542–547. [CrossRef]
26.   Ries, S.K.; Everson, E.H. Protein content and seed size relationships with seedling vigor of wheat cultivars. *Agron. J.* **1973**, *65*, 884–886. [CrossRef]
27.   Ries, S.K.; Ayers, G.; Wert, V.; Everson, E.H. Variation in protein, size and seedling vigor with position of seed in heads of winter wheat cultivars. *Can. J. Plant Sci.* **1976**, *56*, 823–827. [CrossRef]
28.   Wen, D.; Xu, H.; He, M.; Zhang, C. Proteomic analysis of wheat seeds produced under different nitrogen levels before and after germination. *Food Chem.* **2021**, *340*, 127937. [CrossRef]
29.   Ci, D.; Jiang, D.; Wollenweber, B.; Dai, T.; Jing, Q.; Cao, W. Cadmium stress in wheat seedlings: Growth, cadmium accumulation and photosynthesis. *Acta Physiol. Plant* **2010**, *32*, 365–373. [CrossRef]

30. Wang, X.; Shan, X.; Wu, Y.; Su, S.; Li, S.; Liu, H.; Han, J.; Xue, C.; Yuan, Y. iTRAQ-Based quantitative proteomic analysis reveals new metabolic pathways responding to chilling stress in maize seedlings. *J. Proteom.* **2016**, *146*, 14–24. [CrossRef]

31. Moussa, I.D.; Chtourou, H.; Karray, F.; Sayadi, S.; Dhouib, A. Nitrogen or phosphorus repletion strategies for enhancing lipid or carotenoid production from *Tetraselmis marina*. *Biorese. Technol.* **2017**, *238*, 325–332. [CrossRef]

32. Stumpf, B.; Yan, F.; Honermeier, B. Nitrogen fertilization and maturity influence the phenolic concentration of wheat grain (*Triticum aestivum*). *J. Plant Nutr. Soil Sci.* **2015**, *178*, 118–125. [CrossRef]

33. Kaur, N.; Singh, B.; Kaur, A.; Yadav, M.P.; Singh, N.; Ahlawat, A.K.; Singh, A.M. Effect of growing conditions on proximate, mineral, amino acid, phenolic composition and antioxidant properties of wheatgrass from different wheat (*Triticum aestivum* L.) varieties. *Food Chem.* **2021**, *341*, 128201. [CrossRef] [PubMed]

34. Leoncini, E.; Prata, C.; Malaguti, M.; Marotti, I.; Segura-Carretero, A.; Catizone, P.; Dinelli, G.; Hrelia, S. Phytochemical profile and nutraceutical value of old and modern common wheat cultivars. *PLoS ONE* **2012**, *7*, e45997. [CrossRef] [PubMed]

35. Călinoiu, L.F.; Vodnar, D.C. Whole grains and phenolic acids: A review on bioactivity, functionality, health benefits and bioavailability. *Nutrients* **2018**, *10*, 1615. [CrossRef]

36. Boz, H. p-Coumaric acid in cereals: Presence, antioxidant and antimicrobial effects. *Int. J. Food Sci. Technol.* **2015**, *50*, 2323–2328. [CrossRef]

37. Tomé-Sánchez, I.; Martín-Diana, A.B.; Peñas, E.; Bautista-Expósito, S.; Frias, J.; Rico, D.; González-Maillo, L.; Martinez-Villaluenga, C. Soluble phenolic composition tailored by germination conditions accompany antioxidant and anti-Inflammatory properties of wheat. *Antioxidants* **2020**, *9*, 426. [CrossRef]

38. Della Pelle, F.; Scroccarello, A.; Sergi, M.; Mascini, M.; Del Carlo, M.; Compagnone, D. Simple and rapid silver nanoparticles based antioxidant capacity assays: Reactivity study for phenolic compounds. *Food Chem.* **2018**, *256*, 342–349. [CrossRef]

39. Scroccarello, A.; Della Pelle, F.; Neri, L.; Pittia, P.; Compagnone, D. Silver and gold nanoparticles based colorimetric assays for the determination of sugars and polyphenols in apples. *Food Res. Int.* **2019**, *119*, 359–368. [CrossRef]

40. Kumar, S.; Singh, R.; Dhanani, T. Rapid estimation of bioactive phytochemicals in vegetables and fruits using near infrared reflectance spectroscopy. In *Fruit and Vegetable Phytochemicals: Chemistry and Human Health*, 2nd ed.; Wiley Blackwell: Hoboken, NJ, USA, 2017; pp. 781–802.

41. Sytar, O.; Brücková, K.; Kovar, M.; Živčák, M.; Hemmerich, I.; Brestič, M. Nondestructive detection and biochemical quantification of buckwheat leaves using visible (VIS) and near-infrared (NIR) hyperspectral reflectance imaging. *J. Cent. Eur. Agric.* **2017**, *18*, 864–878. [CrossRef]

42. Xiong, C.; Liu, C.; Pan, W.; Ma, F.; Xiong, C.; Qi, L.; Chen, F.; Lu, X.; Yang, J.; Zheng, L. Non-destructive determination of total polyphenols content and classification of storage periods of Iron Buddha tea using multispectral imaging system. *Food Chem.* **2015**, *176*, 130–136. [CrossRef]

43. Zhang, G.; Li, P.; Zhang, W.; Zhao, J. Analysis of multiple soybean phytonutrients by near-infrared reflectance spectroscopy. *Anal. Bioanal. Chem.* **2017**, *409*, 3515–3525. [CrossRef]

44. García-García, M.D.C.; Martín-Expósito, E.; Font, I.; Martínez-García, B.D.C.; Fernández, J.A.; Valenzuela, J.L.; Gómez, P.; Río-Celestino, M.D. Determination of quality parameters in mangetout (*Pisum sativum* L. ssp. *arvense*) by using Vis/Near-Infrared Reflectance Spectroscopy. *Sensors* **2022**, *22*, 4113. [CrossRef] [PubMed]

45. Galieni, A.; D'Ascenzo, N.; Stagnari, F.; Pagnani, G.; Xie, Q.; Pisante, M. Past and future of plant stress detection: An overview from remote sensing to positron emission tomography. *Front. Plant Sci.* **2021**, *11*, 1975. [CrossRef] [PubMed]

46. Lichtenthaler, H.K.; Buschmann, C. Extraction of phtosynthetic tissues: Chlorophylls and carotenoids. *Curr. Protoc. Food Anal. Chem.* **2001**, *1*, F4.2.1–F4.2.6. [CrossRef]

47. Oliva, E.; Viteritti, E.; Fanti, F.; Eugelio, F.; Pepe, A.; Palmieri, S.; Sergi, M.; Compagnone, D. Targeted and semi-untargeted determination of phenolic compounds in plant matrices by high performance liquid chromatography-tandem mass spectrometry. *J. Chromatogr. A* **2021**, *1651*, 462315. [CrossRef] [PubMed]

48. Re, R.; Pellegrini, N.; Proteggente, A.; Pannala, A.; Yang, M.; Rice-Evans, C. Antioxidant activity applying an improved ABTS radical cation decolorization assay. *Free Radic. Biol. Med.* **1999**, *26*, 1231–1237. [CrossRef]

49. Rouse, J.W.; Haas, R.H.; Schell, J.A. Monitoring vegetation systems in the Great Plains with ERTS. *NASA Spec. Publ.* **1974**, *351*, 309.

50. Gitelson, A.A.; Merzlyak, M.N. Quantitative estimation of chlorophyll-a using reflectance spectra-experiments with autumn chestnut and maple leaves. *J. Photochem. Photobiol. B.* **1994**, *22*, 247–252. [CrossRef]

51. Gamon, J.A.; Surfus, J.S. Assessing leaf pigment content and activity with a reflectometer. *New Phytol.* **1999**, *143*, 105–117. [CrossRef]

52. Datt, B. Visible/near infrared reflectance and chlorophyll content in Eucalyptus leaves. *Int. J. Remote Sens.* **1999**, *20*, 2741–2759. [CrossRef]

53. Sims, D.A.; Gamon, J.A. Relationships between leaf pigment content and spectral reflectance across a wide range of species, leaf structures, and developmental stages. *Remote Sens. Environ.* **2002**, *81*, 337–354. [CrossRef]

54. Daughtry, C.S.T.; Walthall, C.L.; Kim, M.S.; de Colstoun, E.B.; McMurtrey, J.E. Estimating corn leaf chlorophyll concentration from leaf and canopy reflectance. *Remote Sens. Environ.* **2000**, *74*, 229–239. [CrossRef]

55. Merzlyak, M.N.; Gitelson, A.A.; Chivkunova, O.B.; Rakitin, V.Y.U. Non-destructive optical detection of pigment changes during leaf senescence and fruit ripening. *Physiol. Plant.* **1999**, *106*, 135–141. [CrossRef]

56. Blackburn, G.A. Quantifying chlorophylls and carotenoids at leaf and canopy scales: An evaluation of some hyperspectral approaches. *Remote Sens. Environ.* **1998**, *66*, 273–285. [CrossRef]
57. Gitelson, A.A.; Zur, Y.; Chivkunova, O.B.; Merzlyak, M.N. Assessing carotenoid content in plant leaves with reflectance spectroscopy. *Photochem. Photobiol.* **2002**, *75*, 272–281. [CrossRef]
58. Gitelson, A.A.; Merzlyak, M.N.; Chivkunova, O.B. Optical properties and nondestructive estimation of anthocyanin content in plant leaves. *Photochem. Photobiol.* **2001**, *74*, 38–45. [CrossRef]
59. Gitelson, A.A.; Keydan, G.P.; Merzlyak, M.N. Three-band model for noninvasive estimation of chlorophyll, carotenoids, and anthocyanin contents in higher plant leaves. *Geophys. Res. Lett.* **2006**, *33*, L11402. [CrossRef]
60. Ustin, S.L.; Gitelson, A.A.; Jacquemoud, S.; Schaepman, M.; Asner, G.P.; Gamon, J.A.; Zarco-Tejada, P. Retrieval of foliar information about plant pigment systems from high resolution spectroscopy. *Remote Sens. Environ.* **2009**, *113*, S67–S77. [CrossRef]
61. R Core Team. R: A Language and Environment of Statistical Computing. R Foundation. 2020. Available online: https://www.rproject.org (accessed on 8 July 2022).

Article

In Vitro Oxidative Stress Threatening Maize Pollen Germination and Cytosolic Ca²⁺ Can Be Mitigated by Extracts of Emmer Wheatgrass Biofortified with Selenium

Alberto Marco Del Pino, Beatrice Falcinelli, Roberto D'Amato, Daniela Businelli, Paolo Benincasa * and Carlo Alberto Palmerini

Department of Agricultural, Food, and Environmental Sciences, University of Perugia, 06121 Perugia, Italy; alberto.delpino@unipg.it (A.M.D.P.); beatricefalcinelli90@gmail.com (B.F.); roberto.damato@unipg.it (R.D.); daniela.businelli@unipg.it (D.B.); carlo.palmerini@unipg.it (C.A.P.)
* Correspondence: paolo.benincasa@unipg.it

Abstract: In this work, we studied the effects of in vitro oxidative stress applied by H_2O_2 to maize pollen germination and cytosolic Ca^{2+}, taken as an experimental model to test the biological activity of extracts of emmer (*Triticum turgidum* L. spp. *dicoccum* (Schrank ex Shubler) Thell.) wheatgrass obtained from grains sprouted with distilled water, or salinity (50 mM) or selenium (45 mg L^{-1} of Na_2SeO_3). Wheatgrass extracts were obtained in two ways: by direct extraction in methanol, which represented the free phenolic fraction of extracts (Ef), and by residual content after alkaline digestion, which made it possible to obtain extracts with the bound fraction (Eb). Comparative tests on maize pollen were carried out by differently combining H_2O_2 and either wheatgrass extracts or pure phenolic acids (4-HO benzoic, caffeic, p-coumaric and salicylic). The cytosolic Ca^{2+} of maize pollen was influenced by either H_2O_2 or pure phenolic acids or Ef, but not by Eb. The negative effect of H_2O_2 on maize pollen germination and cytosolic Ca^{2+} was mitigated by Ef and, slightly, by Eb. The extent of the biological response of Ef depended on the sprouting conditions (i.e., distilled water, salinity or selenium). The extracts of Se-biofortified wheatgrass were the most effective in counteracting the oxidative stress.

Keywords: *Triticum dicoccum*; sprouting; salinity; calcium homeostasis; phenolic acid; hydrogen peroxide

Citation: Del Pino, A.M.; Falcinelli, B.; D'Amato, R.; Businelli, D.; Benincasa, P.; Palmerini, C.A. In Vitro Oxidative Stress Threatening Maize Pollen Germination and Cytosolic Ca²⁺ Can Be Mitigated by Extracts of Emmer Wheatgrass Biofortified with Selenium. *Plants* **2022**, *11*, 859. https://doi.org/10.3390/plants11070859

Academic Editor: Filippo Maggi

Received: 26 February 2022
Accepted: 22 March 2022
Published: 24 March 2022

Publisher's Note: MDPI stays neutral with regard to jurisdictional claims in published maps and institutional affiliations.

1. Introduction

Wheatgrass (1–2 weeks old seedlings of *Graminaceae* species) is recognized as a great source of phytochemicals, i.e., the secondary metabolites of plants having antioxidant activity and related benefits in human health [1,2]. Wheatgrass from *Triticum* species has been extensively studied, and recent trends on healthy diets have been promoting the rediscovery of ancient species, often grown in low-input systems and organic farming [3]. In particular, emmer (*Triticum turgidum* L. spp. *dicoccum* (Schrank ex Schübler) Thell.) has been found to be rich in phenolic compounds, especially phenolic acids (PAs) [4,5], which would also contribute to the abiotic stress tolerance of this species [6,7].

Phytochemicals may have complementary and/or overlapping mechanisms of action [8] and this may explain why the biological activity of wheatgrass is not easy to study and may be different from that expected from each purified molecule it contains. Moreover, phenolic compounds, similarly to other phytochemicals, exist in either free forms or bound forms (covalently conjugated through ester bonds to cell wall components such as cellulose, pectin and polysaccharides), which have different fates and healthy properties after ingestion [9,10].

A body of recent literature demonstrates that the production of phytochemicals in sprouts and wheatgrass of many species can be boosted by appropriate elicitation techniques [1,3,11], with changes also occurring in the proportion between free and bound phenolics. As far as emmer sprouts and wheatgrass are concerned, the application of moderate salt stress was found to increase the total content of phenolic compounds, especially the free ones [6]. On the other hand, bio-fortification with selenium (Se), which has not been tested yet on emmer wheatgrass, has been found to increase phenolic acids and antioxidant activity in sprouts of maize [12] and rice [13], confirming that Se is a powerful protecting agent against abiotic stresses [14].

Although sprouts and wheatgrass of many cereal species have been characterized for the kinds and amounts of phenolic compounds, very few studies have dealt with testing the effects of these matrices on biological systems. In this regard, the measurement of cytosolic Ca^{2+} and germination of maize pollen has proven to be an effective, simple and cheap tool for evaluating the effect of a plant matrix in a biological system [15,16]. In fact, calcium (Ca^{2+}) plays an important role in the signal transduction, growth and development of plants [17,18]. Ca^{2+} homeostasis is maintained by keeping the cytosolic ion concentration below 0.1 μM. When cells are stimulated, raising cytosolic Ca^{2+} levels above 200 nM, they activate a molecular signal to trigger downstream responses [19–21]. Furthermore, pollen represents a good experimental model, because it can be easily labeled with the FURA 2AM fluorescent probe and kept in suspension, which is a strategy that is not always possible with plant cells. This model is also useful to evaluate the occurrence of oxidative stress, since the increase in reactive oxygen species (ROS) in the cells would alter molecular signals including cytosolic Ca^{2+} [22–24]. Oxidative stress is a harmful process that can negatively affect several cellular structures, such as membranes, lipids, proteins, lipoproteins and DNA, and this is the case for both plant [25] and animal cells [26]. With regard to the study of oxidative stress, a body of literature reports the use of hydrogen peroxide (H_2O_2) in in vitro experiments. This is because, among the three primary ROS (superoxide anion, hydroxyl radical and hydrogen peroxide), only H_2O_2 has a half-life long enough (few seconds) to be used for inducing oxidative stress in vitro [27,28]. Recently, Del Pino et al. [16] demonstrated that extracts of emmer wheatgrass grown with distilled water, or in the presence of salinity (i.e., 50 mM) or Se (45 mg L^{-1}), affected cytosolic Ca^{2+} and maize pollen germination. In that case, maize pollen was germinated in optimal conditions and the effects of emmer extracts were evaluated considering only the extracts of free phenolic compounds (i.e., those obtained by extraction with methanol).

The present work follows up the work by Del Pino et al. [16], by using the same biological model (maize pollen germination and cytosolic Ca^{2+}) to test the effect of either free or bound extracts of emmer wheatgrass grown with distilled water, or salinity or selenium, and to evaluate the protective effect against the oxidative stress caused in pollen by the application of H_2O_2.

2. Results

2.1. Total Selenium Content in Extracts of Emmer Wheatgrass

The Se content in Ef_{Se} and Eb_{Se} was 10 and 2.4 times higher, respectively, than in Ef_c, Ef_s, Eb_c and Eb_s (Table 1).

Table 1. Total selenium content in the free and bound extracts of emmer wheatgrass grown with distilled water as a control (Ef_c), or in the presence of salinity as NaCl 50 mM (Ef_s) or selenium as 45 mg L^{-1} of Na_2SeO_3 (Ef_{Se}). All analyses were performed in triplicate.

Free Extracts	Se Content (ppb)	Bound Extracts	Se Content (ppb)
Efc	108 ± 10	Ebc	105 ± 8
Efs	111 ± 13	Ebs	109 ± 11
EfSe	1155 ± 25	EbSe	240 ± 15

2.2. Cytosolic Ca^{2+} of Maize Pollen

The concentration of cytosolic Ca^{2+} of pollen $[Ca^{2+}]_{cp}$ increased with each kind of free extract, both in the absence and in the presence of $CaCl_2$ in the incubation medium;

however, Ef$_s$ caused the greatest increase (Figure 1). On the contrary, none of the bound extracts affected $[Ca^{2+}]_{cp}$ (data not shown).

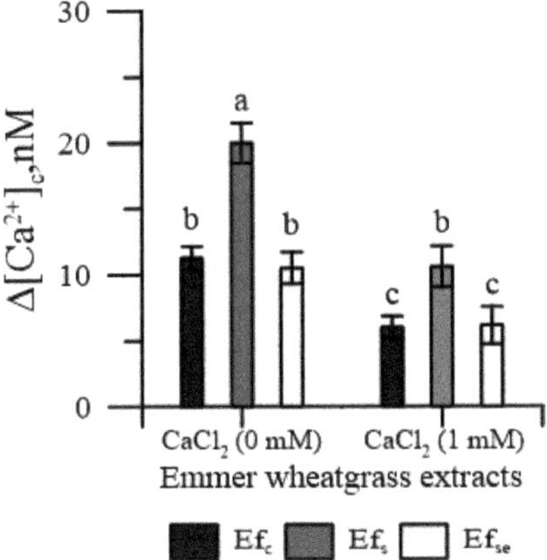

Figure 1. Effects of 50 mg of free extracts (Ef) of emmer wheatgrass obtained with distilled water as a control (Ef$_c$), or with NaCl (Ef$_s$) 50 mM or Na$_2$SeO$_3$ 0.45 mg L^{-1} (Ef$_{Se}$), on the cytosolic Ca^{2+} of maize pollen. The measurements were carried out in the absence of Ca^{2+} and in the presence of Ca^{2+} (1 mM CaCl$_2$) in the incubation medium. Data are expressed as means ± SEM from four independent tests. Different letters indicate significant differences for $p < 0.05$.

In the absence of CaCl$_2$ in the incubation medium, salicylic and 4-HO benzoic acids increased $[Ca^{2+}]_{cp}$, while coumaric and caffeic acids reduced it (Figure 2). With the addition of CaCl$_2$ in the incubation medium, an increase in $[Ca^{2+}]_{cp}$ was observed with all PAs, although it was slighter for p-coumaric and caffeic acids.

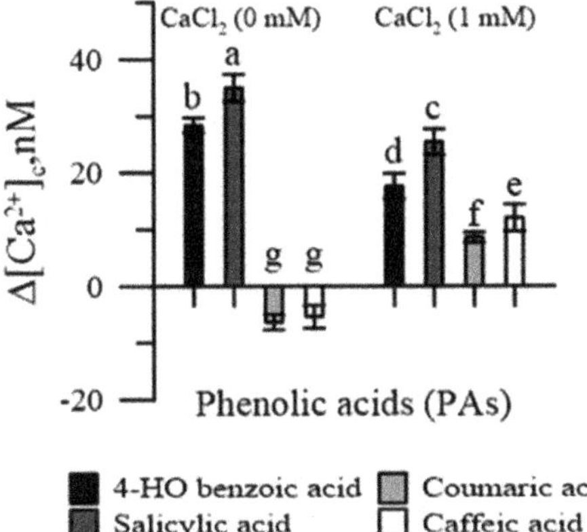

Figure 2. Effects of 4-HO benzoic, caffeic, p-coumaric and salicylic acids on cytosolic Ca^{2+} of maize pollen, in Ca^{2+}-free conditions and in the presence of CaCl$_2$. Data are expressed as means ± SEM from five independent tests. Different letters indicate significant differences for $p < 0.05$.

Hydrogen peroxide increased $[Ca^{2+}]_{cp}$; the higher the dose, the higher the increase (Figure 3).

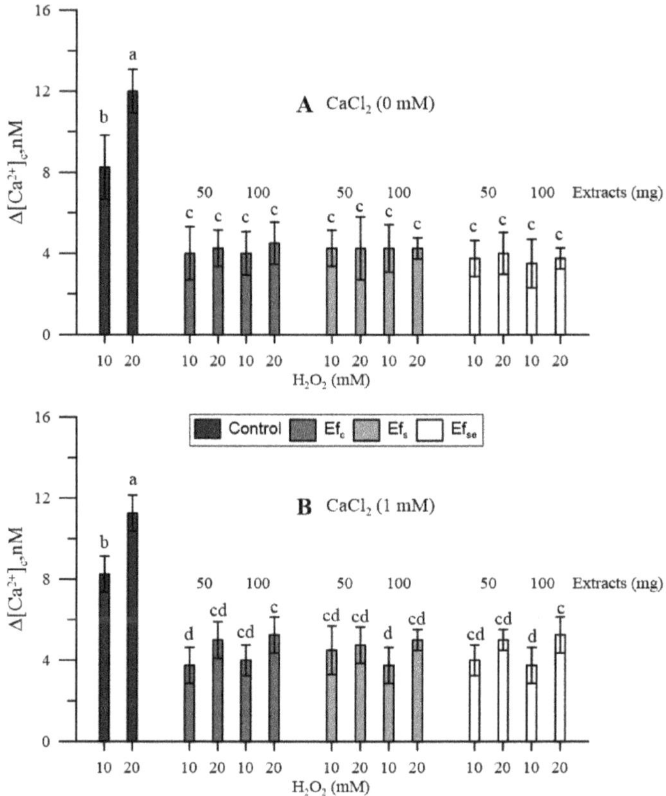

Figure 3. Effects of H$_2$O$_2$ (10 and 20 mM) on the cytosolic Ca^{2+} of maize pollen pre-treated with 50 or 100 mg of free extracts of emmer wheatgrass grown with distilled water as a control (Ef$_c$), or in the presence of salinity as NaCl 50 mM (Ef$_s$) or selenium as 45 mg L^{-1} of Na$_2$SeO$_3$ (Ef$_{Se}$). Maize pollen was (**A**) in the absence or (**B**) in the presence of 1 mM CaCl$_2$ in the growing medium. All analyses were performed in triplicate. Data are expressed as means ± SEM from four independent tests. Different letters indicate significant differences for $p < 0.05$.

Since the bound extracts had not affected the [Ca^{2+}]$_{cp}$, only free extracts (Ef) were tested for their effect against oxidative conditions. All three types of free extracts antagonized the effects of H$_2$O$_2$ on cytosolic Ca^{2+}. The effect did not substantially change in the absence (Figure 3A) or presence (Figure 3B) of CaCl$_2$ in the incubation medium, although, in the presence of CaCl$_2$, the highest H$_2$O$_2$ caused further and generally non-significant increases in [Ca^{2+}]$_{cp}$ (Figure 3B).

2.3. Germination of Maize Pollen

Both Ef and Eb reduced the germination rate of maize pollen. The magnitude of this effect was different between treatments: (Ef$_s$ and Eb$_s$) > Eb$_{Se}$ > (Ef$_c$ and Eb$_c$) > Ef$_{Se}$ (Figure 4).

Considering the Ef and Eb groups, the lowest inhibiting effect on pollen germination was caused by Ef$_{Se}$ (−25%) and Eb$_{Se}$ (−45%). No substantial dose-dependent effect was observed in both Ef and Eb.

The p-coumaric and caffeic acid severely depressed pollen germination (−81% and −76%, respectively), while the effect was less marked with 4-HO benzoic acid (−32%) and salicylic acid (−15%) (Figure 5).

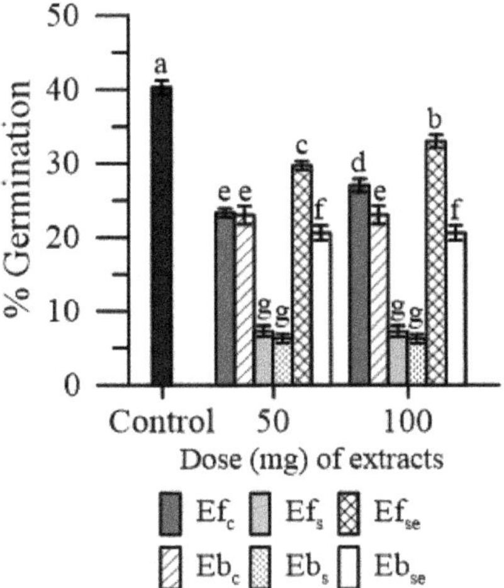

Figure 4. Germination of maize pollen in the presence of 50 or 100 mg of free (Ef) and bound (Eb) extracts of emmer wheatgrass grown with distilled water as a control (Ef$_c$, Eb$_c$), or in the presence of salinity as NaCl 50 mM (Ef$_s$, Eb$_s$) or selenium as 45 mg L^{-1} of Na$_2$SeO$_3$ (Ef$_{Se}$, Eb$_{Se}$). Data are expressed as means ± SEM from five independent tests. Different letters indicate significant differences for $p < 0.05$.

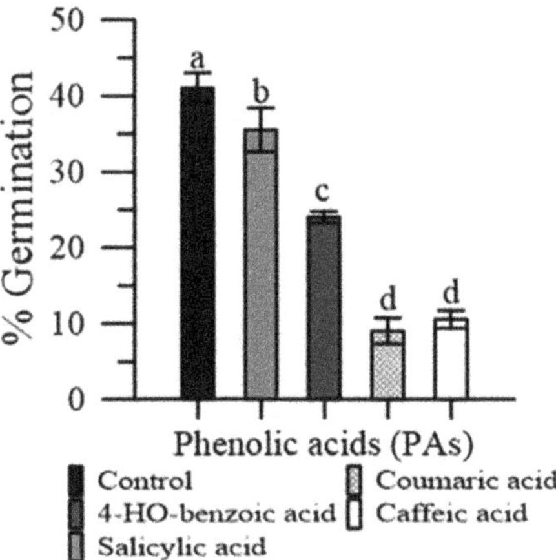

Figure 5. Germination of maize pollen grains in the presence of pure PAs: 4-HO benzoic, caffeic, p-coumaric and salicylic. Data are expressed as means ± SEM from four independent tests. Different letters indicate significant differences for $p < 0.05$.

In the pollen that was not pre-treated with wheatgrass extracts, the oxidative stress induced, with H$_2$O$_2$, a marked reduction in the germination rate (−78% and −92% with 10 and 20 mM H$_2$O$_2$, respectively) (Figure 6).

A similar situation was observed in the case of pollen pre-treated with extracts of wheatgrass grown in distilled water, while the situation was even worse in the case of wheatgrass grown under salinity. On the contrary, extracts of wheatgrass biofortified with Se allowed an appreciable recovery of the germination percentage, especially in the case of free extracts (Ef$_{Se}$) and milder oxidative stress (10 mM H$_2$O$_2$). For all three kinds of wheatgrass, free extracts allowed better germination performances that bound extracts. No relevant dose-dependent effects were observed for both free and bound extracts.

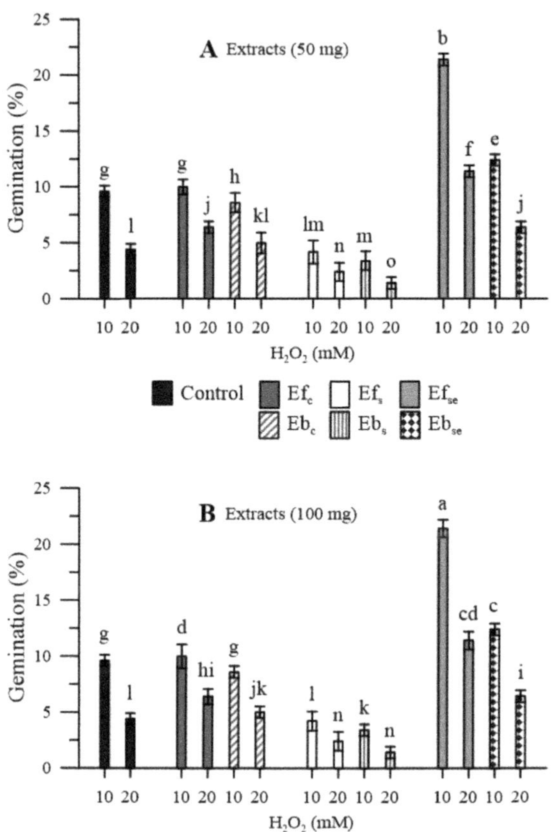

Figure 6. Germination of maize pollen during oxidative stress induced by 10 and 20 mM of H_2O_2 as affected by pre-treating the pollen with (**A**) 50 mg or (**B**) 100 mg of free (Ef) and bound (Eb) extracts of emmer wheatgrass grown with distilled water as a control (Ef_c, Eb_c) or in the presence of salinity as NaCl 50 mM (Ef_s, Eb_s) or selenium as 45 mg L^{-1} of Na_2SeO_3 (Ef_{Se}, Eb_{Se}). Data are expressed as means ± SEM from five independent tests. Different letters indicate significant differences for $p < 0.05$.

3. Discussion

The high phytochemical content and antioxidant activity of sprouts and wheatgrass are well documented, whereas the supposed benefits for living organisms have rarely been ascertained. Maize pollen germination and cytosolic Ca^{2+} were used in this work as a model to evaluate the actual effect of emmer wheatgrass and its role in mitigating the oxidative stress caused by H_2O_2. Of course, evidence obtained with this model cannot be used to speculate on the effect of wheatgrass on other kind of cells, least of all animal cells. Anyway, the use of this model represents a means to observe integrated effects of the wheatgrass matrix instead of focusing on single compounds. The results obtained in this experiment give rise to some reasoning.

In the absence of oxidative stress, all three types of free extracts of emmer wheatgrass (i.e., either obtained with distilled water, or salinity or selenium) had a negative effect on both the cytosolic Ca^{2+} and germination of maize pollen (Figures 1 and 4), while bound extracts affected only pollen germination. Moreover, the wheatgrass obtained under salinity had the most negative effect (Figure 1).

It is not easy to explain why cytosolic Ca^{2+} was affected only by free extracts, while pollen germination was affected by both free and bound extracts. Assuming that phenolic compounds of extracts might have a role in pollen performances, it is worth pointing out that bound forms are the most represented in emmer wheatgrass [4–6] but, differently from free phenolics, they are not readily available to exert their activity.

Moreover, other undetected phytochemicals could take over pollen germination, as this represents a complex biological event to which many effectors, in addition to cytosolic Ca^{2+}, may play a role [22,29,30]. We only examined changes in cytosolic Ca^{2+} because germination is activated by Ca^{2+} signals [29,30]. The specific analytical methods used for the identification of PAs in free and bound extracts could not detect the presence of other active molecules [31]. Finally, the severity of the method used to extract bound phenolics might further explain the low activity of these forms [10].

We actually tested four of the most represented PAs in emmer wheatgrass (Figures 2 and 5), two hydroxy-cinnamic (caffeic and p-coumaric) and two hydroxy-benzoic (4-HO benzoic and salicylic) [4–6]. All the four PAs perturbed the cytosolic Ca^{2+} homeostasis (Figure 2), the two hydroxybenzoic acids by having an agonist activity, and the two hydroxycinnamic by having a chelating activity. This would suggest that these PAs might be responsible, among other compounds, for the effect of free extracts. It is worth noting that in the presence of external $CaCl_2$ 1 mM, the concentration of cytosolic Ca^{2+} was lower than in the absence of external $CaCl_2$. This was expected, because the entrance of Ca^{2+} from outside the cell depends on the Ca^{2+} depletion in intracellular stores. Since, in normal conditions, the Ca^{2+} concentration in the cytosol is in the order of nM, while the Ca^{2+} concentration in the extracellular medium was 1 mM, any agent altering the cytosolic ion concentration could affect the molecular mechanisms of homeostasis. In our case, phenolic acids increased cytosolic Ca^{2+} (due to agonist or chelating activity) and depleted the internal stores, causing a higher entrance of the extracellular Ca^{2+} to restore the basal conditions. In particular, the chelating activity of hydroxy-cinnamic acids could explain their higher inhibitory effect on germination. The worsening of pollen performances observed with extracts of wheatgrass grown under salinity (Figures 1 and 4) could be a consequence of the increased content of free phenolics caused by this elicitor, as observed by Stagnari et al. [6] for emmer wheatgrass obtained with 50 mM NaCl in the growing medium.

In the presence of oxidative stress, all three types of free extracts antagonized the effects of H_2O_2 on cytosolic Ca^{2+} (Figure 3). There is a well-known and documented bidirectional relationship between ROS, which can modulate calcium-dependent cellular networks, and calcium signaling, which plays a key role in ROS assembly [22,23,32]. ROS behave as Ca^{2+} agonists, stimulate ion mobilization from internal reserves and activate the entry of Ca^{2+} from the extracellular medium [23,24,33]. Although both H_2O_2 and free extracts affected cytosolic Ca^{2+}, it is rational to assume that the two agents acted with different molecular mechanisms. Free extracts had a transient effect on the cytosolic Ca^{2+} of the pollen, which was then restored, while H_2O_2 caused a prolonged increase in cytosolic Ca^{2+} over time, a depletion of pollen Ca^{2+} reserves and a persistent loss of Ca^{2+} homeostasis. In fact, free extracts mitigated the effects of H_2O_2 on cytosolic Ca^{2+}, allowing the recovery of Ca^{2+} homeostasis and Ca^{2+} signal function. Bound extracts had no effect. Actually, previous works demonstrated that the highest antioxidant activity in emmer wheatgrass is associated with free phenolics [5], while bound phenolics were found to have low antioxidant activity in both rice [34] and in rapeseed sprouts [35].

Only the free extract of the Se-biofortified wheatgrass allowed an appreciable recovery of pollen germination and only at the lowest level of oxidative stress (10 mM H_2O_2) (Figure 6). A very slight effect was observed also with bound extracts of Se-biofortified wheatgrass. It is reasonable to assume that the positive effect of the Se-enriched extracts was mainly due to their higher Se contents. In fact, biofortification was successful, as demonstrated by the higher Se content in both the free and bound extracts (Table 1). Selenium is known as a protective agent in oxidative stress [36,37]. The protective effect on the homeostasis of cytosolic Ca^{2+} is one of the positive effects of Se in plants [22,23,28]. Indeed, some studies conducted on maize and olive pollen subjected to oxidative stress reported that Se restored Ca^{2+} homeostasis and improved pollen germination [15,38]. On the other hand, a recent study by Benincasa et al. [12] reported that endogenous Se is a promoter of phenolic compounds and antioxidant activity mainly in combination with other factors, such as salinity. However, the effect of Se-biofortification on phenolic content may be contradictory, depending on the plant species and on the Se dose, as reviewed by D'Amato et al. [14]. Since 100 mg of extract gave no further benefit in the recovery of pollen germination, we can deduce that the amount of Se contained in 50 mg of extract was enough to achieve the maximum effect in terms of mitigation of the stress caused by H_2O_2 (saturation effect). Further research with lower doses of Se-biofortified wheatgrass extracts (e.g., including treatments with 10, 20, 30 and 40 mg, besides the treatments with 0 and 50 mg) is needed to ascertain the minimum rate required to achieve the maximum stress mitigation.

4. Materials and Methods

4.1. Reagents

Hydrogen peroxide (30% *w/v*) and nitric acid (65%, *w/v*) were purchased from Supra-pur Reagents Merck (Darmstadt, Germany). 4-hydroxybenzoic acid (4-OHBA), caffeic acid (CA), p-coumaric acid (p-CA) and salicylic acid (SA), were purchased from Sigma Aldrich (St. Louis, MO, USA). All standards were prepared as a stock solution at 5 mg mL^{-1} in methanol and stored at $-20\,^\circ$C in the dark. FURA 2-AM (FURA-2-pentakis (acetoxymethyl ester)), Triton X-100 (t-octylphenoxypolyethoxyethanol), EGTA (ethylene glycol-bis (β-aminoethyl ether) $-$N, N, N$'$, N$'$-tetracetic acid), sodium selenite (Na_2SeO_3), NaCl, KCl, $MgCl_2$, Hepes, dimethyl sulfoxide (DMSO) and $CaCl_2$ were purchased from Sigma-Aldrich (St. Louis, MO, USA). Other reagents (reagent grade) were obtained from common commercial sources.

4.2. Emmer Wheatgrass Production

Emmer wheatgrass was obtained by germinating grains and growing seedlings with distilled water, or salinity or selenium as described in Del Pino et al. [16]. Briefly, grains of emmer (*Triticum turgidum* L. spp. *dicoccum* (Schrank ex Shubler) Thell., cv. Zefiro) were incubated in plastic trays (20 g per tray) containing sterile cotton and filter paper wetted with distilled water as a control (C), or with a solution containing NaCl 50 mM (S) or 45 mg L^{-1} of Na_2SeO_3 (Se), according to a completely randomized block design with four replicates (trays). To maintain the air circulation while preventing dehydration, the trays were covered by a drilled top and then they were incubated in a growth chamber at 20 $^\circ$C in the dark. After germination, trays were moved in a light–dark regime of 10:14 h, with a light intensity at 200 μmol photons m^{-2} s^{-1}. Distilled water was periodically added to trays to restore initial tray weights, considering the change in seedling biomass as negligible, and thus, approximately keeping the initial NaCl and Se concentrations of these treatments [6,7]. Wheatgrass from C treatment was collected 8 days after sowing (DAS), while wheatgrass of S and Se treatments were collected when they reached the same seedling growth stage as in C (9 DAS for S and 11 DAS for Se treatments), because either S or Se slowed seedling growth compared to C. Only shoots were harvested, and replicates of each treatment were re-grouped two by two for the chemical analysis, performed in triplicate. Samples were stored at $-20\,^\circ$C until extraction.

4.3. Preparation of Emmer Wheatgrass Extracts

Emmer wheatgrass extracts were obtained, using methanol as a solvent, as described in Del Pino et al. [16] and according to the method of Krygier et al. [39], with slight modifications. Two grams of frozen wheatgrass were mixed with 20 mL of MeOH and homogenized on ice using an Ultraturrax three times, alternating 30 s homogenization and 30 s pause to prevent the material from heating. The solution was then kept in agitation for 24 h and centrifuged at 5000 rpm for 10 min. The supernatant (free fraction) was recovered and evaporated to dryness using a rotary evaporator. The dry extracts were suspended in 2 mL of methanol. This represented the extract of free phenolics (Ef).

The remaining solid residue was mixed with 10 mL of NaOH (5 N) for 1 h and then HCl (5 M) was added until pH = 2. Samples were mixed with 10 mL of ethyl acetate, vortexed and centrifuged at 3000 rpm for 10 min and the supernatant was then recovered (bound fraction). This extraction was performed three times, and the supernatants were pooled and evaporated to dryness using a rotary evaporator. The dry residue was suspended in 2 mL of methanol. This represented the extract of bound phenolics (Eb).

Preliminary tests were carried out on the wheatgrass extracts (Ef and Eb) to evaluate the most suitable concentration for the measurement of cytosolic Ca^{2+} and the germination of maize pollen. Based on the preliminary tests mentioned above, the treatments tested in this study were: free (Ef) and bound (Eb) extracts from wheatgrass grown in distilled water (Ef_c and Eb_c), and with a solution of 50 mM NaCl (Ef_s and Eb_s) or 45 mg L^{-1} Na_2SeO_3 (Ef_{Se} and Eb_{Se}).

4.4. Determination of Total Selenium in Wheatgrass Extracts

Measurements of the total selenium content were performed in all emmer wheatgrass extracts (Ef_c, Ef_s, Ef_{Se} and Eb_c, Eb_s, Eb_{Se}) following the method of D'Amato et al. [40]. Wheatgrass extracts (200 µg) were microwave-digested (ETHOS one high-performance microwave digestion system; Mile-stone Inc., Sorisole, Bergamo, Italy) with 8 mL of ultrapure concentrated nitric acid (65% w/w) and 2 mL of hydrogen peroxide (30% w/w). The heating program for the digestion procedure was 30 min at 1000 W and 200 °C. After cooling down, the digests were diluted with water up to 20 mL and passed through 0.45-µm filters. The samples were analyzed by ICP-MS (Agilent 7900, Agilent Technologies, Santa Clara, CA, USA) with an Octopole Reaction System (ORS). Total Se standard solutions were prepared by diluting the corresponding stock solutions (Se standard 1000 mg L^{-1} for AAS TraceCert, 89498, Sigma-Aldrich, Milan, Italy) with HPLC-grade water. Results were expressed as micrograms per kilograms. This method was accurately validated with a recovery test (n = 3) by adding a Se standard solution (4 mg L^{-1}) into the mixture of Se-enriched sample and nitric acid prior to digestion in tubes and after appropriate dilution.

4.5. Measurement of Cytosolic Ca^{2+} with the Addition of Emmer Wheatgrass Extracts, Pure Phenolic Acids and H_2O_2 for an Oxidative Stress Induced In Vitro

Cytosolic Ca^{2+} levels were determined spectrofluorometrically using the FURA-2AM probe according to Del Pino et al. [16]. Aliquots (100 mg) of maize pollen, stored in the dark at 5 °C until use, were suspended in 10 mL of PBS and hydrated for 2 days at 25 °C. Hydrated pollens were harvested by centrifugation at 1000 g for 4 min and then resuspended in 2 mL Ca^{2+}-free HBSS buffer (120 mM NaCl, 5.0 mM KCl, $MgCl_2$ 1 mM, 5 mM glucose, 25 mM Hepes, pH 7.4). Pollen suspensions were incubated in the dark with FURA-2 (2 µL of a 2 mM solution in DMSO) for 120 min, and then centrifuged at 1000 g for 4 min. Pollens were then harvested and suspended in ~10 mL of Ca^{2+}-free HBSS containing 0.1 mM EGTA, which was included to rule out or, at least, minimize a potential background due to contaminating ions (so as to obtain a suspension of 1 × 106 pollen granules hydrated per mL).

Fluorescence was measured in a PerkinElmer LS 50 B spectrofluorometer (ex. 340 and 380 nm, em. 510 nm), set with 10 and a 7.5 nm slit widths in the excitation and emission windows, respectively. Fluorometric readings were normally taken after 300–400 s. In detail, the determination of cytosolic Ca^{2+} started after placing the pollen suspension labelled with FURA 2AM in the cuvette and lasted for 100 s.

After determining the basal cytosolic Ca^{2+} content of the pollen, the following agents were added, singularly or in different combinations according to specific purposes. In detail, aliquots (50 mg) of each of the three free (Ef_c, Ef_s, Ef_{Se}) and three bound (Eb_c, Eb_s, Eb_{Se}) extracts, and aliquots of pure phenolic acids (0.250 mg of 4-HO benzoic, caffeic, p-coumaric, and salicylic acid) were used in the assay to determine the effect of wheatgrass extract and of most representative wheatgrass PAs on the cytosolic Ca^{2+} of maize pollen labelled with the FURA 2AM fluorescent probe. The measurements were carried out in the absence of Ca^{2+} (Ca^{2+}-free) and in the presence of Ca^{2+} (1 mM $CaCl_2$) in the incubation medium.

Oxidative stress was induced in vitro with hydrogen peroxide at 10 and 20 mM, and in the absence of Ca^{2+} (Ca^{2+}-free) or presence of Ca^{2+} (1 mM $CaCl_2$) in the incubation medium, and in the absence or presence of two doses (50 and 100 mg) of each of the three free extracts.

After the addition of each test agent, changes in cytosolic calcium were monitored for another 200–300 s. Cytosolic calcium concentrations of pollen ($[Ca^{2+}]_{cp}$) were calculated following Grynkiewicz et al. [41]. The extent of the determined variations was expressed as $\Delta[Ca^{2+}]_{cp}$, nM.

4.6. Germination of Maize Pollen Grains with the Addition of Emmer Wheatgrass Extracts, Pure Phenolic Acids and H_2O_2 for an Oxidative Stress Induced In Vitro

Fresh pollen samples from each plot were hydrated in a humid chamber at room temperature for 30 min [42], and then transferred to 6-well culture Corning plates (1 mg of pollen per plate) containing 3 mL of an agar-solidified growing medium composed of 1.2% agar, 10%, sucrose, 0.03% boric acid and 0.15% calcium chloride (pH 5.5) [43]. Pollen suspensions were incubated for 24–48 h in a growth chamber at 27 °C with gentle shaking to ensure homogeneous distribution of the samples in the wells.

Two doses (50 or 100 mg) of free and bound (E_f and E_b) extracts, and aliquots of pure phenolic acids (0.250 mg of 4-HO benzoic, caffeic, p-coumaric and salicylic acid) were applied to test, in vitro, the germination on maize pollen grains.

Oxidative stress was induced in vitro with hydrogen peroxide at 10 and 20 mM, and in the absence or presence of two doses (50 and 100 mg) of each of the three free or bound extracts.

Germinated and non-germinated pollen grains were counted under a 10x magnification microscope. Germination rates were calculated based on three replicates, each of which consisted of 100 grains. Germination of grains was confirmed when the pollen tube had grown longer than the grain's diameter [43].

4.7. Statistical Analysis

Statistical evaluations were performed using the OriginPro software [44]. Variance assessments included homogeneity analysis using the Levene test and normality analysis using the D'Agostino–Pearson test. Significance of differences was assessed by the Fisher's least significant differences test, after the analysis of variance according to the randomized complete split-plot with five (Figures 2, 4 and 6) or four replicates (Figures 1 and 3), and with randomized complete one-way design with four replicates for Figure 5. The results obtained are expressed as mean values $\pm$ standard error of the mean (SEM). Differences with $p < 0.05$ were considered statistically significant.

5. Conclusions

Maize pollen germination and cytosolic Ca^{2+} were confirmed to be a good model to test the integrated effects of the wheatgrass matrix. Cytosolic Ca^{2+} was affected only by free extracts of emmer wheatgrass, while the germination of maize pollen was affected by both free and bound extracts. Based on the effects observed by applying pure phenolic acids, it is reasonable to assume that these compounds may partly contribute to the effects of the extracts, but other compounds are likely involved, which deserve to be further studied. The extracts, mainly the free ones, were able to counteract the perturbation of Ca^{2+} homeostasis caused by H_2O_2 and to slightly mitigate its depressive effect on pollen germination. The mitigation effect depended on the wheatgrass growing conditions (i.e., distilled water, salinity or selenium). The best results were obtained with Se-biofortified wheatgrass, likely due to the Se itself rather than to its boosting effect on phenolic compounds.

Author Contributions: Conceptualization, C.A.P. and P.B.; methodology, C.A.P., A.M.D.P. and P.B.; formal analysis, A.M.D.P., B.F. and R.D.; investigation, A.M.D.P. and C.A.P.; resources, P.B., C.A.P. and R.D.; data curation, A.M.D.P. and C.A.P.; writing—original draft preparation A.M.D.P., C.A.P., P.B. and B.F.; writing—review and editing, P.B., C.A.P., A.M.D.P. and D.B.; supervision, C.A.P. and P.B.; project administration, P.B.; funding acquisition, P.B. and D.B. All authors have read and agreed to the published version of the manuscript.

Funding: This research was supported by the Project "Ricerca di Base 2020" of the Department of Agricultural, Food and Environmental Sciences of the University of Perugia (Principal Investigator: Paolo Benincasa).

Institutional Review Board Statement: Not applicable.

Informed Consent Statement: Not applicable.

Data Availability Statement: The data that support the findings of this study are available from the last author (C.A.P.) on reasonable request.

Acknowledgments: We gratefully acknowledge Silvano Locchi for technical assistance in the production of emmer sprouts and maize pollen.

Conflicts of Interest: The authors declare no conflict of interest.

References

1. Benincasa, P.; Falcinelli, B.; Lutts, S.; Stagnari, F.; Galieni, A. Sprouted grains: A comprehensive review. *Nutrients* **2019**, *11*, 421.
2. Tufail, T.; Saeed, F.; Afzaal, M.; Suleria, H.A.R. Wheatgrass juice: A nutritional weapon against various maladies. In *Human Health Benefits of Plant Bioactive Compounds*; Apple Academic Press: Palm Bay, FL, USA, 2019; pp. 245–253.
3. Galieni, A.; Falcinelli, B.; Stagnari, F.; Datti, A.; Benincasa, P. Sprouts and microgreens: Trends, opportunities, and horizons for novel research. *Agronomy* **2020**, *10*, 1424.
4. Benincasa, P.; Galieni, A.; Manetta, A.C.; Pace, R.; Guiducci, M.; Pisante, M.; Stagnari, F. Phenolic compounds in grains, sprouts and wheatgrass of hulled and non-hulled wheat species. *J. Sci. Food Agric.* **2015**, *95*, 1795–1803. [PubMed]
5. Benincasa, P.; Tosti, G.; Farneselli, M.; Maranghi, S.; Bravi, E.; Marconi, O.; Falcinelli, B.; Guiducci, M. Phenolic content and antioxidant activity of einkorn and emmer sprouts and wheatgrass obtained under different radiation wavelengths. *Ann. Agric. Sci.* **2020**, *65*, 68–76.
6. Stagnari, F.; Galieni, A.; D'Egidio, S.; Falcinelli, B.; Pagnani, G.; Pace, R.; Pisante, M.; Benincasa, P. Effects of sprouting and salt stress on polyphenol composition and antiradical activity of einkorn, emmer and durum wheat. *Ital. J. Agron.* **2017**, *12*, 293–301.
7. Falcinelli, B.; Benincasa, P.; Calzuola, I.; Gigliarelli, L.; Lutts, S.; Marsili, V. Phenolic content and antioxidant activity in raw and denatured aqueous extracts from sprouts and wheatgrass of einkorn and emmer obtained under salinity. *Molecules* **2017**, *22*, 2132.
8. Holst, B.; Williamson, G. Nutrients and phytochemicals: From bioavailability to bioefficacy beyond antioxidants. *Curr. Opin. Biotechnol.* **2008**, *19*, 73–82.
9. Acosta-Estrada, B.A.; Gutiérrez-Uribe, J.A.; Serna-Saldívar, S.O. Bound phenolics in food, a review. *Food Chem.* **2014**, *152*, 46–55.
10. Shahidi, F.; Yeo, J. Insoluble-bound phenolics in food. *Molecules* **2016**, *21*, 1216.
11. Liu, H.; Kang, Y.; Zhao, X.; Liu, Y.; Zhang, X.; Zhang, S. Effects of elicitation on bioactive compounds and biological activities of sprouts. *J. Funct. Foods* **2019**, *53*, 136–145.
12. Benincasa, P.; D'Amato, R.; Falcinelli, B.; Troni, E.; Fontanella, M.C.; Frusciante, S.; Guiducci, M.; Beone, G.M.; Businelli, D.; Diretto, G. Grain endogenous selenium and moderate salt stress work as synergic elicitors in the enrichment of bioactive compounds in maize sprouts. *Agronomy* **2020**, *10*, 735. [CrossRef]
13. D'Amato, R.; Fontanella, M.C.; Falcinelli, B.; Beone, G.M.; Bravi, E.; Marconi, O.; Benincasa, P.; Businelli, D. Selenium biofortification in rice (*Oryza sativa* L.) sprouting: Effects on Se yield and nutritional traits with focus on phenolic acid profile. *J. Agric. Food Chem.* **2018**, *66*, 4082–4090. [CrossRef] [PubMed]
14. D'Amato, R.; Regni, L.; Falcinelli, B.; Mattioli, S.; Benincasa, P.; Dal Bosco, A.; Pacheco, P.; Proietti, P.; Troni, E.; Santi, C.; et al. Current knowledge on selenium biofortification to improve the nutraceutical profile of food: A comprehensive review. *J. Agric. Food Chem.* **2020**, *68*, 4075–4097. [CrossRef] [PubMed]
15. Del Pino, A.M.; Guiducci, M.; D'Amato, R.; Di Michele, A.; Tosti, G.; Datti, A.; Palmerini, C.A. Selenium maintains cytosolic Ca^{2+} homeostasis and preserves germination rates of maize pollen under H_2O_2-induced oxidative stress. *Sci. Rep.* **2019**, *9*, 13502. [CrossRef]
16. Del Pino, A.M.; Falcinelli, B.; D'Amato, R.; Businelli, D.; Benincasa, P.; Palmerini, C.A. Extracts of emmer wheatgrass grown with distilled water, salinity or selenium differently affect germination and cytosolic Ca^{2+} of maize pollen. *Agronomy* **2021**, *11*, 633.
17. Thor, K. Calcium-Nutrient and Messenger. *Front. Plant Sci.* **2019**, *10*, 440–446. [CrossRef]
18. Fox, T.C.; Guerinot, M.L. Molecular biology of cation transport in plants. *Annu. Rev. Plant Physiol. Plant Mol. Biol.* **1998**, *49*, 669–696.
19. Dodd, A.N.; Kudla, J.; Sanders, D. The language of calcium signaling. *Annu. Rev. Plant Biol.* **2010**, *61*, 593–620. [CrossRef]

20. Ordenes, V.R.; Reyes, F.C.; Wolff, D.; Orellana, A. A thapsigargin sensitive Ca^{2+} pump is present in the pea Golgi apparatus membrane. *Plant Physiol.* **2002**, *129*, 1820–1828.
21. Kudla, J.; Becker, D.; Grill, E.; Hedrich, R.; Hippler, M.; Kummer, U.; Parniske, M.; Romeis, T.; Schumacher, K. Advances and current challenges in calcium signaling. *New Phytol.* **2018**, *218*, 414–431.
22. Steinhorst, L.; Kudla, J. Calcium—A central regulator of pollen germination and tube growth. *Biochim. Biophys. Acta* **2013**, *1833*, 1573–1581. [CrossRef] [PubMed]
23. Görlach, A.; Bertram, K.; Hudecova, S.; Krizanova, O. Calcium and ROS: A mutual interplay. *Redox Biol.* **2015**, *6*, 260–271. [CrossRef]
24. Orrenius, S.; Gogvadze, V.; Zhivotovsky, B. Calcium and mitochondria in the regulation of cell death. *Biochem. Biophys. Res. Commun.* **2015**, *460*, 72–81. [CrossRef]
25. Demidchik, V. Mechanisms of oxidative stress in plants: From classical chemistry to cell biology. *Environ. Exp. Bot.* **2015**, *109*, 212–228. [CrossRef]
26. Pizzino, G.; Irrera, N.; Cucinotta, M.; Pallio, G.; Mannino, F.; Arcoraci, V.; Squadrito, F.; Altavilla, D.; Bitto, A. Oxidative stress: Harms and benefits for human health. *Oxid. Med. Cell. Longev.* **2017**, *2017*, 8416763. [CrossRef] [PubMed]
27. Dickinson, B.C.; Chang, C.J. Chemistry and biology of reactive oxygen species in signaling or stress responses. *Nat. Chem. Biol.* **2011**, *7*, 504–511. [CrossRef] [PubMed]
28. Collin, F. Chemical basis of reactive oxygen species reactivity and involvement in neurodegenerative diseases. *Int. J. Mol. Sci.* **2019**, *20*, 2407. [CrossRef] [PubMed]
29. Michard, E.; Alves, F.; Feijò, J.A. The role of ion fluxes in polarized cell growth and morphogenesis: The pollen tube as an experimental paradigm. *Int. J. Dev. Biol.* **2009**, *53*, 1609–1622. [CrossRef] [PubMed]
30. Cheung, A.Y.; Wu, H.M. Structural and signaling networks for the polar cell growth machinery in pollen tubes. *Annu. Rev. Plant Biol.* **2008**, *59*, 547–572. [CrossRef]
31. Kim, K.-H.; Tsao, R.; Yang, R.; Cui, S.W. Phenolic acid profiles and antioxidant activities of wheat bran extracts and the effect of hydrolysis conditions. *Food Chem.* **2006**, *95*, 466–473. [CrossRef]
32. Proietti, P.; Marinucci, M.T.; Del Pino, A.M.; D'Amato, R.; Regni, L.; Acuti, G.; Chiaradia, E.; Palmerini, C.A. Selenium maintains Ca^{2+} homeostasis in sheep lymphocytes challenged by oxidative stress. *PLoS ONE* **2018**, *13*, e0201523. [CrossRef] [PubMed]
33. Brini, M.; Calì, T.; Ottolini, D.; Carofoli, E. Intracellular calcium homeostasis and signaling. In *Metallomics and the Cell*; Metal Ions in Life Sciences; Banci, L., Ed.; Springer: Berlin/Heidelberg, Germany, 2013; Volume 12, pp. 119–168.
34. Ti, H.; Li, Q.; Zhang, R.; Zhang, M.; Deng, Y.; Wei, Z.; Chi, Z.; Zhang, Y. Free and bound phenolic profiles and antioxidant activity of milled fractions of different indica rice varieties cultivated in southern China. *Food Chem.* **2014**, *159*, 166–174. [CrossRef] [PubMed]
35. Falcinelli, B.; Sileoni, V.; Marconi, O.; Perretti, G.; Quinet, M.; Lutts, S.; Benincasa, P. Germination under moderate salinity increases phenolic content and antioxidant activity in rapeseed (*Brassica napus* var *oleifera* Del.) sprouts. *Molecules* **2017**, *22*, 1377. [CrossRef] [PubMed]
36. Rayman, M.P. Food-chain selenium and human health: Emphasis on intake. *Br. J. Nutr.* **2008**, *100*, 254–268. [CrossRef]
37. Prins, C.N.; Hantzis, L.J.; Quinn, C.F.; Pilon-Smits, E.A.H. Effects of selenium accumulation on reproductive functions in *Brassica juncea* and *Stanleya pinnata*. *J. Exp. Bot.* **2011**, *62*, 5633–5640. [CrossRef]
38. Del Pino, A.M.; Regni, L.; D'Amato, R.; Tedeschini, E.; Businelli, B.; Proietti, P.; Palmerini, C.A. Selenium-enriched pollen grains of *Olea europaea* L.: Ca^{2+} signaling and germination under oxidative stress. *Front. Plant Sci.* **2019**, *10*, 1611. [CrossRef] [PubMed]
39. Krygier, K.; Sosulski, F.; Hogge, L. Free, esterified, and insoluble-bound phenolic acids. Extraction and purification procedure. *J. Agric. Food Chem.* **1982**, *30*, 330–334. [CrossRef]
40. D'Amato, R.; De Feudis, M.; Guiducci, M.; Businelli, D. *Zea mays* L. grain: Increase in nutraceutical and antioxidant properties due to se fortification in low and high water regimes. *J. Agric. Food Chem.* **2019**, *67*, 7050–7059. [CrossRef]
41. Grynkiewicz, G.; Poenie, M.; Tsien, R.Y. A new generation of Ca^{2+} indicators with greatly improved fluorescence properties. *J. Biol. Chem.* **1985**, *260*, 3440–3450. [CrossRef]
42. Rejo´n, J.D.; Zienkiewicz, A.; Rodríguez-García, M.I.; Castro, A.J. Profiling and functional classification of esterases in olive (*Olea europaea*) pollen during germination. *Ann. Bot.* **2012**, *110*, 1035–1045. [CrossRef]
43. Martins, E.S.; Davide, L.M.C.; Miranda, G.J.; de Oliveira Barizon, J.; de Assis Souza, F.; de Carvalho, R.P.; Gonçalves, M.C. In vitro pollen viability of maize cultivars at different times of collection. *Ciênc. Rural* **2017**, *47*, e20151077. [CrossRef]
44. *OriginPro*, version 2019b; OriginLab Corporation: Northampton, MA, USA, 2019.

Article

Potential Importance of Molybdenum Priming to Metabolism and Nutritive Value of *Canavalia* spp. Sprouts

Mohammad K. Okla [1,*,†], Nosheen Akhtar [2,*,†], Saud A. Alamri [1], Salem Mesfir Al-Qahtani [3], Ahmed Ismail [4], Zahid Khurshid Abbas [5], Abdullah A. AL-Ghamdi [1], Ahmad A. Qahtan [1], Walid H. Soufan [6], Ibrahim A. Alaraidh [1], Samy Selim [7] and Hamada AbdElgawad [8,9]

[1] Botany and Microbiology Department, College of Science, King Saud University, Riyadh 11451, Saudi Arabia; saualamri@ksu.edu.sa (S.A.A.); abdaalghamdi@ksu.edu.sa (A.A.A.-G.); aqahtan@ksu.edu.sa (A.A.Q.); ialaraidh@ksu.edu.sa (I.A.A.)
[2] Department of Biological Sciences, National University of Medical Sciences, Rawalpindi 46000, Pakistan
[3] Biology Department, University College of Taymma, Tabuk University, Tabuk 71411, Saudi Arabia; salghtani@ut.edu.sa
[4] Pharmacognosy Department, Faculty of Pharmacy, Fayoum University, Fayoum 63514, Egypt; Ais03@fayoum.edu.eg
[5] Department of Biology, Faculty of Science, University of Tabuk, Tabuk 71491, Saudi Arabia; Znourabbas@ut.edu.sa
[6] Department of Plant Production, Faculty of Food and Agriculture Sciences, King Saud University, Riyadh 11451, Saudi Arabia; waoufan@ksu.edu.sa
[7] Department of Clinical Laboratory Sciences, College of Applied Medical Sciences, Jouf University, Sakaka 72388, Saudi Arabia; sabdulsalam@ju.edu.sa
[8] Integrated Molecular Plant Physiology Research, Department of Biology, University of Antwerp, 2020 Antwerpen, Belgium; hamada.abdelgawad@uantwerpen.be
[9] Botany and Microbiology Department, Faculty of Science, Beni-Suef University, Beni-Suef 62511, Egypt
* Correspondence: malokla@ksu.edu.sa (M.K.O.); nosheenakhtar@numspak.edu.pk (N.A.)
† These authors contributed equally and share first authorship.

Citation: Okla, M.K.; Akhtar, N.; Alamri, S.A.; Al-Qahtani, S.M.; Ismail, A.; Abbas, Z.K.; AL-Ghamdi, A.A.; Qahtan, A.A.; Soufan, W.H.; Alaraidh, I.A.; et al. Potential Importance of Molybdenum Priming to Metabolism and Nutritive Value of *Canavalia* spp. Sprouts. *Plants* **2021**, *10*, 2387. https://doi.org/10.3390/plants 10112387

Academic Editors: Beatrice Falcinelli and Angelica Galieni

Received: 22 September 2021
Accepted: 26 October 2021
Published: 5 November 2021

Publisher's Note: MDPI stays neutral with regard to jurisdictional claims in published maps and institutional affiliations.

Abstract: Molybdenum ions (Mo) can improve plants' nutritional value primarily by enhancing nitrogenous metabolism. In this study, the comparative effects of seed priming using Mo were evaluated among sproutings of *Canavalia* species/cultivars, including *Canavalia ensiformis var. gladiata* (CA1), *Canavalia ensiformis var. truncata Ricker* (CA2), and *Canavalia gladiata var. alba Hisauc* (CA3). Mo impacts on growth, metabolism (e.g., nitrogen and phenolic metabolism, pigment and total nutrient profiles), and biological activities were assayed. Principal component analysis (PCA) was used to correlate Mo-mediated impacts. The results showed that Mo induced photosynthetic pigments that resulted in an improvement in growth and increased biomass. The N content was increased 0.3-fold in CA3 and 0.2-fold in CA1 and CA2. Enhanced nitrogen metabolism by Mo provided the precursors for amino acids, protein, and lipid biosynthesis. At the secondary metabolic level, phenolic metabolism-related precursors and enzyme activities were also differentially increased in *Canavalia* species/cultivars. The observed increase in metabolism resulted in the enhancement of the antioxidant (2,2-diphenyl-1-picryl-hydrazyl-hydrate (DPPH) free radical scavenging, 2,2-azino-bis(3-ethylbenzothiazoline-6-sulfonic acid) (ABTS), ferric reducing antioxidant power (FRAP)) and antidiabetic potential (Glycemic index (GI) and inhibition activity of α-amylase, and α-glucosidase) of species. The antioxidant activity increased 20% in CA3, 14% in CA1, and 8% in CA2. Furthermore, PCA showed significant variations not only between Mo-treated and untreated samples but also among *Canavalia* species. Overall, this study indicated that the sprouts of *Canavalia* species have tremendous potential for commercial usage due to their high nutritive value, which can be enhanced further with Mo treatment to accomplish the demand for nutritious feed.

Keywords: *Canavalia*; molybdenum; seed priming; nitrogen assimilation

1. Introduction

Developing countries are facing an increasing demand for protein and other nutrient-rich foods. In this context, legumes can contribute as the most valuable source of nutrients and provide high-quality dietary proteins [1]. Legume plants have desirable characteristics such as an abundance of carbohydrates, the ability to lower serum cholesterol, high fiber, a high concentration of polyunsaturated fatty acids, and long shelf life. In addition to B complex vitamins such as folate, thiamin, and riboflavin, minerals, and fiber, legumes are also major sources of proteins and calories [2]. Furthermore, it is evidenced that sprouts are the richest source of proteins and other compounds of nutritional value compared to un-sprouted plants [3]. Moreover, sprouts have also been associated with a variety of biologically active constituents with potential health benefits. During germination, metabolic enzymes are activated, which can lead to the release of some amino acids and peptides, and the synthesis or use of them can form new proteins. As a consequence, the nutritional and medicinal value might be enhanced by sprouting in legumes. Research has to be geared to exploit the sprouting of legumes and enhance their nutrition values to meet the nutritional requirements of the increasing population.

The genus of *Canavalia* is considered the third largest family among flowering plants [1]. It comprises approximately 50 species of tropical vines widely distributed in tropical and subtropical regions all over the world [4]. This genus was used traditionally as a food due to its nutritional significance. Sridhar and Seena envisaged a comparative account of nutritional and functional properties of *Canavalia* species [5]. *Canavalia gladiata* and *Canavalia ensiformis* are the common legume species having the potential to be a rich protein source, like edible legumes [5]. Pharmacological effects of *Canavalia gladiata* are reported for cancer [6], allergies [7], antioxidants [8], and inflammation [9]. *Canavalia gladiata* in complex with *Arctium lappa* extract is proposed to develop as a functional food for stimulating immunity [10]. Similarly, the seeds of *Canavalia ensiformis* are a source of proteins with biotechnological importance including ureases [11] and proteases [12]. Processed seeds of *Canavalia ensiformis* are reported for enhanced antioxidant activity [13]. Hence, *Canavalia* species are of high medical importance, and with proper seed priming with micronutrients and using other treatments nutritional and pharmacological values can be enhanced.

Micronutrients are vital for plant growth because they act as a cofactor of the enzyme, take part in redox reactions, and have several other important functions [14]. Furthermore, despite addition to the soil, micronutrient application using seeds improves the stand formation, advances phenological events, and increases yield and micronutrient grain contents [14]. Like different micronutrients, molybdenum (Mo) is very vital and essential for plants' physiological functions. In plants with inadequate Mo, nitrates accrue in leaves, which then do not assimilate into proteins. In legumes, Mo serves an additional function: to help root nodule bacteria to fix atmospheric nitrogen (N). Studies revealed that the application of Mo in beans, via seeds, increases the mass of the nodule [15]; moreover, its foliar spray increases nodule formation, nitrogen contents, the grain number, the grain mass, and overall productivity [16]. The enhanced benefits may be attributed to the acceleration of N absorption and assimilation processes via BNF (biological nitrogen fixation) [14]. Mo is a crucial element for more than 40 enzymes, four of which have been found in plants, including nitrate reductase (NR) and nitrogenase synthesis [17,18]. Thus, plants receiving increased Mo will have increased N productivity levels based on the higher activities of the abovementioned enzymes. Mo bound to a unique pterin compound, named Mo cofactor, binds to diverse apoproteins and aids in anchoring the Mo center at the correct position within the holo-enzyme for its correct interaction with other components of the electron transport chain [17]. Thus, Mo supply can strengthen plant metabolism at different growth stages through an improved enzymatic and non-enzymatic antioxidant defense system and also enhance other pharmacological properties [19]. Moreover, Mo is also an essential mineral for the human body as it is part of important metabolic enzymes such as xanthine oxidase sulfite oxidase, aldehyde oxidase, and mitochondrial amidoxime reducing component.

With this background, we hypothesized that Mo application can enhance the nutritional and pharmacological value of *Canavalia* species by improving their N content and resulting in primary and secondary metabolite production. Thus, the present study aimed to evaluate the impact of Mo seed priming on three *Canavalia* species/cultivar sprouts. To the best of our knowledge, no study has been conducted to assess the influence of Mo seed priming on *Canavalia ensiformis var. gladiata, Canavalia ensiformis var. truncata Ricker,* and *Canavalia gladiata var. alba* Hisauc. Herein, we evaluated the impacts of Mo treatment on growth, N content, and phenolic metabolism as well as on the concentrations of several phytochemical compounds. We further examined the role of Mo in the enhancement of pharmacological properties of *Canavalia* species. Overall, our study contributed to an understanding of the biochemical basis of Mo that induces high growth and tissue quality of different *Canavalia* species/cultivars.

2. Results

2.1. Biomass and Pigment Content

To quantify the biomass of the molybdenum (Mo)-treated cultivar/species of *Canavalia*, the validated method was employed. An increase in fresh weight was observed in all the sproutings of the three studied species/cultivars, as shown in Figure 1A. Significant differences were observed among Mo-treated and non-treated groups. The difference was almost negligible within the sproutings of different species. Next, different types of chlorophylls such as chlorophyll a, chlorophyll b, and chlorophyll ab were quantified in both Mo-treated and untreated plants. The results indicated that exogenous Mo application increased chlorophyll content in the leaves of all the studied *Canavalia* species, as shown in Figure 1B. In untreated sprouts of *Canavalia ensiformis var. gladiata* (CA1), *Canavalia ensiformis var. truncata Ricker* (CA2), and *Canavalia gladiata var. alba Hisauc* (CA3), chlorophyll ab was present in a higher amount as compared to chlorophyll a and b, which was significantly increased after seed priming with Mo, as shown in Figure 1B. Further, the impact of Mo on the production of carotenoids was studied. Concentrations of α-carotene, β- carotene, and lycopene were quantified. A significant increase in the production of T-carotene and B-carotene was observed in all the Mo-treated species/cultivars, while the concentration of lycopene remained the same, as shown in Figure 1B.

2.2. Nutritional Value

The variations in the concentration of nutrients (lipids, proteins, and fibers) and phytochemicals (alkaloids, flavonoids, saponins, and glycosides) were quantified in both Mo-exposed and unexposed seeds of *Canavalia*. The quantity of total lipids was significantly increased in CA1, a non-significant change was observed in CA2, while a decrease was recorded in CA3. Furthermore, a significant increase in fiber content was recorded in CA1, a decrease was found in CA3, while there was no change in the fiber content of CA2. Upward trends in total proteins, flavonoids, saponins, and glycosides were noticed (Table 1).

2.3. Nitrogen and Amino Acid Metabolism

The outcome of Mo treatment on the nitrogen (N) content was studied in *Canavalia* species. The results showed a slight increase in N production in the sproutings of three species, as shown in Figure 2A. Moreover, the quantification of nitrogen reductase (NR) activity showed its increase in CA3 and decrease in CA1 and CA2 in Mo-treated plants, as shown in Figure 2B. Next, we studied the effect of Mo on two pathways of glutamate synthesis, i.e., glutamate-dehydrogenase (GDH) and glutamine synthetase (GS)/glutamate synthase (GOGAT) pathways. The activities of GDH, GOGAT, and GS were measured (Figure 2C–E). The results showed an increase in GDH and GOGAT activity in the three species. The rise in GDH activity was significant in CA3, while it was non-significant in other species. Inversely, GS activity was decreased after Mo treatment and the differences were highly significant in the three species/cultivars. Further, the impact of Mo treatment on dihydrodipicolinate synthase (DHDPS) and cystathionine γ- synthase (CGS) activity was explored. The experiment showed that the activity of DHDPS significantly increased in CA3 and non-significantly in CA1 and CA2 (Figure 2F). Moreover, a notable increase in the activity of CGS was measured in sprouts of CA3 (Figure 2G). Furthermore, amino acids, i.e., asparagine, glutamine, glycine, glutamic acid, isoleucine, arginine, tyrosine, lysine, serine, alanine, proline, histidine, leucine, isoleucine, valine, cystine, threonine, tryptophan, and methionine, were quantified in Mo-treated *Canavalia* species and data were compared to untreated controls. The changes in each amino acid content due to Mo exposure are shown in Table 2. Overall, an increase in amino acid production was detected in the three species. The quantity of cystine, isoleucine, leucine, glycine, alanine, and proline was increased significantly in CA1; asparagine, glutamine, glutamic acid, alanine, proline, cystine, and lysine were enhanced in CA2; and asparagine, glutamine, glycine, histidine, methionine, cystine, isoleucine, tyrosine, lysine, threonine, and tryptophan were enhanced in CA3. Variable trends were noticed among different sprouts. The concentrations of alanine and proline were found to be improved in CA1 and CA2, while they declined in CA3. Levels of histidine, arginine, valine, threonine, and tryptophan increased in CA1 and CA3, whereas they decreased in CA2. Leucine was increased in CA1 and reduced in CA3 and CA2. All results are enumerated in Table 2.

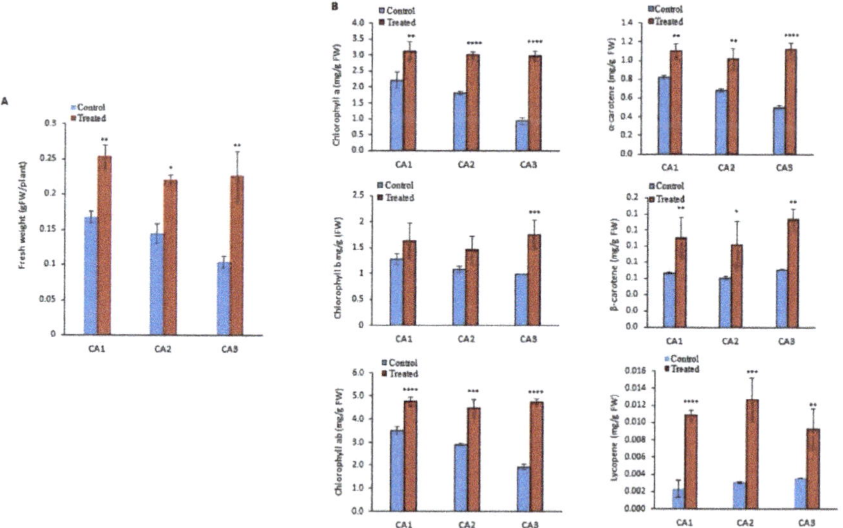

Figure 1. Impacts of molybdenum (Mo) seed priming on different *Canavalia* species. (**A**): Total biomass (gFW). (**B**): Chlorophyll a, b, and ab content, α-carotene content, β-carotene content, and lycopene content were measured and expressed as mg/gFW. Control; without treatment, Treated; Mo treatment. CA1; *Canavalia ensiformis var. gladiata*, CA2; *Canavalia ensiformis var. truncata Ricker*, CA3; *Canavalia gladiata var. alba Hisauc*. α-carotene, or β-carotene, Beta-carotene. The bars above mean indicate ± standard deviation (S.D) of three independent replicates ($n = 3$). Asterisks (*) show the level of significance * $p < 0.05$, ** $p < 0.01$, *** $p < 0.001$, **** $p < 0.0001$.

2.4. Impact of Mo Treatment on Phenolic Level and Metabolism

2.4.1. Total Phenolic Content

The impact of Mo on the total phenolic content of *Canavalia* species was evaluated. The total phenolic content was expressed as gallic acid equivalent (GAE). The results indicated that the production of phenols was increased significantly in CA3. A non-significant increase was observed in CA2, whereas a decrease was noticed in CA1, as given in Table 1.

Table 1. Effect of molybdenum treatment on the nutritional value of *Canavalia* species/cultivar sprouts. Total lipids (µg/gFW), total proteins (µg/gFW), fibers (µg/gFW), alkaloids (µg/gFW), phenolics (µg/gFW), flavanoids (µg/gFW), saponins (µg/gFW), and glycosides (µg/gFW) were measured. Experiments were executed in triplicate and the data are presented as mean ± standard deviation (SD); Level of significance * $p < 0.05$. Mo, Molybdenum.

	C. ensiformis var. gladiata (CA1)		*C. ensiformis var. truncata Ricker* (CA2)		*C. gladiata var. alba Hisauc* (CA3)	
	Control	**Mo-Treated**	**Control**	**Mo-Treated**	**Control**	**Mo-Treated**
Total lipids	57.009 ± 5.12	98.085 ± 0.72 *	81.201 ± 7.52	108.250 ± 28.92	92.659 ± 18.44	70.837 ± 3.35
Total proteins	189.607 ± 2.93	219.279 ± 14.73	146.753 ± 40.14	188.859 ± 7.88	120.820 ± 24.16	160.784 ± 14.59
Fibers	9.172 ± 1.65	12.241 ± 0.04 *	14.543 ± 1.04	14.542 ± 3.46	24.364 ± 7.42	18.242 ± 3.75
Alkaloids	2.324 ± 0.77	2.626 ± 0.41	2.669 ± 0.02	2.594 ± 0.47	2.788 ± 1.24	2.437 ± 0.78
Phenolics	1.868 ± 0.11	1.663 ± 0.41	2.323 ± 0.40	2.732 ± 0.22	0.931 ± 0.06	2.630 ± 0.71 *
Flavonoids	0.240 ± 0.00	0.523 ± 0.24 *	0.273 ± 0.07	0.486 ± 0.03	0.132 ± 0.012	0.613 ± 0.13 *
Saponins	1.386 ± 0.23	2.284 ± 0.39 *	2.400 ± 0.32	2.998 ± 0.30	1.420 ± 0.33	2.348 ± 0.30
Glycosides	2.532 ± 0.45	3.712 ± 0.46	3.194 ± 0.37	4.948 ± 0.08	2.708 ± 0.68	3.749 ± 0.09

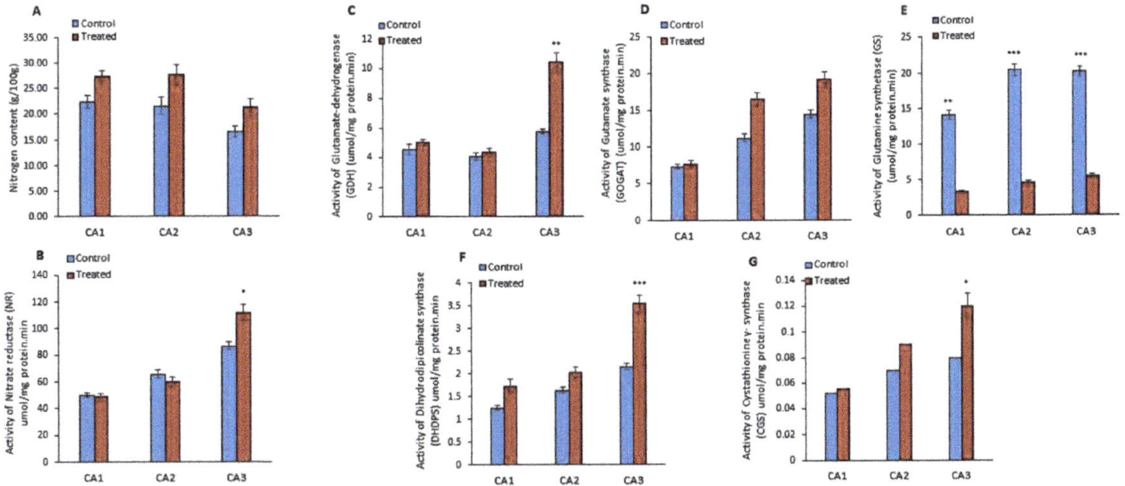

Figure 2. Effects of molybdenum (Mo) treatment on nitrogen assimilation. (**A**): Nitrogen (N) content (g/100gFW). (**B**): Nitrate reductase (umol/mg protein. min) activity. (**C**): Glutamte dehydrogenase (GDH) activity (µmol/mg protein. min). (**D**): Gutamate synthase (GOGAT) activity (µmol/mg protein. min). (**E**): Glutamine synthetase (GS) activity (µmol/mg protein. min). (**F**): Dihydrodipicolinate synthase (DHDPS) activity (µmol/mg protein. min). (**G**): Cystathionine γ- synthase (CGS) activity (µmol/mg protein. min). Control; without treatment, Treated; Mo treatment. CA1; *Canavalia ensiformis var. gladiata*, CA2; *Canavalia ensiformis var. truncata Ricker*, CA3; *Canavalia gladiata var. alba Hisauc*. Experiments were carried out in triplicate and the data are expressed as the mean ± standard deviation; Level of significance * $p < 0.05$, ** $p < 0.01$, *** $p < 0.001$.

Table 2. Effect of molybdenum treatment on amino acid content (μmol/ gFW) of *Canavalia* species/cultivar sprouts. Experiments were carried out in triplicate and the data are expressed as mean ± standard deviation (S.D); Level of significance, * $p < 0.05$; Mo, Molybdenum.

	C. ensiformis var. gladiata (CA1)		*C. ensiformis var. truncata Ricker* (CA2)		*C. gladiata var. alba Hisauc* (CA3)	
	Control	Mo-Treated	Control	Mo-Treated	Control	Mo-Treated
Asparagine	1.22 ± 0.05	1.41 ± 0.06	0.84 ± 0.08 *	1.261 ± 0.03	1.03 ± 0.07	1.7 ± 0.11 *
Glutamine	3.3 ± 0.19	3.98 ± 0.37	2.62 ± 0.21	5.060 ± 0.4 *	3.81 ± 0.22	6.29 ± 0.55 *
Glutamic acid	3.04 ± 0.39	2.42 ± 0.93	1.5 ± 0.23	4.040 ± 0.2 *	2.75 ± 0.22	4.92 ± 0.65
Serine	0.8 ± 0.03	1.13 ± 0.08	0.89 ± 0.05	1.301 ± 0.7	1 ± 0.06	1.15 ± 0.06
Glycine	0.46 ± 0.01	0.77 ± 0.02 *	0.6 ± 0.03	0.6 ± 0.00	0.58 ± 0.05	0.89 ± 0.01 *
Arginin	0.25 ± 0.01	0.34 ± 0.01	0.3 ± 0.01	0.25 ± 0.00	0.32 ± 0.02	0.34 ± 0.03
Alanine	0.49 ± 0.17	0.92 ± 0.05 *	0.49 ± 0.22	0.834 ± 0.01 *	1.5 ± 0.59	1.13 ± 0.7
Proline	0.3 ± 0.17	0.67 ± 0.13 *	0.32 ± 0.23	0.698 ± 0.03 *	1.33 ± 0.19	0.96 ± 0.04
Histidine	0.52 ± 0.02	0.71 ± 0.07	0.41 ± 0.07	0.366 ± 0.08	0.33 ± 0.02	0.58 ± 0.05 *
Valine	0.48 ± 0.03	0.51 ± 0.04	0.29 ± 0.05	0.28 ± 0.0	0.30 ± 0.03	0.46 ± 0.04
Methionine	0.4 ± 0.01	0.5 ± 0.08	0.41 ± 0.01	0.544 ± 0.08	0.42 ± 0.02	0.85 ± 0.05 *
Cystine	0.32 ± 0.08	0.72 ± 0.11 *	0.74 ± 0.07	0.944 ± 0.1 *	0.65 ± 0.08	1.2 ± 0.05 *
Isoleucine	0.41 ± 0.08	1.09 ± 0.11 *	0.72 ± 0.03	0.757 ± 0.2	0.64 ± 0.06	1.35 ± 0.13 *
Leucine	0.66 ± 0.12	1.44 ± 0.14 *	0.75 ± 0.11	0.707 ± 0.11	1.21 ± 0.11	0.9 ± 0.1
Tyrosine	0.33 ± 0.01	0.39 ± 0.02	0.3 ± 0	0.396 ± 0.08	0.28 ± 0.01	0.45 ± 0.0 *
Lysine	0.61 ± 0.03	1 ± 0.2	0.62 ± 0.03	1.075 ± 0.05 *	0.48 ± 0.04	0.91 ± 0.04 *
Threonine	0.55 ± 0.04	0.71 ± 0.02	0.67 ± 0.04	0.76 ± 0.01	0.47 ± 0.03	0.77 ± 0.01 *
Tryptophan	0.2 ± 0.01	0.29 ± 0.02	0.21 ± 0.01	0.38 ± 0.01 *	0.18 ± 0.02	0.4 ± 0.02 *

2.4.2. Phenolic Compounds

Further, we evaluated the quantity of individual phenolic compounds in the Mo-treated group and compared them with untreated controls. Different phenols, i.e., gallic acid, caffeic acid, *p*-coumaric acid, chicoric acid, rosmarinic acid, protocatechuic acid, quercetin, naringenin, kaempferol, luteolin, apigenin, naringenin, rutin, and chlorogenic acid, were quantified using HPLC. The results revealed that Mo impacted the synthesis of different phenols in an interesting pattern among different species of *Canavalia*. Gallic acid, *p*-coumaric acid, rosmarinic acid, naringenin, and chlorogenic acid were enhanced significantly in CA1; caffeic acid, *p*-coumaric acid, chicoric acid, rosmarinic acid, naringenin, and apigenin were increased significantly in CA2; and gallic acid, chicoric acid, kaempferol, and chlorogenic acid were raised significantly in CA3. Enhanced production of caffeic acid, rosmarinic acid, and luteolin was observed in all three species after exposure to Mo. Interestingly, differential patterns were observed for other phenols (Table 3). Gallic and chlorogenic acid production was induced by Mo in CA1, while a reduction pattern was observed in CA2 and CA3. On the other hand, the biosynthesis of chicoric acid, protocatechuic acid, quercetin, and naringenin was decreased in CA1 and increased in CA2 and CA3. Moreover, *p*-coumaric acid and kaempferol were raised in CA1 and CA2 and reduced in CA3. Interestingly, apigenin synthesis was not affected by Mo treatment in CA2, while a reduction was found in CA2 and CA3—values are given in Table 3.

Table 3. Effect of molybdenum treatment on phenolic compounds of *Canavalia* species/cultivar sprouts. Values are expressed as μmol/gFW. Experiments were performed in triplicate and the data are shown as mean ± standard deviation (S.D); Asterisks (*) show the level of significance, * $p < 0.05$; Mo, Molybdenum.

	C. ensiformis var. gladiata (CA1)		*C. ensiformis var. truncata* Ricker (CA2)		*C. gladiata var. alba* Hisauc (CA3)	
	Control	Mo-Treated	Control	Mo-Treated	Control	Mo-Treated
Gallic acid	0.88 ± 0.03	1.44 ± 0.1 *	0.72 ± 0.11	0.58 ± 0.1	1.2 ± 0.09	0.63 ± 0.22 *
Caffeic acid	1.71 ± 0.3	2.36 ± 0.34	1.782 ± 0.38	3.15 ± 0.26 *	3.218 ± 0.18	3.27 ± 0.37
p-Coumaric acid	2.3 ± 0.31	3.44 ± 0.41 *	2.22 ± 0.44	3.32 ± 0.27 *	3.935 ± 0.21	3.42 ± 0.22
Chicoric acid	0.79 ± 0.03	0.55 ± 0.18	0.83 ± 0.04	1.34 ± 0.13 *	0.93 ± 0.07	1.22 ± 0.13 *
Rosmarinic acid	0.26 ± 0.01	0.37 ± 0.03 *	0.27 ± 0.05	0.41 ± 0.02 *	0.483 ± 0.06	0.48 ± 0.03
Protocatechuic acid	1.52 ± 0.21	1.48 ± 0.19	1.53 ± 0.1	1.60 ± 0.12	2.24 ± 0.28	1.94 ± 0.1
Quercetin	0.114 ± 0.02	0.14 ± 0.03	0.09 ± 0.01	0.105 ± 0.02	0.131 ± 0.02	0.141 ± 0.02
Naringenin	0.15 ± 0.01	0.98 ± 0.02	0.78 ± 0.01	1.56 ± 0.01 *	1.58 ± 0.01	1.23 ± 0.01
Kaempferol	0.72 ± 0.02	0.615 ± 0.1	1.10 ± 0.16	1.104 ± 0.17	1.094 ± 0.11	1.91 ± 0.11 *
Luteolin	0.48 ± 0.23	0.34 ± 0.22	0.426 ± 0.11	0.42 ± 0.18	0.48 ± 0.16	0.40 ± 0.2
Apigenin	0.35 ± 0.12	0.3 ± 0.03	0.004 ± 0.03	0.015 ± 0.01 *	0.017 ± 0.03	0.015 ± 0.01
Naringenin	0.59 ± 0.02	1.272 ± 0 *	0.861 ± 0.2	0.97 ± 0	1.66 ± 0	1.87 ± 0.1
Rutin	0.945 ± 0	1.10 ± 0.014	0.108 ± 0.22	0.104 ± 0.08	1.44 ± 0.2	1.47 ± 0.28
Chlorogenic acid	0.08 ± 0.06	1.41 ± 0.28 *	0.84 ± 0.45	1.04 ± 0.24	1.04 ± 0.36	1.71 ± 0.29 *

2.4.3. Phenolic Metabolism

Various parameters were determined to assess the endogenous metabolism of phenols in treated and untreated groups. For this purpose, the concentrations of phenylalanine, L-phenylalanine aminolyase, cinnamic acid, 3-deoxy-d-arabino heptulosonate-7-phosphate synthase (DAHPS), and shikimic acid were quantified (Figure 3). An inverse relation of DAHPS and shikimic acid was observed in the treated group of CA1 and CA2 (Figure 3A,B). The DAHPS activity was increased significantly in CA3, while it was decreased in CA1. A significant rise in the concentration of shikimic acid was observed in both CA2 and CA1. Phenylalanine was reduced in CA1, CA2, and CA3, but the change was not significant (Figure 3C). In the case of L-phenylalanine aminolyase, we found that its activity was increased in CA2 and CA3, though decreased in CA1, after the application of Mo (Figure 3D). Cinnamic acid was decreased in CA1 and CA2, but in the third species an upward trend was observed upon Mo exposure. Cinnamic acid metabolism was not affected in CA3, while a decreasing pattern was found in CA1 and CA2 (Figure 3D).

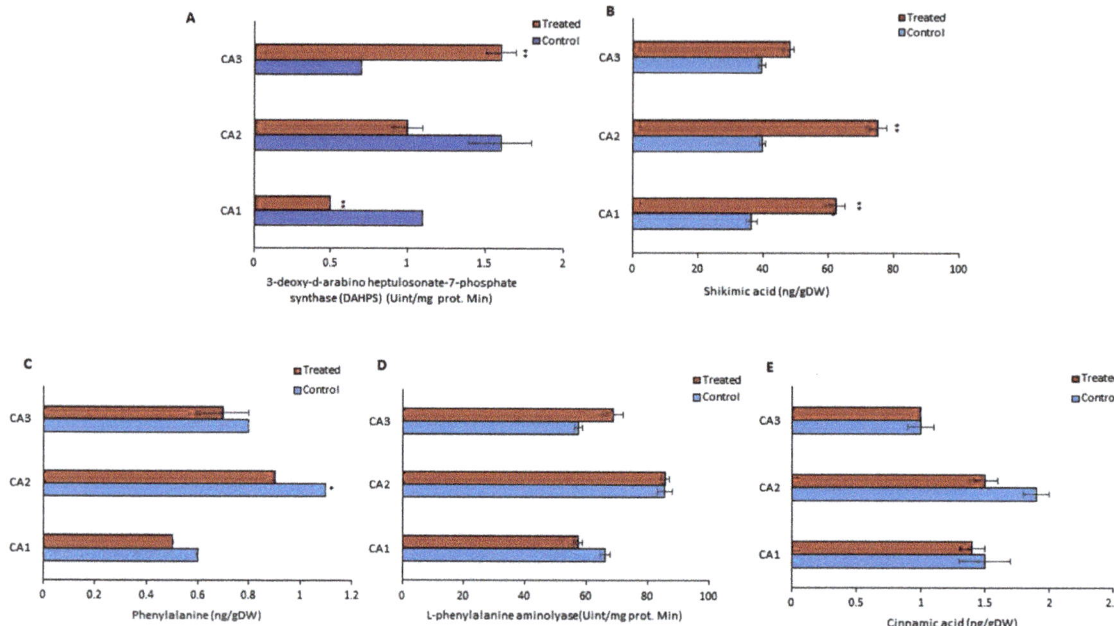

Figure 3. Effects of molybdenum (Mo) treatment on phenol synthesis pathway. (**A**): 3-deoxy-d-arabino heptulosonate-7-phosphate synthase (DAHPS) activity. (**B**): Shikimic acid level. (**C**): Phenylalanine. (**D**): L-phenylalanine aminolyase activity. (**E**): Cinnamic acid content. Control; without treatment, Treated; Mo treatment. CA1; *Canavalia ensiformis var. ladiate*, CA2; *Canavalia ensiformis var. ladiate Ricker*, CA3; *Canavalia ladiate var. alba Hisauc*. The bars above mean indicate ± standard deviation (S.D) of three independent replicates ($n = 3$). Asterisks (*) show the level of significance, * $p < 0.05$, ** $p < 0.01$.

2.5. Impact of Mo Treatment on Biological Activity

Figure 4 shows the antioxidant activity, quantified by the 2,2-diphenyl-1-picryl-hydrazyl-hydrate (DPPH) free radical scavenging, 2,2-azino-bis(3-ethylbenzothiazoline-6-sulfonic acid) (ABTS), ferric reducing antioxidant power (FRAP) methods, for the three plant species/cultivars of *Canavalia* after Mo exposure. The antioxidant activity, measured for each of the species, corresponds to an extract and Mo concentration. Regarding the antioxidant activity quantified by ABTS, the three species exhibited different antioxidant potentials, which was further increased after Mo exposure (Figure 4A). The antioxidant capacity pattern in controls was observed as CA1 > CA2 > CA3. Mo treatment induced a significant increase in CA3. Further, the values obtained from the FRAP assay also revealed the dramatic increase in antioxidant activity after Mo exposure, and a significant increase was recorded in both CA1 and CA3. Next, for the antioxidant activity quantified by DPPH, an increasing trend was obtained in all three species in the Mo-treated group, but the difference was non-significant. The glycemic index (GI), α-amylase inhibition activity, and α-glucosidase inhibition activity were assessed to explore the impact of Mo on the antidiabetic potential of *Canavalia* species. Through GI, a significant increase in antidiabetic potential was recorded in CA3, but in CA1 a decrease was noted. Further, the α-amylase and α-glucosidase inhibition potentials of *Canavalia* species were also increased. A-amylase inhibition was highly significant by CA2. A non-significant decrease was observed in CA3, as shown in Figure 4B. The evaluated percentage values of the inhibition of α-glucosidase revealed that Mo positively enhanced the inhibition potential of the three species/cultivars of *Canavalia*.

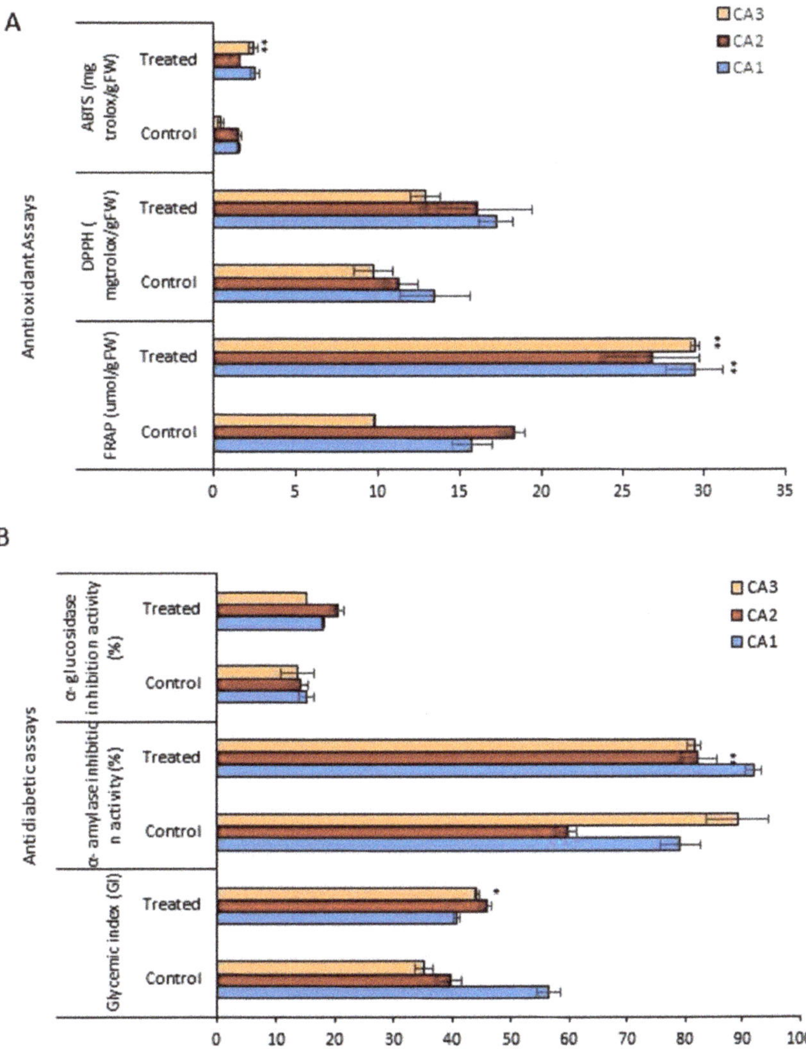

Figure 4. Effects of molybdenum (Mo) treatment on antioxidant and antidiabetic properties of *Canavalia* species (**A**): Antioxidant property, i.e., FRAP, ABTS, and DPPH. (**B**): Glycemic index (GI), α-amylase inhibition activity, and α-glucosidase inhibition activity. Control; without treatment, Treated; Mo treatment. CA; *Canavalia ensiformis var. gladiata*, CA2; *Canavalia ensiformis var. truncata Ricker*, CA3; *Canavalia gladiata var. alba*. The bars above means indicate ± standard deviation (S.D) of three independent replicates (*n* = 3). Asterisks (*) show the level of significance according to analysis of variance (ANOVA), * *p* < 0.05, ** *p* < 0.01.

2.6. Principle Component Analysis (PCA)

R software was used to perform principal component analysis (PCA). For this, Mo-treated and untreated species of *Canavalia* were chosen to analyze the interrelationship of amino acid content, phenolics, antioxidant activities, and antidiabetic activities. Here, the 1st principal component (PC1) showed 36% of the variance, while the 2nd principal component showed 28% variance between untreated and Mo-treated *Canavalia* species, as shown in Figure 5. PC1 vs. PC2 showed significant differences among Mo-treated and untreated plant species as well as between CA1 and CA3. Both PCs displayed positive correlations among most of the parameters. PC1 was highly and positively correlated with many phenols, such as rutin, caffeic acid, kaempferol, rosmarinic acid, naringenin, and coumaric acid, etc., and amino acids including alanine, proline, cystine, and methionine, etc., whereas PC2 was positively related to histidine and valine. The results showed that all phenolic acids, as well as the total amount of phenols, differed with Mo treatment, whereas the treatments and species of *Canavalia* were well separated in the map of PCA. Moreover, the results of antioxidant activities also showed a positive correlation, assessed by ABTS and FRAP assays.

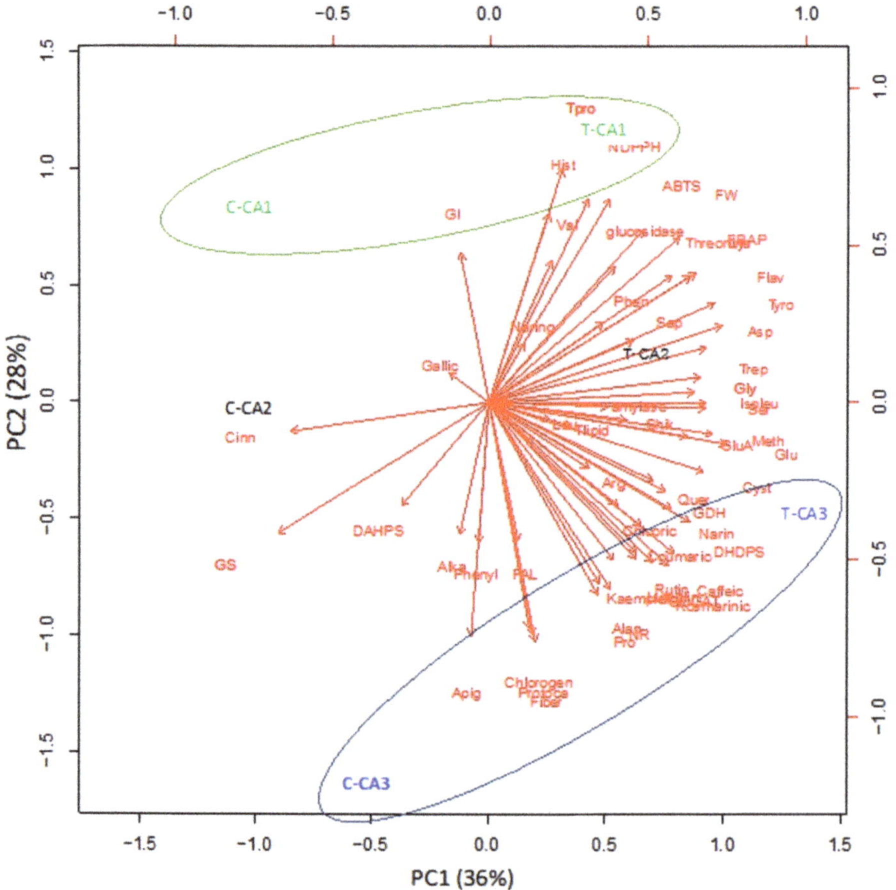

Figure 5. Principle component analysis (PCA) of 3 *Canavalia* species after seed priming with molybdenum (Mo). CA; *Canavalia ensiformis var. gladiata*, CA2; *Canavalia ensiformis var. truncata Ricker*, CA3; *Canavalia gladiata var. alba Hisauc*.

3. Discussion

The present study was conducted to explore the effects of seed priming using Mo on different endogenous chemical parameters of three species/cultivars of *Canavalia*. For this study, 0.1% ammonium molybdate was used and effects were monitored in sproutings of *Canavalia ensiformis var. gladiata* (CA1), *Canavalia ensiformis var. truncata Ricker* (CA2), and *Canavalia gladiata var. alba Hisauc* (CA3). We used the seed priming method as several studies indicated that seed treatment is a more efficient method for Mo than soil application [20,21]. Kumar Rao et al. reported that seed priming with Mo (0.5 g L^{-1} solution of sodium molybdite) increased the yield to 27%, compared to Mo soil application, in a pot study on chickpea [22]. However, genetic variations led to interesting differential patterns among different species/cultivars in our study, which we have discussed. Similarly, the results also revealed that Mo has not targeted the biosynthetic pathways uniformly.

3.1. Increased Canavalia Biomass Production by Improving Photosynthetic Pigments

Micronutrients are crucial for plant growth and development. Their deficiency contributes to a reduction in growth and changes in photosynthesis due to variations in pigment synthesis. Increased chloroplast deformation, the over-production of antioxidant enzymes, and increased production of proteins are the most common signs of stress-related responses in plants. Mo, a micronutrient, acts as a cofactor for several enzymes, thus helping to promote plant growth and biomass [23], and its exposure can lead to dramatic effects on *Canavalia* species. In the current study, the biomass of sproutings, in both Mo-treated and untreated groups, was quantified. Our findings revealed that Mo treatment significantly increased the fresh weight of sproutings in all the studied *Canavalia* species/cultivars. Alam et al. also observed that Mo application proportionately enhanced the weight of nodules in hairy vetch roots [24]. This observation might be attributed to the fact that plants require micronutrients for biosynthetic pathways and plant growth [25].

Next, the impact of Mo on the photosynthesis process was assessed, for which pigments such as chlorophyll and carotenoids play an important role. Chlorophyll, which includes chlorophyll a, chlorophyll b, and chlorophyll ab, is a green pigment for photosynthesis. Carotenoids are also important pigments that exhibit antioxidant and provitamin A roles and are used extensively as natural and safe colorants for food and cosmetics [26]. In our study, exogenous Mo increased chlorophyll content in *Canavalia*. Indeed, Mo stabilizes the structure of chloroplasts, enhances the volume and number of chloroplast grana, and increases the synthesis of chlorophyll [27]. In agreement with the current results, it has been demonstrated that micronutrient Mo application enhanced chlorophyll content in Mung Bean (*Vigna radiata* L.) when its seeds were primed with Mo [28].

3.2. Improved Nitrogen (N) Content and Amino Acid Metabolism

Amino acids are the main part of the cellular structure. Their synthesis costs in terms of energy are expected to play an important role in energy allocation. It was experimentally shown that the source of nitrogen, predominantly nitrate and/or ammonium, affects amino acid and protein levels, as well as the rate of growth and, consequently, overall biomass [29]. Nitrogen assimilation by plants directly takes part in the synthesis and conversion of amino acids through the reduction of nitrate. In our study, Mo application improved N metabolism and as a result increased total protein content and amino acids in the *Canavalia* sprouts. Similar to our results, the study of Toledo et al. showed that Mo foliar application increased nitrogenase and nitrate reductase activities, which resulted in an increase in N accumulated in the soybean leaves [30]. Among the molybdo-enzymes, nitrate reductase (NR) represents the cytosolic key enzyme of nitrogen assimilation that reduces nitrate to nitrite [31]. We have observed that the Mo priming of seeds significantly increased NR expression in CA3, which is in line with the study of Camp et al. In that study, Mo treatment resulted in a rise in total grain N in soybean seeds [32]. Interestingly, this increase was not observed in the sprouting of CA1 and CA2, which might be due to different genetic backgrounds. Mechanistically, N production via Mo priming led to the modulation of many biochemical reactions of *Canavalia* species, which was observed in different experiments of the study. Normally, all N taken up by plants is first reduced to NH4+ because it is the only reduced N form available to plants for assimilation into N-carrying amino acids [33]. Ammonium is then integrated into glutamine and glutamate. It was discovered that there are two pathways for glutamate synthesis. The glutamine synthetase (GS)/glutamate synthase (GOGAT) pathway is believed to assimilate ammonia at normal intracellular concentrations, while GDH (glutamate-dehydrogenase) plays a key role in the assimilation of ammonia into amino acids [34]. In our study, an increase in GDH and GOGAT production was detected in the three species, which was significant in CA3. GS production decreased in sprouts of the treated group and the change was significant in all three species/cultivars. The results are in line with the study of Imran et al., where Mo application up-regulated the expressions of GOGAT genes in winter wheat (*Triticum aestivum* L.) under a sole NH_4^+ source [35,36]. Moreover, dihydrodipicolinate synthase (DHDPS) and cystathionine γ-synthase (CGS) are the first committed step in the biosynthesis of lysine and methionine, respectively, which occurs naturally in plants [37,38]. We observed an increase in the activity of DHDPS and CGS, which consequently increased the production of lysine and methionine in Mo-treated *Canavalia* species. The increase was significant in CA3, which might be due to the more prominent NR activity in CA3.

The amino acids in plants are not only essential components of protein synthesis, but also serve as precursors for a wide range of secondary metabolites that are important for plant growth as well as for human nutrition and health. The enhanced quantity of most of the amino acids in the study evidenced that there was an additional availability of amino acids, probably due to the greater biological fixation of N enhanced by Mo for the enzyme nitrogenase and the production of a greater quantity of ureides. Ureides are transported from the nodules to the aerial part and later converted into amino acids. The process resulted in a greater source of amino acids and, consequently, a higher protein content, as observed in our study. Conversely, Toledo et al. did not observe an increased protein content with applications of 30 and 60 g ha^{-1} Mo applied by leaf spray or with 24 g ha^{-1} per seed treatment [30], which might be due to differences in species and/or different underlying biosynthetic mechanisms. Similar to our study, Oliveira et al. found a linear increase in the protein content of soybean when they applied doses of 0 to 800 g ha^{-1} Mo via foliar application [39]. In the current study, most of the amino acids were significantly increased in CA3 as compared to CA1 and CA2. This increase correlates to enhanced N metabolism and a significant increase in the biomass of CA3.

3.3. Improved Nutritional Value

Furthermore, the quantification of nutrients in sprouts showed a substantial increase in the lipid content of CA1, which might be due to increased synthesis and/or decreased degradation of lipids. Previous documents show that increased photosynthetic capacity due to Mo supply contributes to the accumulation of carbohydrates, such as fructose and glucose. We observed that lipids, fibers, and saponins were significantly increased in CA1. Flavonoids were increased reasonably in CA3. No significant change was observed in CA1 in all the studied parameters. Different impacts of Mo were observed in different species/cultivars, which might be due to different genetic backgrounds and the variable uptake of Mo.

3.4. Secondary Phenolic Production Improvement as a Basis for High Bioactivity

Phenolics play essential roles in plant development; these aromatic benzene ring compounds with one or more hydroxyl groups are produced by plants mainly for protection against stress [40]. In our study, the Mo priming of *Canavalia* seeds significantly increased the phenolic content of CA3 sprouts; a non-significant increase was found in CA2 and, surprisingly, a slight decrease was noticed in CA1. This might be due to differences in endogenous molecular pathways of different species. This link to the inherent levels of mechanisms may also lead to differences in the enhancing patterns of the individual phenolic compounds after Mo treatment, which we observed in our study. A similar trend was observed in DAHPS, the first enzyme of the shikimate pathway, which converts PEP and E4-P into 3-dehydroquaianate. Further, to assess the impact of Mo seed priming and enhanced phenolic production on the pharmacological properties of plants, we performed antioxidant and antidiabetic assays. FRAP, ABTS, and DPPH assays were used to evaluate the antioxidant capacity of species samples spectrophotometrically. An increase in antioxidant potential was observed after Mo treatment in all species. In the FRAP assay, the increase was significant in both CA1 and CA3. CA3 also expressed a significant change in antioxidant activity, measured via ABTS assay. This correlates to the above findings that more impacts of Mo on synthetic pathways were observed in CA3. Antioxidant activity has recently become a target for product development in the pharmaceutical and cosmetics industries [41]. *Canavalia* species are of medicinal importance due to their potential antioxidant properties. In a study, *Canavalia gladiata* extract, at the concentration of 2 mg/mL, showed an antioxidant effect comparable to that of ascorbic acid of the control group [42]. Plant growth and antioxidant capacity are greatly dependent upon N availability. Higher N improves the stress tolerance of plants via enhancement of the antioxidant ability and inhibition of lipid peroxidation [43]. Mo primarily improves the nitrogen fixation to the plant and increases its antioxidant potential, which we observed in our study. This also might be the reason for the enhanced antidiabetic potential of *Canavalia* species assessed by GI, α-amylase inhibition activity, and α-glucosidase inhibition activity. Terpenoids and flavonoids of *Canavalia gladiata* are reported to play a role in lowering glucose levels and possessing antioxidant potential [44]. The Mo-mediated enhanced concentrations of terpenoids and flavonoids might have led to the increased antioxidant and antidiabetic activities of *Canavalia* sprouts in our study.

4. Materials and Methods

4.1. Plant Material and Growth Conditions

Seeds of three *Canavalia* species: *Canavalia ensiformis var. gladiata* (CA1), *Canavalia ensiformis var. truncata Ricker* (CA2), and *Canavalia gladiata var. alba Hisauc* (CA3), were collected from the Agricultural Research Center (Giza, Egypt). Healthy seeds with a similar shape and size were washed with distilled water and dripped for 1 h in 5 gL^{-1} sodium hypochlorite, and then they were washed thoroughly with distilled water. According to a preliminary experiment, regarding the effect of ammonium molybdate (Mo) priming for 10 h at six concentrations, 0 (distilled water) and 0.025, 0.05, 0.075, 0.1, and 1.5% ammonium molybdate, 0.1% Mo-treated plants showed the highest biomass accumulation, thus 0.1% was selected to study the effect of Mo-priming on sprouts and tissue chemical composition of *Canavalia species*. About 250 seeds were primed for 10 h with distilled water or Mo (0.1%) at room temperature (24 °C). For sprouting processes, the seeds were distributed on trays filled with vermiculite and irrigated with Milli-Q water every two days. Then, the seeds were evenly transferred to trays covered with vermiculite and moistened with 150 mL of aquaponic water (mineral-rich water that was collected from a catfish tank and adjusted to a target pH of 7). The growth conditions were 150 μmol PAR m^{-2} s^{-1}, 23/18.5 °C air temperature, 63% humidity, and 16/8 h day/night photoperiod. The sprout tissues from each treatment were harvested after 9 days and weighed (Figure 6), then they were frozen in liquid nitrogen and kept at -80 °C for biochemical analyses. Each experiment was replicated at least two times, and for all assays, 3 to 5 replicates were used, and each replicate corresponded to a group of sprouts and mature plants harvested from a certain tray.

4.2. Preparation of Extracts

For sample preparation, we used an ETA 0067 grinder with grinding stones, VIPO mini grinder, followed by homogenization by Vibrom S2 (Jebavý, Trebechovice p. O., Czech Republic), and a cryogenic grinder accompanied by liquid nitrogen. Supercritical fluid extraction (SFE) using SE-1 (SeKo-K, Brno, Czech Republic) extractor and a steam distillatory apparatus according to CSN 58 0110 and CSN 6571 were successively applied for extraction and subsequent determination of the total content of *Canavalia* oils. Approximately 500 mg of each sample was transferred into an extraction column for SFE extraction. The extraction was performed at 40 MPa for 60 min, and extractor and restrictor temperatures were 80 and 120 °C, respectively. The extract was further trapped into a hexane layer inside a trapping vessel.

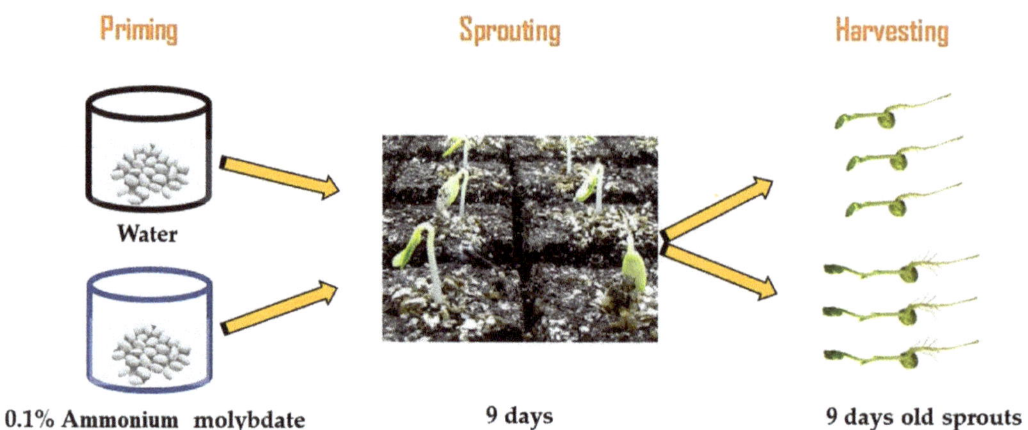

Figure 6. Flow chart of the experimental design.

4.3. Pigment Analysis

To extract pigments, samples were homogenized in acetone and then centrifuged (4 °C, 14,000× g, 20 min). HPLC (Shimadzu SIL10-ADvp, Tokyo, Japan, reversed-phase, at 4 °C) was used to analyze the obtained solution (Almuhayawi et al., 2020). Carotenoids were separated (silica-based C18 column, Waters Spherisorb, 5 μm ODS1, 4.6 × 250 mm), whereas they were eluted by (A) acetonitrile: methanol: water (81:9:10) and (B) methanol: ethyl acetate (68:32). The pigment was quantified using a diode-array detector (Shimadzu SPDM10Avp, Japan) at four wavelengths (420, 440, 462, and 660 nm). Pigments were identified by comparing the standard mixture to the relative retention time of each pigment and the concentrations were calculated using the peak area of the corresponding standard.

4.4. Determination of Nitrogen Content and Metabolism

For total nitrogen (N) content, fine ground sprout samples (0.2 g) were digested in H_2SO_4–H_2O at 260 °C, and the nitrogen levels were assessed with a CN element analyzer (NC-2100, Carlo Erba Instruments, Milan, Italy). Glutamine synthetase (GS, EC 6.3.1.2), glutamate synthase (EC 1.4.7.1), and glutamine 2-oxoglutarate aminotransferase (GOGAT, EC 1.4.7.1) enzyme activities were measured by monitoring the reduction of NADH at A_{340}. GDH determining 2-oxoglutarate-dependent NADH oxidation. GS activity was detected by assessing the formation of γ-glutamyl hydroxamate at A340. GOGAT activity was estimated by following the glutamine-dependent NADH oxidation at A340. The activity of the nitrate reductase (NR, EC 1.7.1.1) enzyme was measured by following the nitrite-dependent NADH oxidation (A_{340}) [45]. Protein concentrations were determined according to Lowry et al. [46].

4.5. Amino Acid Analysis

About 3 mg of each *Canavalia* sample was hydrolyzed with 6 M HCl (6 h, 150 °C), and the acidic suspension was evaporated by rotary evaporation (RE500 Yamato Scientific America Inc., Santa Clara, CA, USA) and redissolved in 2 mL of sodium citrate buffer (pH 2.2). For the derivation step, phthalaldehyde (OPA) (7.5 mM) was mixed with samples in citrate buffer (OPA reagent containing β-mercaptoethanol and Brij 35). The HPLC (Shimadzu SPDM10Avp, Tokyo, Japan) analysis was evaluated using internal and external standards with the aid of fifteen amino acid reference standards (0.05 μmoles mL^{-1} amino acid) for retention time detection of each single amino acid. In the meantime, the internal standard (0.05 μmoles mL^{-1} α aminobutyric) was added to both the plant sample and the reference. Reversed-phase C18 column (100 × 4.6 mm × 1/4″ Microsorb 100-3 C18, Agilent Technologies, Santa Clara, CA, USA) was used and gradient elution was performed by mobile phase consisting of 0.1 M sodium acetate and methanol (9:1). Measurement was at a wavelength of 360 and 455 nm. Star Chromatography software (Varian version 5.51) was applied for amino acid peak integration and final calculation was carried out to express values as μmol/gFW.

4.6. Determination of Total Carbohydrates, Protein, Lipids, and Fibers

Carbohydrate evaluation was processed for each *Canavalia* sample (eCO$_2$-treated and control plants) by Nelson's method [47], while the concentration of protein was measured for each frozen *Canavalia* sample (0.2 g FW). Detection of total lipids was performed using Folch's method [48], whereas the samples were subjected to homogenization in a chloroform/methanol mixture (2:1), followed by centrifugation at 3000× g for 15 min and concentration of chloroform extract containing lipids via a rotary evaporator (RE500 Yamato Scientific America Inc.); after that, the produced pellets were re-dissolved in a mixture of toluene/ethanol (4/1 v/v) and then mixed with saline solution, re-concentrated again, and weighed to calculate the total lipid content (μg/gFW). Fibers were also extracted and evaluated according to the AOAC method [49], starting with sample gelatinization using α-amylase (30 min, pH 6, 100 °C), then enzymatic digestion by protease (30 min, pH 7.5, 60 °C). Thereafter, starch and proteins were removed by amyloglucosidase (30 min, pH 6 and 0 °C). Finally, fibers were precipitated using ethanol, washed, and the yielded residue was weighed and expressed as μg/gFW.

4.7. Determination of Phenolics and Their Precursors and Related Enzymes

The determination of phenolics and their precursor metabolites was carried out using an ultra-performance liquid chromatography system (Waters Acquity UPLC, Boston, MA, USA) coupled with a quadrupole mass spectrometer (Waters Xevo TQ, Manchester, UK) provided with an ESI source according to Wang et al.'s method [50]. Phenolic and flavonoid levels were identified by comparing the standard mixture of different phenols and flavonoids to the relative retention time. The concentration of each compound (μmol/gFW) was calculated using the peak area of the corresponding standard.

In addition, deoxy-d-arabino 201 heptulosonate-7-phosphate synthase (DAHPS) activity (umol/mg protein. min) was analyzed according to Wang et al., (2020). This enzyme catalyzes the reactions in cinnamic and shikimic acid pathways that are involved in phenylpropanoid biogenesis and therefore in the biosynthesis of coumarins. Samples were first homogenized in 3 mL 50 mM Tris-HCL buffer (pH = 7.5). The assay mixture contained 0.1 mM erythrose-4-phosphate, 0.2 mM phosphoenolpyruvate, and 0.1 mM MnSO$_4$/0.1 mM CoCl2. In total, the reaction was initiated by enzyme addition and terminated by 25% (w/v) trichloroacetic acid addition. The activity of DAHPS was measured at 549 nm. Regarding PAL, it was extracted in 1 mL of 200 mM sodium borate buffer (pH 8.8) and then evaluated by measuring the production of trans-cinnamic acid at 290 nm.

4.8. Measurement of Antioxidant Capacity

The determination of antioxidant activity was performed through ferric reducing antioxidant potential, 2,2-diphenyl-1-picryl-hydrazyl-hydrate (DPPH) free radical scavenging, 2,2-azino-bis(3-ethylbenzothiazoline-6-sulfonic acid) (ABTS), and ferric reducing antioxidant power (FRAP) assays according to the reported method [51]. In the FRAP method, about 0.2 g of *Canavalia* samples was firstly extracted in ethanol (80%) and centrifuged at 14,000 rpm for 20 min. Afterward, 0.1 mL of tested extracts was added to 0.25 mL of FRAP reagent (20 mM FeCl$_3$ in 0.25 M acetate buffer, pH 3.6) at room temperature, the absorbance was further measured at 517 nm, and antioxidant activity is expressed as μmol/ gFW. In the DPPH assay, the reaction mixture was composed of 3.9 mL of 200 μM DPPH (prepared in ethanol) and 0.1 mL of samples, incubated in the dark for half an hour at room temperature (35 ± 2 °C), and the absorbance was further detected at 517 nm. The percentage of inhibition was calculated versus a control. ABTS (2,2′ -azino-bis(3-ethylbenzothiazoline-6-sulfonic acid) scavenging activity was measured by mixing ABTS with potassium persulphate (2.4 mM), then the reaction was performed for 12 in the dark, and absorbance was measured at 734 nm and antioxidant activity was calculated as trolox/ gFW.

4.9. Anti-Diabetic Activity

4.9.1. Determination of In Vitro Glycemic Index (GI)

In vitro starch hydrolysis assays were used for the evaluation of GI [52]. Sprouts were powdered and incubated with pepsin (100 mg/mL) in a reaction buffer (HCl-KCl buffer, pH 1.5), incubated for 1 h at 40 °C with shaking, and then the mixture was diluted in phosphate buffer (pH 6.9), adding α-amylase and incubation at 37 °C. Approximately 1mL of aliquots was taken every 30 min and boiled for 20 min for the sake of amylase enzyme inactivation. Residual starch was digested to glucose by adding 60 µL amyloglucosidase together with 0.4 M of sodium acetate buffer (pH 4.75), and the reaction mixture was previously incubated at 60 °C for 50 min. Aliquots (0.6-mL) were incubated with 1.2 mL glucose oxidase/peroxidase at 37 °C for 35 min, followed by absorbance of the mixture at 500 nm. Starch digestion rate (SDR) was evaluated as a hydrolyzed starch percentage at different times (0, 30, 60, 90, 120, and 180 min). The area under the hydrolysis curve (AUC, 0–180 min) was enumerated.

4.9.2. α-Glucosidase Inhibition Assay

The percentage of α-glucosidase inhibition activity was measured, as already reported [53]. The assay principle is about measuring the amount of para-nitrophenolate released by para-nitrophenyl glucopyranoside. The hydroethanolic extract of seeds and sprouts was mixed with α-glucosidase (2 U/mL) and incubated at 37 °C for 5 min. Then, 1 mM of para-nitrophenylglucopyranoside was added, dissolved in a phosphate buffer of 50 mM (pH 6.8) to the reaction buffer, and incubated for 20 min at 37 °C. The reaction was shut down by the addition of sodium carbonate (1M). The activity of α-glucosidase was measured at 405 nm. The percentage of inhibition was calculated.

4.9.3. α-Amylase Inhibition Assay

α-Amylase inhibition was evaluated by mixing starch (2 mg) with 0.5 M Tris-HCl buffer (pH 6.9) and 0.01 M CaCl2, boiling for 5 min, cooling at room temperature, and then incubating for 5 min at 37 °C. Then, the α-amylase (U/mL) was added and incubated again at 37 °C for 10 min. After that, 500 µL 0.1% 3,5-dinitro salicylic acid was added and incubated for 10 min at 100 °C. After cooling, the absorbance was determined at 540 nm for α-amylase inhibition calculation. The percentage of inhibition was calculated.

4.10. Statistical Analyses

The $p < 0.05$ values were used to illustrate the statistical importance among the groups. The statistical analysis was conducted using both SPSS and MS Excel. Replication of each experiment was performed twice. Three to five replicates were used for all assays and each of the replicates corresponded to a group of control plants and Mo-treated plants. The PCA was carried out on R software.

5. Conclusions

The results demonstrated the benefits of Mo seed priming of three *Canavalia* species/cultivars to improve growth and N nutrition. Mo enrichment significantly triggered the photosynthetic pigments and enhanced the biomass of sprouts. In CA1, the lipids, fibers, flavonoids, and saponins were significantly increased, and in CA3 a significant rise was observed in phenolics and flavonoids. In CA3, higher differences were observed in N assimilation processes and amino acids as compared to CA1 and CA2, indicating that Mo-mediated impacts were greater in CA3. Further, Mo treatment affected the biosynthesis pathway of phenols, revealed by increased phenolic compounds in the sprouts. The raised levels of phenols and flavonoids were attributed to enhanced antioxidant acidity in CA1 and CA3. Differential patterns revealed that Mo-mediated impacts were not uniform in the *Canavalia* species/cultivars, which might be due to genetic variations and differential uptake. Overall, the findings from this research may facilitate a better understanding of the relationship of Mo application with yield and provide useful information for increasing the nutritional and pharmacological value of *Canavalia* species.

Author Contributions: Conceptualization, M.K.O. and H.A.; methodology, S.A.A., S.M.A.-Q.; software, N.A., A.A.A.-G., I.A.A. and H.A.; validation, M.K.O., A.I. and H.A.; formal analysis, Z.K.A., N.A. and S.S.; investigation, A.A.Q., W.H.S. and A.A.A.-G.; resources, M.K.O.; data curation, S.S., H.A. and N.A.; writing—original draft preparation, N.A.; writing—review and editing, N.A. and H.A. All authors have read and agreed to the published version of the manuscript.

Funding: This research was funded by a grant from the Research Group Program (RG-1439-63), King Saud University, Riyadh, Saudi Arabia.

Institutional Review Board Statement: Not applicable.

Informed Consent Statement: Not applicable.

Data Availability Statement: Data presented in this study are available on reasonable request.

Acknowledgments: The authors extend their appreciation to the Dean of Scientific Research, King Saud University for funding this work through the research group project number RG-1439-63. We also thank the Deanship of Scientific Research and RSSU at King Saud University for their support.

Conflicts of Interest: The authors declare no conflict of interest.

References

1. Michael, K.; Sogbesan, O.; Onyia, L. Effect of processing methods on the nutritional value of *Canavalia ensiformis* jack bean seed meal. *J. Food Process. Technol.* **2018**, *9*, 766.
2. Gebrelibanos, M.; Tesfaye, D.; Raghavendra, Y.; Sintayeyu, B. Nutritional and health implications of legumes. *Int. J. Pharm. Sci. Res.* **2013**, *4*, 1269.
3. Gulewicz, P.; Martínez-Villaluenga, C.; Frias, J.; Ciesiołka, D.; Gulewicz, K.; Vidal-Valverde, C. Effect of germination on the protein fraction composition of different lupin seeds. *Food Chem.* **2008**, *107*, 830–844. [CrossRef]
4. Yamashiro, A.; Yamashiro, T. Utilization on extrafloral nectaries and fruit domatia of *Canavalia lineata* and *C. cathartica* (Leguminosae) by ants. *Arthropod-Plant Interact.* **2008**, *2*, 1–8. [CrossRef]
5. Sridhar, K.R.; Seena, S. Nutritional and antinutritional significance of four unconventional legumes of the genus *Canavalia*—A comparative study. *Food Chem.* **2006**, *99*, 267–288. [CrossRef]
6. Jeon, K.S.; Na, H.J.; Kim, Y.M.; Kwon, H.J. Antiangiogenic activity of 4-O-methylgallic acid from *Canavalia gladiata*, a dietary legume. *Biochem. Biophys. Res. Commun.* **2005**, *330*, 1268–1274. [CrossRef] [PubMed]
7. Kim, O.K.; Chang, J.Y.; Nam, D.E.; Park, Y.K.; Jun, W.; Lee, J. Effect of *Canavalia gladiata* Extract Fermented with Aspergillus oryzae on the Development of Atopic Dermatitis in NC/Nga Mice. *Int. Arch. Allergy Immunol.* **2015**, *168*, 79–89. [CrossRef]
8. Byun, J.S.; Han, Y.S.; Lee, S.S. The effects of yellow soybean, black soybean, and sword bean on lipid levels and oxidative stress in ovariectomized rats. *Int. J. Vitam. Nutr. Res.* **2010**, *80*, 97–106. [CrossRef]
9. Han, S.S.; Hur, S.J.; Lee, S.K. A comparison of antioxidative and anti-inflammatory activities of sword beans and soybeans fermented with Bacillus subtilis. *Food Funct.* **2015**, *6*, 2736–2748. [CrossRef]
10. Lyu, Y.R.; Jung, S.-j.; Lee, S.-W.; Yang, W.-K.; Kim, S.-H.; Jung, I.C.; Kim, K.-H.; Kim, H.-Y.; Yang, Y.J.; Lee, Y.; et al. Efficacy and safety of CAEC (*Canavalia gladiata* arctium lappa extract complex) on immune function enhancement: An 8 week, randomised, double-blind, placebo-controlled clinical trial. *J. Funct. Foods* **2020**, *75*, 104259. [CrossRef]
11. Jashni, M.K.; Mehrabi, R.; Collemare, J.; Mesarich, C.H.; de Wit, P.J. The battle in the apoplast: Further insights into the roles of proteases and their inhibitors in plant–pathogen interactions. *Front. Plant Sci.* **2015**, *6*, 584. [CrossRef] [PubMed]
12. Mosolov, V.; Valueva, T. Participation of proteolytic enzymes in the interaction of plants with phytopathogenic microorganisms. *Biochemistry* **2006**, *71*, 838–845. [CrossRef]
13. Sowndhararajan, K.; Siddhuraju, P.; Manian, S. Antioxidant activity of the differentially processed seeds of Jack bean (*Canavalia ensiformis* L. DC). *Food Sci. Biotechnol.* **2011**, *20*, 585–591. [CrossRef]
14. Farooq, M.; Wahid, A.; Siddique, K.H. Micronutrient application through seed treatments: A review. *J. Soil Sci. Plant Nutr.* **2012**, *12*, 125–142. [CrossRef]
15. Matoso, S.C.; Kusdra, J.F. Nodulation and growth of bean in response to application of molybdenum and rhizobia inoculation. *Rev. Bras. Eng. Agrícola Ambient.* **2014**, *18*, 567–573. [CrossRef]
16. De Albuquerque, H.C.; Pegoraro, R.F.; Vieira, N.M.; de JesusFerreira Amorim, I.; Kondo, M.K. Nodulor capability and agronomic characteristics of common bean plants subjected to fragmented molybdenum and nitrogen fertilization. *Rev. Ciência Agronômica* **2012**, *43*, 214.

17. Mendel, R.R.; Hänsch, R. Molybdoenzymes and molybdenum cofactor in plants. *J. Exp. Bot.* **2002**, *53*, 1689–1698. [CrossRef]
18. Almeida, F.F.D.; Araújo, A.P.; Alves, B.J.R. Seeds with high molybdenum concentration improved growth and nitrogen acquisition of rhizobium-inoculated and nitrogen-fertilized common bean plants. *Rev. Bras. Ciência Solo* **2013**, *37*, 367–378. [CrossRef]
19. Hesberg, C.; Hänsch, R.; Mendel, R.R.; Bittner, F. Tandem orientation of duplicated xanthine dehydrogenase genes from Arabidopsis thaliana: Differential gene expression and enzyme activities. *J. Biol. Chem.* **2004**, *279*, 13547–13554. [CrossRef]
20. Donald, C.; Spencer, D. The control of molybdenum deficiency in subterranean clover by pre-soaking the seed in sodium molybdate solution. *Aust. J. Agric. Res.* **1951**, *2*, 295–301. [CrossRef]
21. Johansen, C.; Musa, A.M.; Kumar Rao, J.; Harris, D.; Yusuf Ali, M.; Shahidullah, A.; Lauren, J.G. Correcting molybdenum deficiency of chickpea in the High Barind Tract of Bangladesh. *J. Plant Nutr. Soil Sci.* **2007**, *170*, 752–761. [CrossRef]
22. Kumar Rao, J.; Harris, D.; Johansen, C.; Musa, A. Low cost provision of molybdenum (Mo) to chickpeas grown in acid soils. In Proceedings of the International Fertiliser Association Symposium on Micronutrients', New Delhi, India, 23–25 February 2004.
23. Hadi, F.; Ali, N.; Fuller, M.P. Molybdenum (Mo) increases endogenous phenolics, proline and photosynthetic pigments and the phytoremediation potential of the industrially important plant *Ricinus communis* L. for removal of cadmium from contaminated soil. *Environ. Sci. Pollut. Res.* **2016**, *23*, 20408–20430. [CrossRef] [PubMed]
24. Alam, F.; Kim, T.Y.; Kim, S.Y.; Alam, S.S.; Pramanik, P.; Kim, P.J.; Lee, Y.B. Effect of molybdenum on nodulation, plant yield and nitrogen uptake in hairy vetch (*Vicia villosa* Roth). *Soil Sci. Plant Nutr.* **2015**, *61*, 664–675. [CrossRef]
25. Gungula, D.; Garjila, Y. The effects of molybdenum application on growth and yield of cowpea in Yola. *J. Agric. Environ. Sci.* **2006**, *1*, 96–101.
26. Zakynthinos, G.; Varzakas, T. Carotenoids: From plants to food industry. *Curr. Res. Nutr. Food Sci.* **2016**, *4*, 38. [CrossRef]
27. Han, Z.; Wei, X.; Wan, D.; He, W.; Wang, X.; Xiong, Y. Effect of molybdenum on plant physiology and cadmium uptake and translocation in rape (*Brassica napus* l.) Under different levels of cadmium stress. *Int. J. Environ. Res. Public Health* **2020**, *17*, 2355. [CrossRef]
28. Hayyawi, N.J.H.; Al-Issawi, M.H.; Alrajhi, A.A.; Al-Shmgani, H.; Rihan, H. Molybdenum Induces Growth, Yield, and Defence System Mechanisms of the Mung Bean (*Vigna radiata* L.) under Water Stress Conditions. *Int. J. Agron.* **2020**, *2020*, 8887329. [CrossRef]
29. Helali, S.M.r.; Nebli, H.; Kaddour, R.; Mahmoudi, H.; Lachaâl, M.; Ouerghi, Z. Influence of nitrate—Ammonium ratio on growth and nutrition of Arabidopsis thaliana. *Plant Soil* **2010**, *336*, 65–74. [CrossRef]
30. Toledo, M.Z.; Garcia, R.A.; Rocha Pereira, M.R.; Fernandes Boaro, C.S.; Lima, G.P.P. Nodulação e atividade da nitrato redutase em função da aplicação de molibdênio em soja. *Biosci. J.* **2010**, *26*, 858–864.
31. Campbell, W.H. Structure and function of eukaryotic NAD(P)H:nitrate reductase. *Cell. Mol. Life Sci. CMLS* **2001**, *58*, 194–204. [CrossRef]
32. Campo, R.J.; Araujo, R.S.; Hungria, M. Molybdenum-enriched soybean seeds enhance N accumulation, seed yield, and seed protein content in Brazil. *Field Crop. Res.* **2009**, *110*, 219–224. [CrossRef]
33. Ruiz, J.M.; Rivero, R.M.; Romero, L. Comparative effect of Al, Se, and Mo toxicity on NO3(-) assimilation in sunflower (*Helianthus annuus* L.) plants. *J. Environ. Manag.* **2007**, *83*, 207–212. [CrossRef]
34. Cammaerts, D.; Jacobs, M. A study of the role of glutamate dehydrogenase in the nitrogen metabolism of Arabidopsis thaliana. *Planta* **1985**, *163*, 517–526. [CrossRef] [PubMed]
35. Imran, M.; Sun, X.; Hussain, S. Molybdenum-induced effects on nitrogen metabolism enzymes and elemental profile of winter wheat (*Triticum aestivum* L.) under different nitrogen sources. *Int. J. Mol. Sci.* **2019**, *20*, 3009. [CrossRef] [PubMed]
36. Bataung, M.; Xianjin, Z.; Guozheng, Y.; Xianlong, Z. Review: Nitrogen assimilation in crop plants and its affecting factors. *Can. J. Plant Sci.* **2012**, *92*, 399–405. [CrossRef]
37. Pearce, F.G.; Hudson, A.O.; Loomes, K.; Dobson, R.C. Dihydrodipicolinate synthase: Structure, dynamics, function, and evolution. *Subcell Biochem.* **2017**, *83*, 271–289.
38. Ravanel, S.; Gakière, B.; Job, D.; Douce, R. The specific features of methionine biosynthesis and metabolism in plants. *Proc. Natl. Acad. Sci. USA* **1998**, *95*, 7805–7812. [CrossRef]
39. Oliveira e Oliveira, C.; Cipriano Pinto, C.; Garcia, A.; Trombeta Bettiol, J.V.; de Sá, M.E.; Lazarini, E. Produção de sementes de soja enriquecidas com molibdênio. *Rev. Ceres* **2017**, *64*. [CrossRef]
40. Bhattacharya, A.; Sood, P.; Citovsky, V. The roles of plant phenolics in defence and communication during Agrobacterium and Rhizobium infection. *Mol. Plant Pathol.* **2010**, *11*, 705–719. [CrossRef]
41. Żymańczyk-Duda, E.; Szmigiel-Merena, B.; Brzezińska-Rodak, M. Natural antioxidants–properties and possible applications. *J. Appl. Biotechnol. Bioeng.* **2018**, *5*, 251–258. [CrossRef]
42. Kim, H.C.; Moon, H.G.; Jeong, Y.C.; Park, J.U.; Kim, Y.R. Anti-oxidant, skin whitening, and antibacterial effects of *Canavalia gladiata* extracts. *Med. Biol. Sci. Eng.* **2018**, *1*, 11–17. [CrossRef]
43. Fu, J.; Huang, B. Effects of foliar application of nutrients on heat tolerance of creeping bentgrass. *J. Plant Nutr.* **2003**, *26*, 81–96. [CrossRef]
44. Anitha, K.; Lakshmi, S.M.; Satyanarayana, S. Antidiabetic, lipid lowering and antioxidant potentiating effect of *Canavalia* species in high fat diet-streptozotocin induced model. *Adv. Tradit. Med.* **2020**, *20*, 609–618. [CrossRef]
45. Hageman, R.; Reed, A. [24] Nitrate reductase from higher plants. In *Methods in Enzymology*; Elsevier: Amsterdam, The Netherlands, 1980; Volume 69.

46. Classics Lowry, O.; Rosebrough, N.; Farr, A.; Randall, R. Protein measurement with the Folin phenol reagent. *J. Biol. Chem.* **1951**, *193*, 265–275. [CrossRef]

47. Clark, R.A.; Delia, J.G. Cognitive complexity, social perspective-taking, and functional persuasive skills in second-to ninth-grade children. *Hum. Commun. Res.* **1977**, *3*, 128–134. [CrossRef]

48. Bligh, E.G.; Dyer, W.J. A rapid method of total lipid extraction and purification. *Can. J. Biochem. Physiol.* **1959**, *37*, 911–917. [CrossRef]

49. Horwitz, W.; Chichilo, P.; Reynolds, H. Official methods of analysis of the Association of Official Analytical Chemists. In *Official Methods of Analysis of the Association of Official Analytical Chemists*; Association of Official Analytical Chemists: Washington, DC, USA, 1970.

50. Wang, Z.; Xiao, S.; Wang, Y.; Liu, J.; Ma, H.; Wang, Y.; Tian, Y.; Hou, W. Effects of light irradiation on essential oil biosynthesis in the medicinal plant *Asarum heterotropoides* Fr. Schmidt var. mandshuricum (Maxim) Kitag. *PLoS ONE* **2020**, *15*, e0237952. [CrossRef] [PubMed]

51. Almuhayawi, M.S.; Hassan, A.H.; Al Jaouni, S.K.; Alkhalifah, D.H.M.; Hozzein, W.N.; Selim, S.; AbdElgawad, H.; Khamis, G. Influence of elevated CO_2 on nutritive value and health-promoting prospective of three genotypes of Alfalfa sprouts (*Medicago sativa*). *Food Chem.* **2021**, *340*, 128147. [CrossRef]

52. Hasan, M.; Hanafiah, M.M.; Aeyad Taha, Z.; AlHilfy, I.H.; Said, M.N.M. Laser irradiation effects at different wavelengths on phenology and yield components of pretreated maize seed. *Appl. Sci.* **2020**, *10*, 1189. [CrossRef]

53. Kazeem, M.; Adamson, J.; Ogunwande, I. Modes of inhibition of α-amylase and α-glucosidase by aqueous extract of *Morinda lucida* Benth leaf. *BioMed Res. Int.* **2013**, *2013*, 527570. [CrossRef]

Article

Effect of Elevated CO_2 on Biomolecules' Accumulation in Caraway (*Carum carvi* L.) Plants at Different Developmental Stages

Hamada AbdElgawad [1,2,*], **Mohammad K. Okla** [3,*], **Saud S. Al-amri** [3], **Abdulrahman AL-Hashimi** [3], **Wahida H. AL-Qahtani** [4], **Salem Mesfir Al-Qahtani** [5], **Zahid Khorshid Abbas** [5,6], **Nadi Awad Al-Harbi** [5], **Ayman Abd Algafar** [3], **Mohammed S. Almuhayawi** [7], **Samy Selim** [8] and **Mohamed Abdel-Mawgoud** [9]

[1]   Botany and Microbiology Department, Faculty of Science, Beni-Suef University, Beni-Suef 62521, Egypt
[2]   Laboratory for Molecular Plant Physiology and Biotechnology, Department of Biology, University of Antwerp, Groenenborgerlaan 171, 2020 Antwerp, Belgium
[3]   Botany and Microbiology Department, College of Science, King Saud University, Riyadh 11451, Saudi Arabia; saualamri@ksu.edu.sa (S.S.A.-a.); aalhashimi@ksu.edu.sa (A.A.-H.); aghafr@ksu.edu.sa (A.A.A.)
[4]   Department of Food Sciences & Nutrition, College of Food & Agriculture Sciences, King Saud University, Riyadh 11451, Saudi Arabia; wahida@ksu.edu.sa
[5]   Biology Department, University College of Taymma, Tabuk University, Tabuk 71491, Saudi Arabia; salghtani@ut.edu.sa (S.M.A.-Q.); Znourabbas@ut.edu.sa or Zahid104@yahoo.com (Z.K.A.); nalharbi@ut.edu.sa (N.A.A.-H.)
[6]   Biology Department, College of Science, Tabuk University, Tabuk 71491, Saudi Arabia
[7]   Department of Medical Microbiology and Parasitology, Faculty of Medicine, King Abdulaziz University, Jeddah 21589, Saudi Arabia; msalmuhayawi@kau.edu.sa
[8]   Department of Clinical Laboratory Sciences, College of Applied Medical Sciences, Jouf University, Sakaka 72388, Saudi Arabia; sabdulsalam@ju.edu.sa
[9]   Department of Medicinal and Aromatic Plants, Desert Research Centre, Cairo 11753, Egypt; Mohamed_drc@yahoo.com
*   Correspondence: hamada.abdelgawad@uantwerpen.be (H.A.); malokla@ksu.edu.sa (M.K.O.)

Citation: AbdElgawad, H.; Okla, M.K.; Al-amri, S.S.; AL-Hashimi, A.; AL-Qahtani, W.H.; Al-Qahtani, S.M.; Abbas, Z.K.; Al-Harbi, N.A.; Abd Algafar, A.; Almuhayawi, M.S.; et al. Effect of Elevated CO_2 on Biomolecules' Accumulation in Caraway (*Carum carvi* L.) Plants at Different Developmental Stages. *Plants* **2021**, *10*, 2434. https://doi.org/10112434

Academic Editors: Angelica Galieni and Beatrice Falcinelli

Received: 17 October 2021
Accepted: 4 November 2021
Published: 11 November 2021

Publisher's Note: MDPI stays neutral with regard to jurisdictional claims in published maps and institutional affiliations.

Abstract: Caraway plants have been known as a rich source of phytochemicals, such as flavonoids, monoterpenoid glucosides and alkaloids. In this regard, the application of elevated CO_2 (eCO_2) as a bio-enhancer for increasing plant growth and phytochemical content has been the focus of many studies; however, the interaction between eCO_2 and plants at different developmental stages has not been extensively explored. Thus, the present study aimed at investigating the changes in growth, photosynthesis and phytochemicals of caraway plants at two developmental stages (sprouts and mature tissues) under control and increased CO_2 conditions (ambient CO_2 (a CO_2, 400 ± 27 µmol CO_2 mol^{-1} air) and eCO_2, 620 ± 42 µmol CO_2 mol^{-1} air ppm). Moreover, we evaluated the impact of eCO_2-induced changes in plant metabolites on the antioxidant and antibacterial activities of caraway sprouts and mature plants. CO_2 enrichment increased photosynthesis and biomass accumulation of both caraway stages. Regarding their phytochemical contents, caraway plants interacted differently with eCO_2, depending on their developmental stages. High levels of CO_2 enhanced the production of total nutrients, i.e., carbohydrates, proteins, fats and crude fibers, as well as organic and amino acids, in an equal pattern in both caraway sprouts and mature plants. Interestingly, the eCO_2-induced effect on minerals, vitamins and phenolics was more pronounced in caraway sprouts than the mature tissues. Furthermore, the antioxidant and antibacterial activities of caraway plants were enhanced under eCO_2 treatment, particularly at the mature stage. Overall, eCO_2 provoked changes in the phytochemical contents of caraway plants, particularly at the sprouting stage and, hence, improved their nutritive and health-promoting properties.

Keywords: high CO_2; caraway plants; sprouting; mature plants; nutritious metabolites; antioxidant; antimicrobial activity

1. Introduction

Sprouts are the young seedlings produced through seed germination; they have become more popular vegetables and functional foods in most western countries [1]. Sprouts are rich in active phenolic compounds, proteins, minerals, vitamins and glucosinolates. Thus, they are considered as healthy foods with higher antioxidant activities [2,3].

Caraway (*Carum carvi* L.), has been known as an important medicinal plant, being cultivated worldwide. Caraway plants have been used as a remedy for various ailments, such as diarrhea, asthma, cholera and hypertension [4]. In addition, caraway fruits have been prescribed in herbal mixtures as a drug for digestive, carminative and lactogenic disorders. Moreover, caraway have been used as antiallergic, antibacterial, anthelmintic [5], antifungal [6], bronchodilator [7] and cholinergic [8] agents. Such versatile biological activities are mainly due to the existence of multiple phyto-constituents including steroptin, cumene, thymene, amino acids such as lysine and threonine, calcium, iron, starch, tannins and dietary fibers [9]. The essential oils of caraway seeds contain thymol, γ-terpinene, *p*-cymene, β-pinene, α-pinene and carvacrol, in addition to monoterpenes such as carvone and limonene, which usually make up 95% of the caraway essential oils [10,11]. Thus, caraway essential oils have largely been used as a raw material in pharmaceutical industries [12].

Thus, in view of the wide traditional uses of caraway plants, several strategies have been approved to improve the growth, bioactivity and nutritive values of *Carum* sprouts [13, 14]. For example, different innovative approaches such as laser light, UV-B irradiation [15, 16] and high atmospheric CO_2 were recommended to enhance plant growth and nutritive values [17,18]. Previous studies have reported that increased atmospheric CO_2 levels significantly influenced the plant chemical composition by changing carbon and nitrogen metabolism [19,20], which in turn increases the availability of phytochemicals required for growth [21]. Moreover, eCO_2 was demonstrated to enhance photosynthesis, leading to biosynthesis of sugars, which could be then broken down in the dark respiration process [20, 22]. Subsequently, these processes are assumed to provide precursors and energy needed for the biosynthesis of active metabolites. Moreover, eCO_2 was found to enhance the biological activities of many plant species, such as peppermint, basil, dill and parsley, as well as lemongrass, brassica and alfalfa sprouts [14,17,23–25]. To the best of our knowledge, the influence of eCO_2 on caraway plants and associated metabolic changes have not been previously investigated. Thus, this study evaluates the effect of eCO_2 treatment on the nutritive values, phytochemicals (i.e., pigments, phenolics, flavonoids, vitamins, mineral profiles and essential oil yields) and antioxidant activities of both caraway sprouts and mature plants. Therefore, this study was performed, not only to enhance the sprouting conditions under eCO_2 but also to obtain insights into how eCO_2 improves the sprouting process, and to compare the significant differences between caraway sprouts and mature plants regarding their response to the effect of eCO_2 on their phytochemical contents and biological activities.

2. Results

2.1. Growth and Biomass Production of Caraway Sprouts and Mature Plants as Affected by eCO_2

The present investigation has clearly shown that eCO_2 treatment has exerted significant increases in biomass (fresh and dry weights) in the examined sprouts and mature plants when compared to controls (Figure 1). Meanwhile, the effect of eCO_2 seemed to be more pronounced on caraway mature plants than the sprouts. On the other hand, eCO_2 significantly enhanced the accumulation of photosynthetic pigments in caraway sprouts and mature plants, whereas significant increments in chlorophyll a, b and (a + b) were obtained for both sprouts and mature plants (Figure 1).

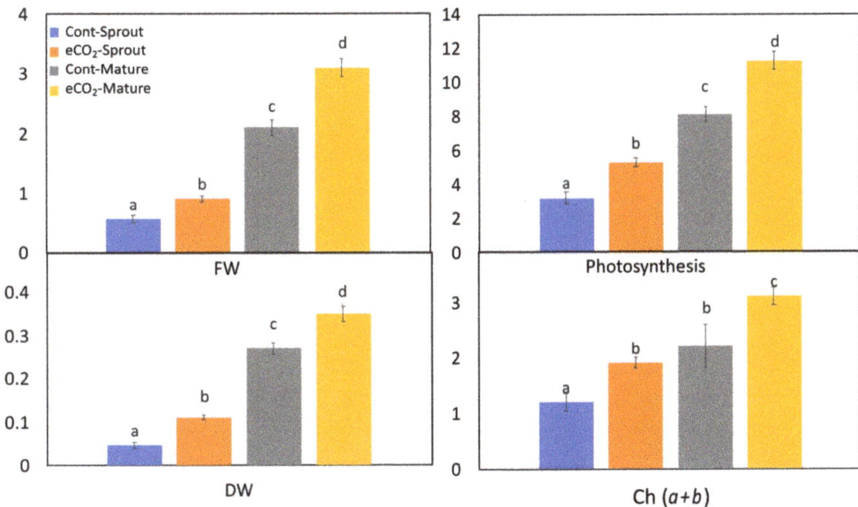

Figure 1. Biomass; fresh weight (FW) (mg/g FW) and dry weight (DW) (mg/g FW), photosynthesis (μmol CO_2 m^{-m} s^{-s}), and pigments content (chlorophyll *a* + *b*) (mg/g FW) of control and eCO$_2$-treated caraway sprouts and mature plants. Data are represented by the means of at least 3 replicates $\pm$ standard deviations. Different small letter superscripts (a, b, c and d) within a row indicate significant differences between control and eCO$_2$-samples. One-way analysis of variance (ANOVA) was performed. Tukey's test was used as the post-hoc test for the separation of means ($p < 0.05$).

2.2. Impact of eCO$_2$ on Minerals and Primary Metabolites of Caraway Sprouts and Mature Plants

In the current study, the impact of eCO$_2$ on total nutrients, i.e., primary (lipid, proteins, sugars and crude fibers) and secondary metabolites (flavonoids, phenols and steroids) was investigated in both caraway sprouts and mature plants. The treatment with eCO$_2$ significantly increased most of the detected total nutrients in sprouts and mature plants, as compared to controls (Figure 2). Moreover, we determined the levels of mineral elements, i.e., macrominerals (Ca, K, P, Mg, N and Na) and microminerals (Cu, Fe and Zn) in both plant stages (Figure 2). Predominantly, higher levels of CO$_2$ increased most of the detected elements in sprouts and mature plants in comparison to controls, except for Na, N and Zn. The predominant macrominerals observed in both stages were K and P, while the lowest amount was reported for Zn.

Regarding their organic acids content, the investigated sprouts and mature plants contained comparable levels of six detected organic acids under control conditions, i.e., malic, succinic and citric acids (predominant), as well as oxalic, isobutyric and fumaric acids (Table 1). eCO$_2$ treatment showed significant increases in most of the organic acids detected in both stages.

The current results also revealed that both untreated caraway sprouts and mature plants had almost the same amino acids profile, whereas glutamine and glutamic acid were detected as the major amino acids in both caraway stages (Table 1). Treatment with eCO$_2$ significantly increased the concentrations of some detected amino acids in sprouts and mature plants, while others were not affected, when compared to controls.

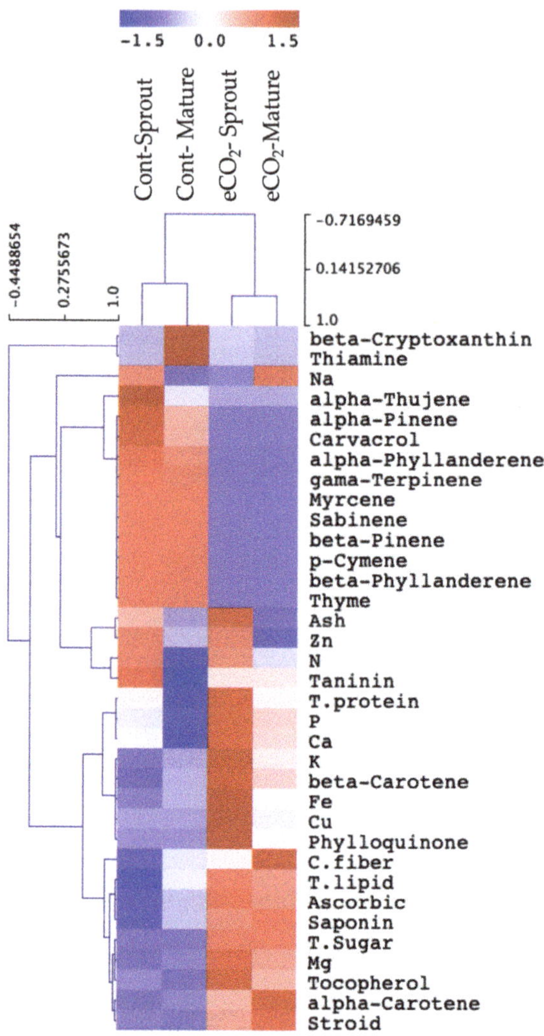

Figure 2. Hierarchical clustering analysis of total nutrients, minerals, vitamins and essential oils of control and eCO$_2$-treated caraway sprouts and mature plants. Data are represented by the means of at least 3 replicates.

Table 1. Organic acids and amino acids of caraway plants at two developmental stages (sprouts and mature tissues) under control or eCO_2 growth conditions. Data are represented by the means of at least 3 replicates $\pm$ standard deviations. Different small letter (a, b, c, and d) within a row indicate significant differences between control and eCO_2-samples. One-way analysis of variance (ANOVA) was performed. Tukey's test was used as the post-hoc test for the separation of means ($p < 0.05$).

Organic Acids	Sprouts		Mature	
(mg/g FW)	Control	eCO_2	Control	eCO_2
Oxalic	1.42 ± 0.14 a	1.99 ± 0.2 b	1.2 ± 0.10 a	1.91 ± 0.2 b
Malic	5.06 ± 0.31 a	6.49 ± 0.5 b	4.1 ± 0.7 a	6.09 ± 0.2 b
Succinic	2.84 ± 0.16 a	4.95 ± 0.43 b	2.3 ± 0.2 a	3.95 ± 0.41 b
Citric	2.45 ± 0.1 a	3.2 ± 0.39 b	2.5 ± 0.4 a	3.1 ± 0.19 b
isobutyric	1.06 ± 0.11 a	1.02 ± 0.01 a	1.16 ± 0.1 a	1.11 ± 0.01 a
Fumaric	0.42 ± 0.03 b	0.45 ± 0.01 b	0.32 ± 0.0 a	0.42 ± 0.0 b
Amino Acids (µg/g FW)				
Lysine	1.57 ± 0.13 a	2.62 ± 0.12 b	1.51 ± 0.11 a	2.12 ± 0.2 b
Histidine	1.61 ± 0.089 a	1.13 ± 0.01 a	1.61 ± 0.09 a	1.13 ± 0.01 a
Alanine	0.54 ± 0.04 a	1.23 ± 0.08 b	0.34 ± 0.041 a	1.7 ± 0.08 b
Arginine	0.98 ± 0.01 a	1.19 ± 0.09 a	0.78 ± 0.03 a	1.39 ± 0.1 b
Isoleucine	0.08 ± 0.00 a	0.13 ± 0.01 a	0.1 ± 0.00 a	0.11 ± 0.01 a
Asparagine	0.52 ± 0.04 a	0.91 ± 0.07 b	0.52 ± 0.05 a	0.77 ± 0.01 a
Ornithine	0.1 ± 0.02 a	0.12 ± 0.02 a	0.11 ± 0.03 a	0.11 ± 0.02 a
Glycine	0.6 ± 0.04 b	0.68 ± 0.06 b	0.2 ± 0.046 a	0.58 ± 0.01 b
Phenylalanine	0.16 ± 0.013 a	0.32 ± 0.01 b	0.11 ± 0.015 a	0.31 ± 0.04 b
Serine	0.18 ± 0.01 a	0.3 ± 0.02 b	0.19 ± 0.01 a	0.26 ± 0.01 b
Proline	0.58 ± 0.04 a	0.57 ± 0.05 a	0.52 ± 0.05 a	0.57 ± 0.05 a
Valine	0.24 ± 0.01 a	0.22 ± 0.03 a	0.34 ± 0.05 b	0.22 ± 0.03 a
Aspartate	0.02 ± 0.001 a	0.02 ± 0 a	0.025 ± 0.0 a	0.02 ± 0.0 a
Cystine	0.01 ± 0.0 a	0.1 ± 0.01 a	0.01 ± 0.0 a	0.1 ± 0.0 a
Leucine	0.16 ± 0 b	0.11 ± 0 a	0.13 ± 0.019 a	0.21 ± 0 b
Methionine	0.01 ± 0.001 a	0.01 ± 0 a	0.02 ± 0.002 b	0.031 ± 0 b
Threonine	0.05 ± 0.0 a	0.08 ± 0.01 b	0.043 ± 0.0 a	0.08 ± 0.01 b
Tyrosine	0.5 ± 0.03 a	0.66 ± 0.05 b	0.42 ± 0.039 a	0.61 ± 0.05 b
Glutamine	46.3 ± 3.5 b	52.79 ± 0.6 c	33.7 ± 4.2 a	51.1 ± 4.2 d
Glutamic acid	33.8 ± 2.5 b	46.2 ± 5.5 c	22.8 ± 5.8 a	33.2 ± 5.1 b

2.3. Secondary Metabolites of Caraway Sprouts and Mature Plants as Influenced by eCO_2

In addition, several essential oils were estimated in control and eCO_2-treated caraway plants, whereby thyme, γ-terpinene, p-cymene, carvacrol and β-pinene were the most prominent essential oils, followed by lower amounts of other components (Figure 2). The results showed that eCO_2 significantly decreased the levels of essential oils in both sprouts and mature plants compared to normally cultivated controls. On the other hand, a considerable amount of several vitamins were determined in caraway sprouts and mature plants (Figure 2). eCO_2 treatment significantly increased almost all investigated vitamins in both caraway stages.

The phenolic profile of caraway sprouts and mature plants was almost similar, whereas gallic acid and quercetin were quantified as the most dominant phenolic acids and flavonoids, respectively, in both stages (Figure 3). eCO_2 caused significant increments in the phenolic and flavonoid profiles of sprouts when compared to controls.

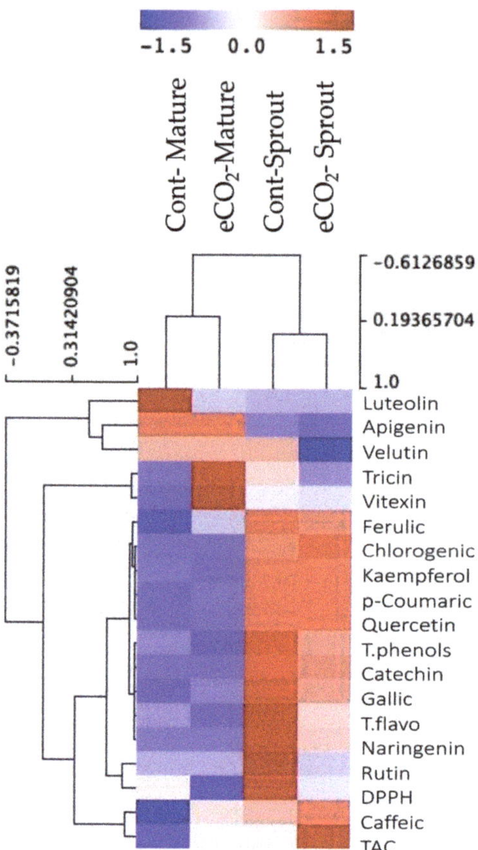

Figure 3. Hierarchical clustering analysis of phenolic compounds and antioxidant capacity of control and eCO$_2$-treated caraway sprouts and mature plants. Data are represented by the means of at least 3 replicates.

2.4. Effect of eCO$_2$ on Antioxidant Activities of Caraway Sprouts and Mature Plants

Antioxidant activities of caraway sprouts and mature plants were estimated in our study (by ferric reducing antioxidant power (FRAP), DPPH• radical, and lipid peroxidation assays), following eCO$_2$ treatment (Figure 3). The results showed that eCO$_2$ had a more enhancing effect on the antioxidant activities of sprouts than the mature plants, when compared to their respective controls.

2.5. Antimicrobial Activity of Caraway Sprouts and Mature Plants under eCO$_2$ Treatment

In the current investigation, the effect of eCO$_2$ on the antimicrobial activity of caraway sprouts and mature plants was determined. The test was conducted against a set of 13 microbes, based on the minimal inhibitory concentration (MIC) of the selected caraway extracts (Table 2). As indicated by the diameter of inhibition zone, the results showed that treatment with eCO$_2$ significantly increased the antimicrobial potency of caraway sprouts and mature plants predominantly against *Staphylococcus epidermidis*, *Staphylococcus saprophyticus*, *Streptococcus salivarius*, *Enterococcus faecalis*, *Salmonella typhimurium*, *Pseudomonas aeruginosa*, *Candida glabrata*, *Serratia marcescens* and *Escherichia coli* (Table 2). Moreover, eCO$_2$ significantly increased the antimicrobial activity of mature caraway plants against *Proteus vulgaris*. The increased antibacterial activity might be attributed to enhanced levels of phytochemicals in eCO_2-treated caraway sprouts and/or mature tissues. The increased levels of photosynthetic pigments might also be involved in antimicrobial effects against some bacteria such as *Vibrio parahaemolyticus*, *Micrococcus luteus*, *Bacillus subtilis* and *Escherichia coli*. However, under eCO_2 treatment, there were significant reductions in the antimicrobial activity of caraway sprouts and mature plants against *Enterobacter aerogenes*. In addition, eCO_2 significantly reduced the antimicrobial activity of caraway sprouts against *Proteus vulgaris* and *Candida albicans*, compared to normally cultivated controls. Meanwhile, eCO_2 had no impact on the antimicrobial activity of both stages against *Aspergillus flavus*, and, also, no change was observed in the antimicrobial activity of eCO_2-treated mature plants against *Candida albicans*. It can be noted that the effect of eCO_2 was more pronounced on the antimicrobial activities of sprouts than their mature plants.

Table 2. Antimicrobial activity of control and eCO_2-treated caraway sprouts and mature plants. The activity is presented as the diameter of inhibition zone (mm). Data are represented by the means of at least 3 replicates $\pm$ standard deviations. Different small letter (a, b, c) within a row indicate significant differences between control and eCO_2-samples. One-way analysis of variance (ANOVA) was performed. Tukey's test was used as the post-hoc test for the separation of means ($p < 0.05$).

Microbial Name	Sprouts		Mature	
	Control (mm)	eCO_2 (mm)	Control (mm)	eCO_2 (mm)
Staphylococcus saprophyticus	15.34 ± 2.2 b	22.48 ± 1.8 c	11.14 ± 1.3 a	16.08 ± 1.29 b
Staphylococcus epidermidis	9.27 ± 1 a	20.13 ± 1.6 b	11.27 ± 1.2 a	22.1 ± 1.6 b
Enterococcus faecalis	14.57 ± 3.4 b	17.9 ± 0.5 c	9.52 ± 3.42 a	16.3 ± 0.56 b
Streptococcus salivarius	7.87 ± 1.15 a	16.14 ± 2.7 b	8.82 ± 1.12 a	13.11 ± 0.1 b
Escherichia coli	6.3 ± 0.9 b	9.68 ± 7.8 c	4.3 ± 0.9 a	9.68 ± 30.2 c
Salmonella typhimurium	11.17 ± 1.9 b	19.91 ± 3.1 c	7.17 ± 1.2 a	12.1 ± 1.1 b
Pseudomonas aeruginosa	20.02 ± 1.7 b	27.15 ± 1.5 c	14.02 ± 1.7 a	17.15 ± 1.1 b
Proteus vulgaris	20.8 ± 1.0 c	18.53 ± 1.82 b	10.8 ± 1.2 a	13.2 ± 1.5 ab
Enterobacter aerogenes	14.92 ± 0.3 c	13.96 ± 0.8 b	12.12 ± 0.9 ab	9.96 ± 0.17 a
Serratia marcescens	5.92 ± 0.2 b	7.68 ± 1.1 c	3.91 ± 0.96 a	4.68 ± 1.6 ab
Aspergillus flavus	13.93 ± 1.1 b	14.85 ± 1.8 b	7.91 ± 1.5 a	8.85 ± 1.0 a
Candida albicans	15.58 ± 5.07 b	15.76 ± 1.2 a	14.51 ± 1.01 a	14.2 ± 0.3 a
Candida glabrata	4.11 ± 0.954 a	8.11 ± 1.4 c	3.41 ± 0.92 a	5.91 ± 0.04 b

3. Discussion

Our study has demonstrated the importance of eCO_2 to enhance the nutritive and health-promoting values of plants, particularly in the context of rising atmospheric CO_2. Depending on their developmental stages, plants might interact differently in response to eCO_2. Preferably, sprouts can be good candidates for the positive impact of eCO_2 on their nutritious metabolites. This can be supported by our previous studies and other research that also discussed the application of eCO_2 to increase the food functional values of many plants and sprouts [2,3,14,16,24]. Thus, the expected availability of eCO_2 in the future could be a promising approach to stimulate the growth and phytochemicals of sprouts.

3.1. eCO_2 Enhanced Biomass Production of Caraway Sprouts and Mature Plants

It has been known that eCO_2 could effectively enhance photosynthesis and plant growth [23]. In our study, the eCO_2-induced biomass accumulation in both sprouts and mature plants could be the result of enhanced photosynthetic activity under eCO_2 treatment, which triggered the biosynthesis of sugars and organic acids, and finally resulted in enhanced growth and biomass accumulation [20,23]. Similar to our results, eCO_2 has reportedly increased photosynthesis and the growth of many plants and sprouts, e.g., tobacco, tomato, peppermint, *Arabidopsis, basil, Hymenocallis littoralis, Isatis indigotica,* broccoli and alfalfa sprouts [20,23–28]. In addition, eCO_2 enhanced the yield of some crops, such as carrot, radish and turnip [22]. Moreover, CO_2 enrichment has been found to increase the levels of sugars and biomass production in dill and parsley [17]. Furthermore, the use of other elicitors, (e.g., laser light) to stimulate the growth and biomass accumulation of buckwheat sprouts has been recently explored [16].

On the other hand, eCO_2 was observed to enhance the photosynthetic pigment content in caraway sprouts and mature plants (Figure 1). It can be noted that caraway mature plants are likely to better respond to eCO_2, thus accumulating higher photosynthetic pigments than the sprouts. Similar to our results, eCO_2 has been recently applied to enhance chlorophyll content (a, b) in alfalfa sprouts [25]. In this regard, the eCO_2-induced chlorophyll content is assumed to enhance much more photosynthesis, which, in turn, leads to more sugar accumulation and, hence, higher biomass production [20].

3.2. eCO_2 Similarly Affected the Primary Chemical Composition of Caraway Sprouts and Mature Tissue

High CO_2 levels could modify the plant chemical composition by changing the carbon and nitrogen metabolism [20]. Herein, the treatment of caraway plants with eCO_2 significantly increased most of the primary metabolites detected in sprouts and mature plants, as compared to controls (Figure 2). Both of the plant stages seemed to equally respond to the effect of eCO_2 on their nutrients. The results are in agreement with those obtained by [23] who reported a significant increase in total lipids, carbohydrates and proteins in peppermint and basil following treatment with eCO_2. Additionally, [29] reported that eCO_2 caused significant increments in the sugars, proteins and phenolic compounds of oregano. Several studies have demonstrated the enhancing effects of high atmospheric CO_2 on increasing the nutritive values, sugar and phytochemical contents of many plants' tissues and sprouts, e.g., dill and parsley, as well as the total proteins, lipids, carbohydrates and fibers of alfalfa sprouts [17,25]. In this regard, the accumulation of sugars, in response to elevated photosynthesis, is expected to provide the energy necessary for the biosynthesis of multiple classes of metabolites [20,26,28].

At individual levels, the present study has shown eCO_2 to increase most of the amino acids and organic acids detected in both caraway stages, whereas the sprouts and mature plants showed a similar response to eCO_2. In accordance with this, it has been observed that some organic acids were accumulated, while others were not affected, in peppermint, basil, parsley and dill treated with eCO_2 [17,23]. Regarding the amino acid contents, both caraway sprouts and mature plants responded in a similar manner to eCO_2 effects on their amino acids (Table 1). Similar to our results, eCO_2 has been shown to enhance almost all amino acids in lemongrass sprouts, broccoli sprouts, dill, basil, parsley and peppermint [14,17,23,24]. Meanwhile, eCO_2 was incapable of increasing the biosynthesis of L-canavanine; a non-protein amino acid in alfalfa sprouts [25]. Such increases in amino acids could be due to the ability of eCO_2 to induce the availability of C skeleton and metabolic energy needed for the biosynthesis of amino acids [28,30]. In addition, the variation in amino acid contents in sprouts might be related to the higher rates of physiological activities during the sprouting process. Some amino acids, reported herein, have been investigated for their health effects, e.g., glutamine has been involved in fighting some neurodegenerative diseases such as Alzheimer's diseases, besides being an essential factor for lymphocyte proliferation and the production of cytokines [31].

3.3. eCO_2 Differentially Affected the Mineral and Secondary Chemical Compostion of Caraway Sprouts and Mature Tissue

Sprouts have been considered as rich sources of vitamins, phenolics and minerals, whose deficiency could risk human health [32]. Therefore, increasing their contents in caraway sprouts and mature plants by using promising approaches, such as eCO_2, might enhance their nutritional and health-promoting values. In this study, the effect of eCO_2 on mineral content was more pronounced in sprouts than mature plants (Figure 2). Matched with our results, alfalfa sprouts have been observed to accumulate more macronutrients (Na, Mg, K, P and Ca) and micronutrients (Fe, Zn, Cu and Mn), when treated with eCO_2 [25]. In addition, significant increments in the levels of Ca, Mg, Zn, K, Cu and Cd were detected following eCO_2 treatment of many plant species such as basil and peppermint [17,23]. Moreover, it has been reported that Ca, Mn and Cu were significantly improved in carrot and radish when treated with eCO_2 [22]. Previous studies also reported enhancing effects of eCO_2 on primary metabolism, nutrient uptake and the levels of minerals in many plant species [28,33,34]. However, eCO_2 did not induce significant effects on most minerals previously detected in dill and parsley [17], besides its negative impact on minerals in many plants [35–37]. Such influence might be ascribed to the dilution effect rather than decreased nutrient absorption [17].

At the antioxidant level, similar responses to eCO_2 were recorded for phenolics and flavonoids, where caraway sprouts seemed to respond better to the eCO_2 effects on phenolic content than the mature plants (Figure 3). Caraway seeds, fruits and sprouts, have been known to contain a diverse amount of flavonoids, monoterpenoid glucosides, lignins, alkaloids and polyacetylenic compounds [38–43]. In the current study, eCO_2 caused significant increments in the phenolic and flavonoid profiles of caraway sprouts. In agreement, eCO_2 has been reported to enhance the phenolic content of alfalfa sprouts, basil, peppermint, parsley, dill and fenugreek [17,23,25,29]. The improved levels of phenols and flavonoids in eCO_2-treated caraway sprouts could be attributed to the abundance of C and N intermediates that could be used for the biosynthesis of these phytochemicals, given that eCO_2 has been reported to affect the metabolism of C and N [28].

The impact of eCO_2 on vitamins and essential oil levels in plants is poorly studied, whereas limited studies have reported improved vitamin C and E contents in eCO_2-treated medicinal plants [23]. Similarly, positive effects of eCO_2 on some vitamins, e.g., vitamin K1, A, B, C and E, were recorded in basil, peppermint, parsley, orange, strawberry and dill as well as alfalfa sprouts [17,23,25,44]. Higher contents of essential oils were reportedly present in caraway, particularly in seeds, whereas the most prominent essential components were carvone and limonene, beside trace amounts of acetaldehyde, camphene, furfural, pinene and phellandrene [45–48]. In our study, the change in essential oil content following treatment with high eCO_2 might be affected by some factors such as the characteristics and tillage of soil, annual precipitation, fertilization amount, breeding, maturation and harvesting [49,50]. It was also previously reported that the composition and the amount of essential oils and vitamins depend on climatic conditions during fruit formation and ripening [51].

3.4. eCO_2 Treatments Induced Higher Biological Activities in Sprouts than in Mature Tissues of Caraway Plants

Caraway plants have been recommended in medicinal uses as antioxidant, antimicrobial, immune-modulatory, antiulcerogenic and antispasmodic as well as antioxidative agents. Such various biological activities could be due to their richness in polyphenolic compounds, flavonoids, essential oils, organic and amino acids, and vitamins [52–56]. Herein, eCO_2 was observed to exert an enhancing effect on the antioxidant activities of both sprouts and mature plants. Such results could indicate that mature plants might be more responsive than sprouts to eCO_2 effects on their antioxidant activities.

In accordance, eCO_2 has previously been found to increase the antioxidant activities of many plants and sprouts, e.g., alfalfa sprouts, lemongrass sprouts, *Medicago lupulina*, basil, peppermint, parsley, dill, fenugreek, strawberry, *Citrus aurantium* and *Labisia pumila* [14,17,23,25,29,44,57–60]. Such eCO_2-induced antioxidant effects could be related to the increased levels of several phenolic compounds accumulated in response to eCO_2. Additionally, in ginger plants treated with eCO_2, a significantly similar improvement in the levels of phenolic compounds and vitamin C was detected, which increased the antioxidant capacity [44]. Previous studies reported that there was a significant link between antioxidant activity and polyphenol and flavonoids contents, as well as vitamins such as α-tocopherol, which collectively increases the radical scavenging activity of the essential oils of plants and vegetables [61–64]. Recently, caraway essential oils have shown a high oxidative stability and antioxidant properties [65]. This could be due to their high contents of unsaturated fatty acids, sterols, α and γ-tocopherols, and phenolic compounds. The improved phenolics, flavonoids and other constituents significantly improved the antioxidant and anti-lipid peroxidation in our studied plants in a similar behavior to other previously reported studies on fenugreek [29]. Moreover, the antioxidant capacity of some minerals detected herein, e.g., Cu, Mn, Zn and Se might be improved following treatment of caraway plants with eCO_2 [66].

In the current investigation, the effect of eCO_2 was more pronounced on the antimicrobial activities of sprouts than their mature plants. Supporting our results, previous studies reported that caraway essential oils play a potential role in the growth inhibition of many bacteria and fungi such as *Vibrio cholera*, *Escherichia coli*, *Staphylococcus aureus*, *Salmonella typhi* and *Mycobacterium tuberculosis* [67,68]. Similar antimicrobial effects were determined for caraway essential oils, particularly carvone, against the growth of bacteria and fungi during potato storage, such as *Fusarium sulphureum* and *Phoma exigua* [69–71]. eCO_2 has also been shown to enhance the antibacterial effects of fenugreek against *Staphylococcus aureus*, *Bacillus subtilis* and *Streptococcus* sp. [29]. Moreover, the antifungal potency of caraway oil was previously reported against *Alternaria alternata*, *Fulvia fulvium*, *Cladosporium cladosporioides*, and *Phoma macdonaldii* [72].

The observed antimicrobial activities can be explained by higher quantities of polyphenolic compounds along with appreciable amounts of vitamins such as α-tocopherol [61–63]. It could also be suggested that eCO_2 increased the antimicrobial activity of eCO_2-treated caraway sprouts and mature plants, which might be related to the improved content of phytochemicals such as phenolic acids, flavonoids, vitamins and mineral contents, which consequently enhance the antibacterial activities, as previously reported for fenugreek [29]. The enhanced levels of photosynthetic pigments might also be involved in antimicrobial effects against *Vibrio parahaemolyticus*, *Micrococcus luteus*, *Bacillus subtilis* and *Escherichia coli*. Similarly, it has been previously demonstrated that high levels of CO_2 increased the antibacterial activities of dill, parsley, peppermint and basil against *Streptococcus* spp. and *Escherichia coli*, the activities that could be supported by their phenolic contents [17,23]. Overall, improving the potential antimicrobial activity of caraway oils and extracts could further support their applications in medicinal purposes, food preservation and the cosmetic industry.

3.5. Principal Component Analysis (PCA) Confirmed the Developmental Stage-Specific Response of Caraway Plants to eCO_2

To better understand the developmental stage responses, we performed PCA of the chemical composition and biological activities of sprouts and mature tissues under control and eCO_2 treatments (Figure 4). We observed a clear separation between the treatment parameters along the PC1, which explains 66% of the total variation. Remarkably, eCO_2 induced the accumulation of minerals and several components as well as antimicrobial and antioxidant activities. There was also a clear separation between the parameters of sprouts and mature stages along PC2 (represents 18% of the total variation). Caraway plants at the mature stage showed high levels of a few organic and amino acids, phenolic acids including methionine, leucine, velutin, valine, isobutyric and aspartic acids and the flavonoid luteolin, as previously discussed [9,52–56]. However, caraway sprouts were richer than mature plants, in many of the essential oils, phenolics, ash, mineral (Zn, N), and amino acids (proline and glycine) as well as antioxidant (DPPH•) and antibacterial activity against *P. aeruginosa*, *S. marcescens*, *P. vulgaris* and *E, aerogenes*. Together, these data showed that caraway plants at different developmental stages were differentially grouped, indicating the specificity of nutritive metabolite accumulation in response to eCO_2 treatment.

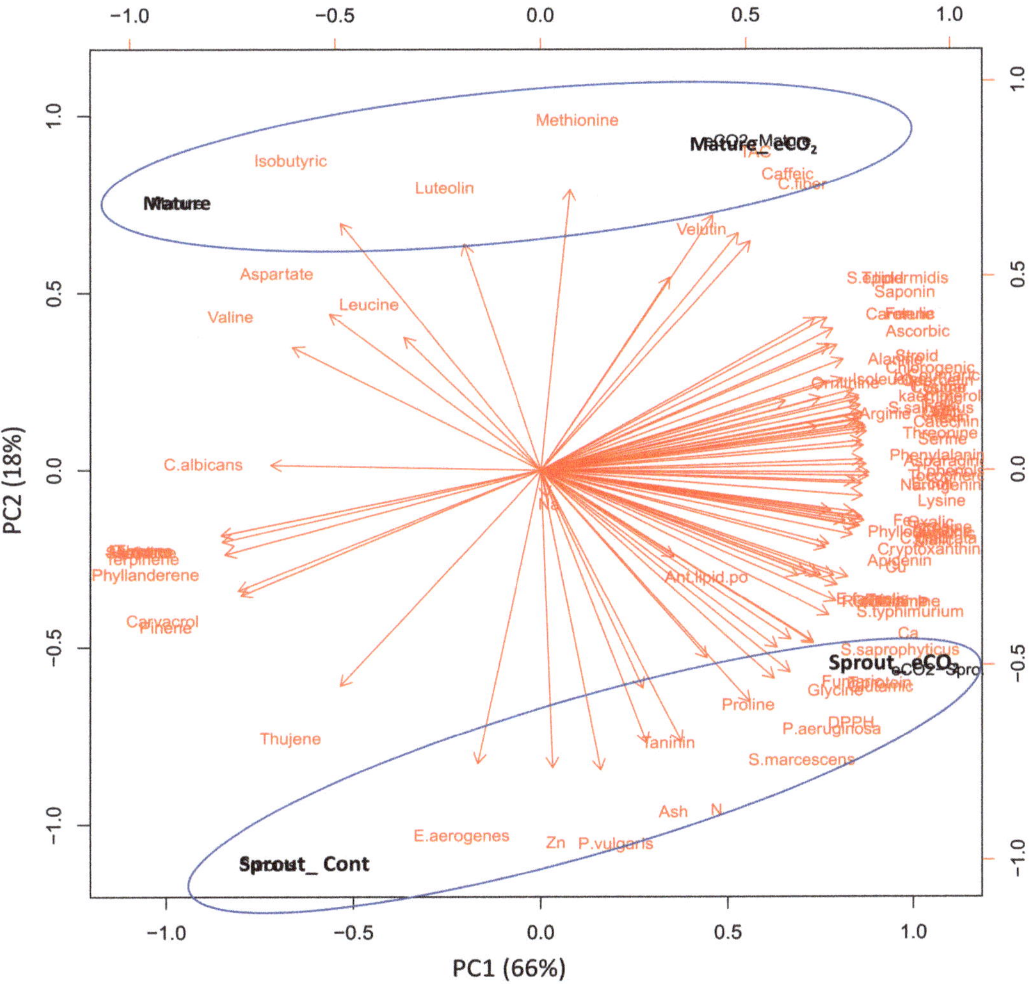

Figure 4. Principal component analysis (PCA) of chemical compositions and biological activities of caraway plants at two developmental stages (sprout and mature tissues) under control or eCO$_2$ growth conditions.

4. Materials and Methods

4.1. Plant Material and Growth Conditions

Seeds of caraway (*Carium carvum* L.) were collected from Agricultural Research Center (Giza, Egypt). The seeds were washed with distilled water and soaked for 1 h in 5 g L^{-1} sodium hypochlorite, then they were kept overnight in distilled water. For sprouting process, the seeds were distributed on trays filled with vermiculite and irrigated with Milli-Q water every two days. For growing plants till mature stage, the seeds were sown in loamy soil and organic compost (50:50%) in pots and the soil water content (SWC) was adjusted to 60%. The growth conditions were adjusted to 25 °C air temperature, a 16/8-h day/night photoperiod using white fluorescent tubes with photosynthetically active radiation (400 µmol m^{-2} s^{-1} and 60% humidity). According to IPCC-SRES B2-scenario prediction of elevated CO$_2$ of the year 2100, the seeds were maintained under two climate conditions, (1) ambient CO$_2$ (a CO$_2$, 400 ± 27 µmol CO$_2$ mol^{-1} air); (2) elevated CO$_2$ (eCO$_2$, 620 ± 42 µmol CO$_2$ mol^{-1} air ppm). The sprouts and mature tissues from each treatment were harvested after 9 and 45 days and weighed, then they were frozen in liquid nitrogen and kept at $-80\ ^\circ$C for biochemical analyses. Each experiment was replicated at least two times, and for all assays, 3 to 5 replicates were used and each replicate corresponded to a group of sprouts and mature plants harvested from a certain tray.

4.2. Determination of Photosynthetic Rate

Photosynthesis (μmol CO_2 m^{-2} s^{-1}) was detected by EGM-4 infrared gas analyzer (PP Systems, Hitchin, UK). Photosynthetic rate was detected from 180 s measurements of net CO_2 exchange (NE).

4.3. Pigment Analysis

Caraway samples were homogenized for 1 min at 7000 rpm in acetone by using a MagNA Lyser (Roche, Vilvoorde, Belgium), then centrifugation was performed at 4 $^\circ$C, 14,000$\times$ g for 20 min. The supernatant was separated then filtered (Acrodisc GHP filter, 0.45 μm 13 mm). Thereafter, analysis of the obtained solution was conducted by using HPLC (Shimadzu SIL10-ADvp, reversed-phase, at 4 $^\circ$C) [16]. Extraction of chlorophyll a and b was performed, then quantified by using a diode array detector (Shimadzu SPDM10Avp) at four wavelengths (420, 440, 462 and 660 nm).

4.4. Preparation of Caraway Extracts

An ETA 0067 grinder with grinding stones, a VIPO grinder and a Vibrom S2 (Jebavý, Trebechovice p. O., CR) cryogenic grinder (liquid nitrogen) were tested for sample homogenization. An SE-1 (SeKo-K, Brno, CR) extractor for supercritical fluid extraction (SFE) and an apparatus for steam distillation according to CSN 58-0110 and CSN 6571 were applied for extraction and subsequent determination of the total content of caraway oils. Approximately 500 mg of exactly weighed fresh sample was transferred into an extraction column for SFE extraction. The extraction was performed for 60 min at 40 MPa, extractor temperature 80 $^\circ$C and restrictor temperature 120 $^\circ$C. The extract was trapped into a hexane layer in a trapping vessel.

4.5. Analysis of Mineral Contents

Macro and microelements were evaluated according to [18]. About 200 mg from both eCO$_2$-treated and non-treated caraway sprouts and mature plants were digested in HNO$_3$/H$_2$O solution (5:1 v/v) in an oven. Then, the concentrations of macrominerals and trace elements were estimated at 25 $^\circ$C by using inductively coupled plasma mass spectrometry (ICP-MS, Finnigan Element XR, and Scientific, Bremen, Germany), where nitric acid (1%) was used as a standard.

4.6. Measurement of Phenolic Acids, Flavonoids and Vitamins

Chromatographic techniques such as HPLC were used to measure the levels of phenolic acids, flavonoids and vitamins in eCO$_2$-treated and non-treated caraway sprouts and mature plants as previously described in [16,73]. Identification of compounds was conducted by comparing the standard mixtures to the relative retention time of each compound from each sample. The peak area of the corresponding standard was used to calculate the concentration of each compound. Phenolic compounds were measured by HPLC. A total of 50 mg of frozen dried samples were mixed with 4:1 v/v acetone–water solution. HPLC system (SCL-10 AVP, Japan), supplied with a LiChrosorb Si-60, 7 μm, 3 $\times$ 150 mm column, diode array detector) was used for determination of flavonoids and phenolic acids, whereas the mobile phase consisted of (90:10) water/formic acid, and (85:10:5) acetonitrile/water/formic acid, at 0.8 mL/min (flow rate), while 3,5-dichloro-4-hydroxybenzoic was used as an internal standard. The concentration of each detected compound was evaluated by using the peak area of the corresponding standard. The contents of thiamine and riboflavin were determined in caraway samples by using UV and/or fluorescence detectors [16]. Separation was conducted on a reverse-phase (C18) column (HPLC, methanol–water).

4.7. Measurement of Antioxidant Capacity

The determination of antioxidant capacity was conducted through ferric reducing antioxidant power (FRAP), DPPH• radical, and lipid peroxidation assays as previously reported [14,74–76].

4.7.1. Ferric Reducing Antioxidant Power (FRAP) Method

About 0.2 g of caraway sample was extracted in ethanol (80%), and then centrifugation was performed (14,000 rpm, 20 min). Afterwards, 0.1 mL of extracts was added to 0.25 mL of FRAP reagent (20 mM $FeCl_3$ in 0.25 M acetate buffer, pH 3.6) at room temperature, and, finally, the absorbance was detected at 517 nm [14].

4.7.2. DPPH• Assay

DPPH· radical has a violet color with a maximum absorbance at 517 nm that becomes colorless in presence of antioxidants. The reaction mixtures consisted of 0.1 mL of samples and 3.9 mL of 200 µM DPPH• (prepared in ethanol), and incubated in dark (30 min, 37 °C). Afterwards, the absorbance was detected (517 nm). The percentage of inhibition was calculated against a control [74].

4.7.3. Lipid Peroxidation Assay

Lipid peroxidation was determined through detection of MDA present in caraway samples of eCO_2-treated and control plants. In this assay, a colorimetric lipid peroxidation (MDA) assay kit (MAK085, 3050 Spruce Street, Saint Louis, MO 63103, USA) was used, whereas the MDA in the sample reacts with thiobarbituric acid (TBA) to generate a MDA-TBA adduct, which can be easily quantified calorimetrically (OD = 532 nm) or fluorometrically (Ex/Em = 532/553 nm). This assay detects MDA levels as low as 1 nmol/well calorimetrically and 0.1 nmol/well fluorometrically [14].

4.8. Determination of Total Carbohydrates, Protein, Lipids and Fibers

Nelson's method was used to measure carbohydrates from each caraway sample (eCO_2-treated and control plants). Concentration of protein was detected for each frozen caraway sample (0.2 g FW) according to Lowry methods [75]. Detection of total lipids was conducted based on Folch method modified by [76], whereas the samples were homogenized in chloroform/methanol (2:1). Afterwards, centrifugation was performed at $3000\times g$ for 15 min. A rotary evaporator was used to evaporate the chloroform phase containing lipids., and then the pellets were redissolved in a mixture of toluene/ethanol (4/1 v/v). A saline solution was mixed with the extract. The extracted lipids were concentrated by a rotary evaporator and then weighed in vials to calculate the total lipid content. Fibers also were extracted from the target samples and evaluated according to AOAC (1990), where α-amylase was used for sample gelatinization (30 min, pH 6, 100 °C), then protease was used for enzymatic digestion (30 min, pH 7.5, 60 °C). Thereafter, amyloglucosidase was used for proteins and starch removal (30 min, pH 6 and 0 °C). Finally, fibers were precipitated with ethanol, and the residue was weighed after washing.

4.9. Analysis of Essential Oils, Organic Acids, and Amino Acids

4.9.1. Analysis of Essential Oils

The steam distillation method, according to the standards CSN 58 0110 and CSN 6571, was used. Depending on the expected content of essential oil, an exactly weighed sample (10–25 g) was transferred into a distillation vessel, and then 400 mL of water and boiling stones were added. The samples were boiled for 4 h. Then, cooling was stopped, and distillation was prolonged for a while until all essential oils were quantitatively transferred into a calibrated tube. Then, the heating was stopped and the volume of the extracted essential oils was measured after 5 min. The extracted or distilled samples were stored in a refrigerator at 1–4 °C (for 2 days), if necessary, and analyzed by GC. A gas chromatograph HP 4890D (Hewlett Packard) with a FID detector was used for determination of limonene-to-carvone ratio in the samples. Separation was performed using an HP-5 (Crosslinked 5% PH ME Siloxane, 15 m × 0.53 mm × 1.5 µm film) column at helium flow rate 2 mL/min, injector temperature 220 °C and detector temperature 240 °C using temperature program 60 °C, 40 °C/min up to 220 °C, 2 min at 220 °C. Portions of 2 µL of each essential oil (dissolved in hexane) were injected into the used analytical column. Resulting chromatograms were treated using CSW (Data Apex, Prague, CR) data station [77]. Identification of oil components was achieved based on their retention indices (RI, determined with reference to a homologous series of normal alkanes) and by comparison of their mass spectral fragmentation patterns (NIST) database (G1036A, revision D.01.00)/Chem-Station data system (G1701CA, version C.00.01.08)].

4.9.2. Organic Acids Analysis

Organic acids were detected in caraway extracts by using HPLC, isocratically, with 0.001 N sulfuric acid, at 210 nm and flow of 0.6 mL min^{-1} [78]. The assay was performed by using liquid chromatographer (Dionex, Sunnyvale, CA, USA) with LED detector Ultimate 3000. The latter cooperated with the following devices: pump (LPG-3400A) EWPS-3000SI autosampler, TCC-3000SD column thermostat and the Chromeleon v.6.8 computer software. Meanwhile, the separation was conducted by using Aminex HPH-87 H (300 × 7.8 mm) column with IG Cation H (30 × 4.6) pre-column of Bio-Red firm, at a temperature of 65 °C.

4.9.3. Amino Acid Analysis

Detection of amino acid in caraway samples was carried out according to [79]. About 3 mg of each sample was hydrolyzed with 6 M HCl (6 h, 150 °C). Afterwards, the acid was evaporated by rotary evaporation (RE500 Yamato Scientific America Inc.), and the samples were redissolved in 2 mL of sodium citrate buffer (pH 2.2). For sample derivation, phthalaldehyde (OPA) (7.5 mM) was added to samples in citrate buffer (OPA reagent contains β-mercaptoethanol and Brij 35). The HPLC method precision was evaluated using external and internal standards. The amino acid reference standards consisted of 15 amino acids (0.05 µmoles mL^{-1} amino acid), which were used for detection of retention times of each amino acid. Meanwhile, the internal standard (0.05 µmoles mL^{-1} α aminobutyric) was added to the reference sample as well as the plant sample. The gradient mobile phase consisted of 0.1 M sodium acetate and methanol (9:1), while C18 column reversed-phase (100 × 4.6 mm × 1/4″ Microsorb 100-3 C18) was used for sample elution. Fluorescence detection was realized using an excitation–emission wavelength of 360 and 455 nm, respectively. For amino acid peak integration, Star Chromatography (Varian version 5.51) was applied.

4.10. Antibacterial Activity of Caraway Extracts

Antibacterial activity of caraway was analyzed by standard dilution in liquid media according to a previously reported methodology [14]. Dimethyl sulfoxide (DMSO) was used to dissolve 100 mg of the essential oil sample. The range of oil dilution from 1 to 20 mg mL^{-1} was prepared in media Mueller-Hinton Broth of Merck. Then, 0.1 mL of 18 h liquid culture of standard strain (*Staphylococcus aureus* ATCC 6538 P) diluted 1:10,000 in the same medium (number of inoculum contained 104–105 bacterial cells in 1 mL), was added to the media. Incubation of the tested samples was conducted in 37 °C for 18 h. The value of MIC (Minimal Inhibitory Concentration) was defined as the lowest concentration of the oil completely inhibiting the growth of standard strain. This value was calculated on antibiotic units (AU), based on that the value of MIC is equivalent to 1 AU. The results were referenced to 1 g of oil. The antibacterial activity of the tested samples was evaluated against *Candida albicans* (ATCC90028), *Candida glabrata* (ATCC90030), *Aspergillus flavus* (ATCC9170), *Staphylococcus saprophyticus* (ATCC 19701), *S. epidermidis* (ATCC 12228), *Enterococcus faecalis* (ATCC

10541), *Streptococcus salivarius* (ATCC25975), *E. coli* (ATCC 29998), *Salmonella typhimurium* (ATCC14028), *Pseudomonas aeruginosa* (ATCC10145), *Proteus vulgaris* (ATCC8427), *Enterobacter aerogenes* (ATCC 13048), *Serratia marcescens* (ATCC99006) and *Salmonella typhimurium* (ATCC14028).

4.11. Statistical Analyses

Statistical analyses were performed using SPSS statistical package (SPSS Inc., Chicago, IL, USA). Replication of each experiment was carried out (two times). Three to five replicates were used for all assays and each replicate corresponded to a group of sprouts and mature plants harvested from a certain tray. One-way analysis of variance (ANOVA) was performed. Tukey's test was used as the post-hoc test for the separation of means ($p < 0.05$). Principal component analysis (PCA) was generated by Multi Experimental Viewer (TM4 software package).

5. Conclusions

This study reported, for the first time, the application of eCO_2 to improve the nutritive value, functionality and health-promoting prospective of caraway sprouts and mature plants. The exposure of caraway sprouts and mature plants to high CO_2 significantly improved the nutritive value, levels of essential minerals, free amino and organic acids, phenolic compounds and other metabolites. In addition, the antimicrobial activity increased; but no change was reported in essential oils' contents under eCO_2 treatment. The impact of eCO_2 was more pronounced on sprouts than mature plants, regarding minerals, vitamins and phenolic compounds, while the mature plants displayed higher antioxidant abilities than that of sprouts, under eCO_2 treatment. Meanwhile, both caraway sprouts and mature plants responded equally to eCO_2 effects on their total nutrients, organic and amino acids as well as antimicrobial activities. Thus, under the current scenario of increasing atmospheric CO_2, eCO_2 could be applied to improve the nutritive value, functionality and health-promoting prospective of caraway sprouts and mature plants, depending on the developmental stage, and the kind of metabolites and bioactivity needed to be modulated under eCO_2 treatment.

Author Contributions: Conceptualization, M.K.O. and H.A.; methodology, S.S.A.-a. and A.A.A.; software, S.M.A.-Q., N.A.A.-H. and H.A.; validation, M.K.O. and H.A.; formal analysis, Z.K.A. and M.A.-M.; investigation, Z.K.A., A.A.-H. and S.M.A.-Q.; resources, M.K.O.; data curation, W.H.A.-Q., M.S.A., S.S., A.A.A., H.A. and N.A.A.-H.; writing—original draft preparation, N.A.A.-H. and M.A.-M.; writing—review and editing, N.A.A.-H. and H.A. All authors have read and agreed to the published version of the manuscript.

Funding: The authors extend their appreciation to the Researchers Supporting Project number (RSP-2021/374) King Saud University, Riyadh, Saud Arabia.

Institutional Review Board Statement: Not applicable.

Informed Consent Statement: Not applicable.

Data Availability Statement: Data presented in this study are available on reasonable request.

Conflicts of Interest: The authors declare that they have no known competing financial interests or personal relationships that could have appeared to influence the work reported in this paper.

References

1. Galieni, A.; Falcinelli, B.; Stagnari, F.; Datti, A.; Benincasa, P. Sprouts and Microgreens: Trends, Opportunities, and Horizons for Novel Research. *Agronomy* **2020**, *10*, 1424. [CrossRef]
2. Silva, L.R.; Pereira, M.J.; Azevedo, J.; Gonçalves, R.F.; Valentão, P.; de Pinho, P.G.; Andrade, P.B. Glycine max (L.) Merr., Vigna radiata L. and Medicago sativa L. sprouts: A natural source of bioactive compounds. *Food Res. Int.* **2013**, *50*, 167–175. [CrossRef]
3. Marton, M.; Mandoki, Z.S.; Csapo-Kiss, Z.S.; Csapo, J. The role of sprouts in human nutrition. A review. *Acta Univ. Sapientiae* **2010**, *3*, 81–117.
4. Kapoor, L.D. *Handbook of Ayurvedic Medicinal Plants*; CRC Press Inc.: Boca Raton, FL, USA, 1990.
5. Tyler, V.E.; Brady, L.R.; Robbers, J.E. *Pharmacognosy*, 9th ed.; Lea Fabiger: Philadelphia, PA, USA, 1988.

6.   Dubey, N.K.; Mishra, A.K. Evaluation of some essential oils against dermatophytes. *Indian Drugs* **1990**, *27*, 529–531.

7.   Boskabady, M.; Shaikhi, J. Inhibitory effect of Carum copticum on histamine (H1) receptors of isolated guinea-pig tracheal chains. *J. Ethnopharmacol.* **2000**, *69*, 217–227. [CrossRef]

8.   Vasudevan, K.; Vembar, S.; Veeraraghavan, K.; Haranath, P.S. Influence of intragastric perfusion of aqueous spice extracts on acid secretion in anesthetized albino rats. *Indian J. Gastroenterol.* **2000**, *19*, 53–56.

9.   Pradeep, K.U.; Geervani, P.; Eggum, B.O. Common Indian spices: Nutrient composition, consumption and contribution to dietary value. *Plant Foods Hum. Nutr.* **1993**, *44*, 137–148. [CrossRef] [PubMed]

10.   Ruszkowska, J. *Main Chemical Constituents of Carum. Caraway-The Genus Carum*; Németh, É., Ed.; Harwood Academic Publishers: Amsterdam, The Netherlands, 1998; pp. 38–60.

11.   Sedláková, J.; Kocourková, B.; Kubáň, V. Determination of essential oils content and composition in caraway (Carum carvi L.). *Czech J. Food Sci.* **2013**, *19*, 31–36. [CrossRef]

12.   Johri, R. Cuminum cyminum and Carum carvi: An update. *Pharmacogn. Rev.* **2011**, *5*, 63–72. [CrossRef] [PubMed]

13.   Hedges, L.J.; Lister, C.E. The nutritional attributes of legumes. *Crop. Food Res. Confid. Rep.* **2006**, *1745*, 50.

14.   Almuhayawi, M.S.; Al Jaouni, S.K.; Almuhayawi, S.M.; Selim, S.; Abdel-Mawgoud, M. Elevated CO_2 improves the nutritive value, antibacterial, anti-inflammatory, antioxidant and hypocholestecolemic activities of lemongrass sprouts. *Food Chem.* **2021**, *357*, 129730. [CrossRef] [PubMed]

15.   Kim, H.J.; Feng, H.; Kushad, M.M.; Fan, X. Effects of Ultrasound, Irradiation, and Acidic Electrolyzed Water on Germination of Alfalfa and Broccoli Seeds and Escherichia coli O157: H7. *J. Food Sci.* **2006**, *71*, M168–M173. [CrossRef]

16.   Almuhayawi, M.S.; Hassan, A.H.A.; Abdel-Mawgoud, M.; Khamis, G.; Selim, S.; Al Jaouni, S.K.; AbdElgawad, H. Laser light as a promising approach to improve the nutritional value, antioxidant capacity and anti-inflammatory activity of flavonoid-rich buckwheat sprouts. *Food Chem.* **2020**, *345*, 128788. [CrossRef]

17.   Saleh, A.M.; Selim, S.; Al Jaouni, S.; AbdElgawad, H. CO_2 enrichment can enhance the nutritional and health benefits of parsley (Petroselinum crispum L.) and dill (Anethum graveolens L.). *Food Chem.* **2018**, *269*, 519–526. [CrossRef] [PubMed]

18.   Saleh, A.M.; Abdel-Mawgoud, M.; Hassan, A.R.; Habeeb, T.H.; Yehia, R.S.; AbdElgawad, H. Global metabolic changes in-duced by arbuscular mycorrhizal fungi in oregano plants grown under ambient and elevated levels of atmospheric CO_2. *Plant Physiol. Biochem.* **2020**, *151*, 255–263. [CrossRef] [PubMed]

19.   Jing, L.; Wang, J.; Shen, S.; Wang, Y.; Zhu, J.; Wang, Y.; Yang, L. The impact of elevated CO_2 and temperature on grain quality of rice grown under open-air field conditions. *J. Sci. Food Agric.* **2016**, *96*, 3658–3667. [CrossRef] [PubMed]

20.   Li, P.; Li, H.; Zong, Y.; Li, F.Y.; Han, Y.; Hao, X. Photosynthesis and metabolite responses of Isatis indigotica Fortune to ele-vated [CO_2]. *Crop J.* **2017**, *5*, 345–353. [CrossRef]

21.   Högy, P.; Wieser, H.; Köhler, P.; Schwadorf, K.; Breuer, J.; Franzaring, J.; Muntifering, R.; Fangmeier, A. Effects of elevated CO_2 on grain yield and quality of wheat: Results from a 3-year free-air CO_2 enrichment experiment. *Plant Biol.* **2009**, *11*, 60–69. [CrossRef]

22.   Azam, A.; Khan, I.; Mahmood, A.; Hameed, A. Yield, chemical composition and nutritional quality responses of carrot, radish and turnip to elevated atmospheric carbon dioxide. *J. Sci. Food Agric.* **2013**, *93*, 3237–3244. [CrossRef] [PubMed]

23.   Al Jaouni, S.; Saleh, A.M.; Wadaan, M.A.M.; Hozzein, W.N.; Selim, S.; AbdElgawad, H. Elevated CO_2 induces a global met-abolic change in basil (Ocimum basilicum L.) and peppermint (Mentha piperita L.) and improves their biological activity. *J. Plant Physiol.* **2018**, *224*, 121–131. [CrossRef] [PubMed]

24.   Almuhayawi, M.; AbdElgawad, H.; Al Jaouni, S.; Selim, S.; Hassan, A.H.A.; Khamis, G. Elevated CO_2 improves glucosinolate metabolism and stimulates anticancer and anti-inflammatory properties of broccoli sprouts. *Food Chem.* **2020**, *328*, 127102. [CrossRef]

25.   Almuhayawi, M.S.; Hassan, A.H.A.; Al Jaouni, S.K.; Alkhalifah, D.H.M.; Hozzein, W.N.; Selim, S.; AbdElgawad, H.; Khamis, G. Influence of elevated CO2 on nutritive value and health-promoting prospective of three genotypes of Alfalfa sprouts (Medicago Sativa). *Food Chem.* **2020**, *340*, 128147. [CrossRef] [PubMed]

26.   Leakey, A.; Ainsworth, F.A.; Bernacchi, C.; Rogers, A.; Long, S.; Ort, D.R. Elevated CO_2 effects on plant carbon, nitrogen, and water relations: Six important lessons from FACE. *J. Exp. Bot.* **2009**, *60*, 2859–2876. [CrossRef] [PubMed]

27.   Li, X.; Zhang, G.; Sun, B.; Zhang, S.; Zhang, Y.; Liao, Y.; Zhou, Y.; Xia, X.; Shi, K.; Yu, J. Stimulated Leaf Dark Respiration in Tomato in an Elevated Carbon Dioxide Atmosphere. *Sci. Rep.* **2013**, *3*, 3433. [CrossRef]

28.   Noguchi, K.; Watanabe, C.K.; Terashima, I. Effects of Elevated Atmospheric CO_2 on Primary Metabolite Levels inArabidopsis thalianaCol-0 Leaves: An Examination of Metabolome Data. *Plant Cell Physiol.* **2015**, *56*, 2069–2078. [CrossRef] [PubMed]

29.   Hozzein, W.N.; Saleh, A.M.; Habeeb, T.H.; Wadaan, M.A.M.; AbdElgawad, H. CO_2 treatment improves the hypocholester-olemic and antioxidant properties of fenugreek seeds. *Food Chem.* **2020**, *308*, 125661. [CrossRef] [PubMed]

30.   Nunes-Nesi, A.; Fernie, A.R.; Stitt, M. Metabolic and Signaling Aspects Underpinning the Regulation of Plant Carbon Nitrogen Interactions. *Mol. Plant* **2010**, *3*, 973–996. [CrossRef]

31.   Kanunnikova, N.P. Role of brain glutamic acid metabolism changes in neurodegenerative pathologies. *J. Biol.* **2012**, *2*, M1–M10.

32.   Zieliński, H.; Frias, J.; Piskuła, M.K.; Kozłowska, H.; Vidal-Valverde, C. Vitamin B 1 and B 2, dietary fiber and minerals content of Cruciferae sprouts. *Eur. Food Res. Technol.* **2005**, *221*, 78–83. [CrossRef]

33.   Watanabe, C.K.; Sato, S.; Yanagisawa, S.; Uesono, Y.; Terashima, I.; Noguchi, K. Effects of elevated CO_2 on levels of primary metabolites and transcripts of genes encoding respiratory enzymes and their diurnal patterns in Arabidopsis thaliana: Possible relationships with respiratory rates. *Plant Cell Physiol.* **2014**, *55*, 341–357. [CrossRef]

34. Asif, M.; Yilmaz, O.; Ozturk, L. Elevated carbon dioxide ameliorates the effect of Zn deficiency and terminal drought on wheat grain yield but compromises nutritional quality. *Plant Soil* **2017**, *411*, 57–67. [CrossRef]
35. Teng, N.; Wang, J.; Chen, T.; Wu, X.; Wang, Y.; Lin, J. Elevated CO_2 induces physiological, biochemical and structural changes in leaves of Arabidopsis thaliana. *New Phytol.* **2006**, *172*, 92–103. [CrossRef]
36. Roberntz, P.; Linder, S. Effects of long-term CO_2 enrichment and nutrient availability in Norway spruce. II. Foliar chemistry. *Trees* **1999**, *14*, 17–27. [CrossRef]
37. LUOMALA, E.; Laitinen, K.; Sutinen, S.; Kellomäki, S.; Vapaavuori, E. Stomatal density, anatomy and nutrient concentra-tions of Scots pine needles are affected by elevated CO_2 and temperature. *Plant Cell Environ.* **2005**, *28*, 733–749. [CrossRef]
38. Ishikawa, T.; Takayanagi, T.; Kitajima, J. Water-Soluble Constituents of Cumin: Monoterpenoid Glucosides. *Chem. Pharm. Bull.* **2002**, *50*, 1471–1478. [CrossRef] [PubMed]
39. Kunzemann, J.; Herrmann, K. Isolation and identification of flavon (ol)-O-glycosides in caraway (Carum carvi L.), fennel (Foeniculum vulgare Mill.), anise (Pimpinella anisum L.), and coriander (Coriandrum sativum L.), and of flavon-C-glycosides in anise. I. Phenolics of spices (author's transl). *Zeitschrift für Lebensmittel-untersuchung und-Forschung.* **1977**, *164*, 194–200.
40. Matsumura, T.; Ishikawa, T.; Kitajima, J. Water-Soluble Constituents of Caraway: Carvone Derivatives and Their Glucosides. *Chem. Pharm. Bull.* **2002**, *50*, 66–72. [CrossRef]
41. Matsumura, T.; Ishikawa, T.; Kitajima, J. Water-soluble constituents of caraway: Aromatic compound, aromatic compound glucoside and glucides. *Phytochemistry* **2002**, *61*, 455–459. [CrossRef]
42. Najda, A.; Dyduch, J.; Brzozowski, N. Flavonoid Content and Antioxidant Activity of Caraway Roots (Carum Carvi L.). *Veg. Crop. Res. Bull.* **2008**, *68*, 127–133. [CrossRef]
43. Nakano, Y.; Matsunaga, H.; Saita, T.; Mori, M.; Katano, M.; Okabe, H. Antiproliferative Constituents in Umbelliferae Plants II. Screening for Polyacetylenes in Some Umbelliferae Plants, and Isolation of Panaxynol and Falcarindiol from the Root of Heracleum moellendorffii. *Biol. Pharm. Bull.* **1998**, *21*, 257–261. [CrossRef]
44. Idso, S.B.; Kimball, B.A.; Shaw, P.E.; Widmer, W.; Vanderslice, J.T.; Higgs, D.J.; Montanari, A.; Clark, W.D. The effect of ele-vated atmospheric CO_2 on the vitamin C concentration of (sour) orange juice. *Agric. Ecosyst. Environ.* **2002**, *90*, 1–7. [CrossRef]
45. De Carvalho, C.C.C.R.; Da Fonseca, M.M.R. Carvone: Why and how should one bother to produce this terpene. *Food Chem.* **2006**, *95*, 413–422. [CrossRef]
46. Sedláková, J.; Kocourková, B.; Lojková, L.; Kubáň, V. Determination of essential oil content in caraway (Carum carvi L.) species by means of supercritical fluid extraction. *Plant Soil Environ.* **2011**, *49*, 277–282. [CrossRef]
47. Oosterhaven, K.; Poolman, B.; Smid, E. S-carvone as a natural potato sprout inhibiting, fungistatic and bacteristatic compound. *Ind. Crop. Prod.* **1995**, *4*, 23–31. [CrossRef]
48. Chizzola, R. Composition of the Essential Oil of Wild Grown Caraway in Meadows of the Vienna Region (Austria). *Nat. Prod. Commun.* **2014**, *9*, 581–592. [CrossRef] [PubMed]
49. Abu-Darwish, M.S.; Ofir, R. Heavy metals content and essential oil yield of Juniperus phoenicea L. in different origins in Jordan. *Environ. Eng. Manag. J.* **2014**, *13*, 3009–3014. [CrossRef]
50. Kolta, R.; Hornok, L. Extraction processes for essential oils. *Cultiv. Process. Med. Plants* **1992**, *4*, 93–104.
51. Van der Mheen, H.J. *Vergelijking en Verloop (Tijdens de Afrijping) van de Zaad en Carvonopbrengst van (Zomer) Karwij en Dille= Comparison and Development (during the Ripening Phase) of the Seed and Carvone Yield of (Summer) Caraway and Dill*; PAGV: Lelystad, The Netherlands, 1994.
52. Al-Essa, M.K.; Shafagoj, Y.A.; Mohammed, F.I.; Afifi, F.U. Relaxant effect of ethanol extract of Carum carvi on dispersed intestinal smooth muscle cells of the guinea pig. *Pharm. Biol.* **2009**, *48*, 76–80. [CrossRef]
53. Joshi, S.G.; Joshi, S.G. *Medicinal Plants*; Oxford and IBH Publishing: New Delhi, India, 2000; Volume 491.
54. Zheng, G.-Q.; Kenney, P.M.; Lam, L.K.T. Anethofuran, Carvone, and Limonene: Potential Cancer Chemoprotective Agents from Dill Weed Oil and Caraway Oil. *Planta Med.* **1992**, *58*, 338–341. [CrossRef]
55. Lado, C.; Then, M.; Varga, I.; Szőke, É.; Szentmihályid, K. Antioxidant property of volatile oils determined by the ferric reducing ability. *Zeitschrift für Naturforschung C* **2004**, *59*, 354–358. [CrossRef] [PubMed]
56. Raphael, T.J.; Kuttan, G. Immunomodulatory Activity of Naturally Occurring Monoterpenes Carvone, Limonene, and Perillic Acid. *Immunopharmacol. Immunotoxicol.* **2003**, *25*, 285–294. [CrossRef] [PubMed]
57. Jaafar, H.Z.; Ibrahim, M.H.; Karimi, E. Phenolics and Flavonoids Compounds, Phenylanine Ammonia Lyase and Antioxidant Activity Responses to Elevated CO2 in Labisia pumila (Myrisinaceae). *Molecules* **2012**, *17*, 6331–6347. [CrossRef]
58. AbdElgawad, H.; Farfan-Vignolo, E.R.; de Vos, D.; Asard, H. Elevated CO_2 mitigates drought and temperature-induced oxidative stress differently in grasses and legumes. *Plant Sci.* **2015**, *231*, 1–10. [CrossRef] [PubMed]
59. Karimi, E.; Oskoueian, E.; Oskoueian, A.; Omidvar, V.; Hendra, R.; Nazeran, H. Insight into the functional and medicinal properties of Medicago sativa (Alfalfa) leaves extract. *J. Med. Plants Res.* **2013**, *7*, 290–297.
60. Wang, S.Y.; Bunce, J.A.; Maas, J.L. Elevated Carbon Dioxide Increases Contents of Antioxidant Compounds in Field-Grown Strawberries. *J. Agric. Food Chem.* **2003**, *51*, 4315–4320. [CrossRef] [PubMed]
61. Szydłowska-Czerniak, A.; Tułodziecka, A.; Szłyk, E. Determination of Antioxidant Capacity of Unprocessed and Processed Food Products by Spectrophotometric Methods. *Food Anal. Methods* **2011**, *5*, 807–813. [CrossRef]
62. Pellegrini, N.; Visioli, F.; Buratti, S.; Brighenti, F. Direct Analysis of Total Antioxidant Activity of Olive Oil and Studies on the Influence of Heating. *J. Agric. Food Chem.* **2001**, *49*, 2532–2538. [CrossRef] [PubMed]

63. Achat, S.; Rakotomanomana, N.; Madani, K.; Dangles, O. Antioxidant activity of olive phenols and other dietary phenols in model gastric conditions: Scavenging of the free radical DPPH• and inhibition of the haem-induced peroxidation of linoleic acid. *Food Chem.* **2016**, *213*, 135–142. [CrossRef]

64. Arabameri, M.; Nazari, R.R.; Abdolshahi, A.; Abdollahzadeh, M.; Mirzamohammadi, S.; Shariatifar, N.; Barba, F.J.; Mousavi Khaneghah, A. Oxidative stability of virgin olive oil: Evaluation and prediction with an adaptive neuro-fuzzy inference system (ANFIS). *J. Sci. Food Agric.* **2019**, *99*, 5358–5367. [CrossRef] [PubMed]

65. Hosseini, S.; Ramezan, Y.; Arab, S. A comparative study on physicochemical characteristics and antioxidant activity of su-mac (Rhus coriaria L.), cumin (Cuminum cyminum), and caraway (Carum carvil) oils. *J. Food Meas. Charact.* **2020**, *14*, 3175–3183. [CrossRef]

66. Fedor, M.; Socha, K.; Urban, B.; Soroczyńska, J.; Matyskiela, M.; Borawska, M.H.; Bakunowicz-Łazarczyk, A. Serum concen-tration of zinc, copper, selenium, manganese, and Cu/Zn Ratio in children and adolescents with myopia. *Biol. Trace Elem. Res.* **2017**, *176*, 1–9. [CrossRef] [PubMed]

67. Sadowska, A.; Obidoska, G. Pharmacological uses and toxicology of caraway. In *Caraway*; CRC Press: Boca Raton, FL, USA, 1999; pp. 198–208. ISBN 0429219679.

68. Toxopeus, H.; Bouwmeester, H. Improvement of caraway essential oil and carvone production in The Netherlands. *Ind. Crop. Prod.* **1992**, *1*, 295–301. [CrossRef]

69. Hartmans, K.J.; Diepenhorst, P.; Bakker, W.; Gorris, L.G. The use of carvone in agriculture: Sprout suppression of potatoes and antifungal activity against potato tuber and other plant diseases. *Ind. Crop. Prod.* **1995**, *4*, 3–13. [CrossRef]

70. Bailer, J.; Aichinger, T.; Hackl, G.; de Hueber, K.; Dachler, M. Essential oil content and composition in commercially available dill cultivars in comparison to caraway. *Ind. Crop. Prod.* **2001**, *14*, 229–239. [CrossRef]

71. Frank, T.; Biert, K.; Speiser, B. Feeding deterrent effect of carvone, a compound from caraway seeds, on the slug Arion lusitanicus. *Ann. Appl. Biol.* **2002**, *141*, 93–100. [CrossRef]

72. Simic, A.; Rančic, A.; Sokovic, M.D.; Ristic, M.; Grujic-Jovanovic, S.; Vukojevic, J.; Marin, P.D. Essential Oil Composition of Cymbopogon winterianus and Carum carvi. and Their Antimicrobial Activities. *Pharm. Biol.* **2008**, *46*, 437–441. [CrossRef]

73. Almuhayawi, M.; Mohamed, M.; Abdel-Mawgoud, M.; Selim, S.; Al Jaouni, S.; AbdElgawad, H. Bioactive Potential of Several Actinobacteria Isolated from Microbiologically Barely Explored Desert Habitat, Saudi Arabia. *Biology* **2021**, *10*, 235. [CrossRef] [PubMed]

74. Müller, L.; Fröhlich, K.; Böhm, V. Comparative antioxidant activities of carotenoids measured by ferric reducing antioxidant power (FRAP), ABTS bleaching assay (αTEAC), DPPH• assay and peroxyl radical scavenging assay. *Food Chem.* **2011**, *129*, 139–148. [CrossRef]

75. Noble, J.E.; Bailey, M.J.A. Quantitation of protein. *Methods Enzymol.* **2009**, *463*, 73–95.

76. Bligh, E.G.; Dyer, W.J. A rapid method of total lipid extraction and purification. *Can. J. Biochem. Physiol.* **1959**, *37*, 911–917. [CrossRef]

77. Habeeb, T.H.; Abdel-Mawgoud, M.; Yehia, R.S.; Khalil, A.M.A.; Saleh, A.M.; AbdElgawad, H. Interactive Impact of Arbuscular Mycorrhizal Fungi and Elevated CO_2 on Growth and Functional Food Value of Thymus vulgare. *J. Fungi* **2020**, *6*, 168. [CrossRef] [PubMed]

78. Sturm, K.; Koron, D.; Stampar, F. The composition of fruit of different strawberry varieties depending on maturity stage. *Food Chem.* **2003**, *83*, 417–422. [CrossRef]

79. Vázquez-Ortiz, F.A.; Caire, G.; Higuera-Ciapara, I.; Hernández, G. High Performance Liquid Chromatographic Determination of Free Amino Acids in Shrimp. *J. Liq. Chromatogr.* **1995**, *18*, 2059–2068. [CrossRef]

Article

Developmental Stages-Specific Response of Anise Plants to Laser-Induced Growth, Nutrients Accumulation, and Essential Oil Metabolism

Mohammad K. Okla [1], Mohamed Abdel-Mawgoud [2,*], Saud A. Alamri [1], Zahid Khorshid Abbas [3], Wahidah H. Al-Qahtani [4], Salem Mesfir Al-Qahtani [5], Nadi Awad Al-Harbi [5], Abdelrahim H. A. Hassan [6], Samy Selim [7], Mohammed H. Alruhaili [8] and Hamada AbdElgawad [9,10]

[1] Botany and Microbiology Department, College of Science, King Saud University, P.O. Box 2455, Riyadh 11451, Saudi Arabia; Okla99@yahoo.com (M.K.O.); Salamri@hotmail.com (S.A.A.)
[2] Department of Medicinal and Aromatic Plants, Desert Research Centre, Cairo 11753, Egypt
[3] Biology Department, College of Science, Tabuk University, Tabuk 71491, Saudi Arabia; Znourabbas@ut.edu.sa
[4] Department of Food Sciences & Nutrition, College of Food & Agriculture Sciences, King Saud University, Riyadh 11451, Saudi Arabia; wahida@ksu.edu.sa
[5] Biology Department, University College of Taymma, Tabuk University, P.O. Box 741, Tabuk 47512, Saudi Arabia; salghtani@ut.edu.sa (S.M.A.-Q.); nalharbi@ut.edu.sa (N.A.A.-H.)
[6] Department of Food Safety and Technology, Faculty of Veterinary Medicine, Beni-Suef University, Beni Suef 62511, Egypt; abdelrahim@vet.bsu.edu.eg
[7] Department of Clinical Laboratory Sciences, College of Applied Medical Sciences, Jouf University, Sakaka 72341, Saudi Arabia; sabdulsalam@ju.edu.sa
[8] Department of Medical Microbiology and Parasitology, Faculty of Medicine, King Abdulaziz University, Jeddah 21589, Saudi Arabia; malruhaili@kau.edu.sa
[9] Botany and Microbiology Department, Faculty of Science, Beni-Suef University, Beni Suef 62521, Egypt; hamada.abdelgawad@uantwerpen.be
[10] Integrated Molecular Plant Physiology Research, Department of Biology, University of Antwerp, 2020 Antwerpen, Belgium
* Correspondence: Mohamed_drc@yahoo.com or mabdelmawgoud35@gmail.com

Citation: Okla, M.K.; Abdel-Mawgoud, M.; Alamri, S.A.; Abbas, Z.K.; Al-Qahtani, W.H.; Al-Qahtani, S.M.; Al-Harbi, N.A.; Hassan, A.H.A.; Selim, S.; Alruhaili, M.H.; et al. Developmental Stages-Specific Response of Anise Plants to Laser-Induced Growth, Nutrients Accumulation, and Essential Oil Metabolism. *Plants* **2021**, *10*, 2591. https://doi.org/10.3390/plants10122591

Academic Editors: Beatrice Falcinelli and Angelica Galieni

Received: 7 October 2021
Accepted: 17 November 2021
Published: 26 November 2021

Publisher's Note: MDPI stays neutral with regard to jurisdictional claims in published maps and institutional affiliations.

Abstract: Compared to seeds and mature tissues, sprouts are well known for their higher nutritive and biological values. Fruits of *Pimpinella anisum* (anise) are extensively consumed as food additives; however, the sprouting-induced changes in their nutritious metabolites are hardly studied. Herein, we investigated the bioactive metabolites, phytochemicals, and antioxidant properties of fruits, sprouts (9-day-old), and mature tissue (5-week-old) of anise under laser irradiation treatment (He-Ne laser, 632 nm). Laser treatment increased biomass accumulation of both anise sprouts and mature plants. Bioactive primary (e.g., proteins and sugars) and secondary metabolites (e.g., phenolic compounds), as well as mineral levels, were significantly enhanced by sprouting and/or laser light treatment. Meanwhile, laser light has improved the levels of essential oils and their related precursors (e.g., phenylalanine), as well as enzyme activities [e.g., O–methyltransferase and 3-Deoxy-D-arabino-heptulosonate-7-phosphate synthase (DAHPS)] in mature tissues. Moreover, laser light induced higher levels of antioxidant and anti-lipidemic activities in sprouts as compared to fruits and mature tissues. Particularly at the sprouting stage, anise was more responsive to laser light treatment than mature plants.

Keywords: anise sprouts; mature plants; He–Ne laser; nutritious metabolites; antioxidant

1. Introduction

Sprouts are considered as rich sources of bioactive metabolites such as vitamins, minerals, and polyphenols, with various biological properties such as antioxidant and antitumor activities [1,2]. When compared to seeds and mature plants, sprouts are known to have low levels of anti-nutritional factors, so they are strongly recommended as popular healthy foods [2].

Recently, the application of different techniques for enhancing the nutritional quality of sprouts and mature plants has been attracting much interest [3–5]. In this regard, the use of laser irradiation as a biophysical elicitor, has proved to increase plant growth and development, induce high content of phytochemicals and enhance plant productivity [4,6]. Laser light has been divided into two types; pulsed laser, which has been used in medical applications; and continuous laser, which was applied in improving crop production [6,7]. Plants could respond physiologically to laser treatment (light, electromagnetism, and temperature) and their responses are based on the ability of plant macromolecules to absorb a specific wavelength of laser light to trigger photosynthetic activity, which leads to increased growth and biomass accumulation [6]. Previous reports have demonstrated the positive impact of laser light on increasing the quality and accumulation of bioactive metabolites of many plants, such as fennel, coriander, white lupine, faba bean, soybean, lemongrass sprouts, and buckwheat sprouts [4,8–11].

Pimpinella species are belonging to the family *Apiaceae*, which are mostly used as condiments or vegetables, in addition to their use in traditional therapy in many parts of the world [12]. Among the most common *Pimpinella* species, *Pimpinella anisum* (anise) has been extensively consumed as food additives and beverages and is also used in the pharmaceutical industry due to its distinct flavor and therapeutic values [13]. Anise fruits (aniseeds) have been used in dry/fresh form in traditional medicine to cure many ailments [14]. Moreover, aniseeds contain essential oils with a wide range of biological activities, such as antimicrobial, antioxidant, carminativum, anti-spasmoliticum, and antidiabetic activities [15–17]. Principally, aniseeds contain *trans*-anethole as a major volatile component [18], in addition to other constituents such as eugenol, anisaldehyde, methylchavicol, coumarins, scopoletin, estragole, umbelliferone, terpene hydrocarbons, estrols, polyenes, and polyacetylenes [17]. The medicinal properties of aniseed are frequently attributed to its content of bioactive molecules such as phenolic components and terpens. Therefore, increasing the bioactive contents of anise sprouts and mature plants by using eco-friendly approaches, such as laser light, could enhance the plant's nutritive and health-promoting values.

To the best of our knowledge, few studies have dealt with the effect of laser light on aniseeds morphological properties and yield [19]. Thus, the present study aimed at investigating the laser-induced effects on growth, minerals, vitamins, essential oils, as well as total bioactive metabolites and biological activities (antioxidant, anti-lipidemic) of anise sprouts and their mature plants under He-Ne laser treatment in comparison with control. At the same time, we also compared the significant differences between anise sprouts and mature plants, regarding their response to the effect of laser light on their bioactive contents and biological activities.

2. Material and Methods

2.1. Plant Material and Experimental Conditions

Fruits of anise were collected and obtained from Agricultural Research Centre, Giza, Egypt. The fruits of anise were soaked in distilled water for 2 h, then they were divided into 2 groups; untreated and laser-treated groups (each one contains 100 seeds). Helium-neon (He-Ne) laser system (equipment whitening, laser II, DMC Equipment Ltd., Vista, CA, USA) was used as a light source. The treated group was irradiated by He–Ne laser (632 nm at 5 mW for 5 min and 500 mJ energy from the embryonic area side, beam diameter 1 mm), whereas the distance between laser light and the fruits was 12 cm. The untreated treated fruits were placed in dark covered with a box, under controlled conditions. All these conditions were chosen after a preliminary experiment and our previous studies [4,11]. The experiment was repeated 3 times. The irradiated fruits were rinsed in distilled water and spread on trays and irrigated for the first three days with Milli-Q water, then placed in the dark. Afterward, the fruits were maintained at 25 °C in a growth chamber under a 16 h light/8 h dark cycle through white fluorescent tubes with 400 μmol m^{-2} s^{-1} photosynthetically active radiation (PAR). Thereafter, 100 mL aquaponic water was poured over the fruits in each tray. Mature plants (five weeks-old) were cultivated at 21/18 °C in a custom-built climate-controlled chamber with a 16/8 h day/night photoperiod (150 μmol PAR m^{-2} s^{-1}, 60% humidity). After 9 days, the sprouts were taken as fresh weight and then stored at $-80\,°C$ for further analyses, where the chemicals and reagents were purchased from Sigma Aldrich. Five biological replicates (each biological replicate was a pooled of ten plants) were used for each measurement

2.2. Determination of Photosynthetic Rate

Photosynthesis (μmol CO_2 m^{-2} s^{-1}) was determined according to [5], by using EGM-4 infrared gas analyzer (PP Systems, Hitchin, UK). The photosynthetic rate was evaluated from 180 s measurements of net CO_2 exchange (NE).

2.3. Pigment Analysis

Anise sprouts and mature plants were homogenized at 7000 rpm for 1 min in acetone by using a MagNALyser (Vilvoorde, Belgium), then centrifuged for 20 min (4 °C, 14,000× g). The supernatant was then filtered (Acrodisc GHP filter, 0.45 μm 13 mm). Afterward, HPLC (Shimadzu SIL10-ADvp, reversed-phase, at 4 °C) was used to analyze the obtained solution [4]. Carotenoids were separated on a silica-based C18 column (Waters Spherisorb, 5 μm ODS1, 4.6 × 250 mm, Agilent Technologies, Santa Clara, CA, USA), whereas two solvents were used; (A) acetonitrile: methanol: water (81:9:10) and (B) methanol: ethyl acetate (68:32). Chlorophyll a,b, and β-carotene were extracted, and then quantified using a diode-array detector (Shimadzu SPDM10Avp, Japan, Tokyo) at four wavelengths (420, 440, 462, and 660 nm).

2.4. Elemental Analysis

Elemental analysis was done, using inductively coupled plasma (ICP-MS, Finnigan Element XR, Scientific, Bremen, Germany). Macro and micro-elements were evaluated according to [5]. About 200 mg from both treated and control non-treated sprouts and mature plants were digested in HNO_3/H_2O solution (5:1 v/v) in an oven. Then, the concentrations of macro-minerals and trace elements were estimated at 25 °C by using inductively coupled plasma mass spectrometry (ICP-MS, Finnigan Element XR, and Scientific, Bremen, Germany), where nitric acid in 1% was used as a standard.

2.5. Vitamins Analysis

Tocopherols and ascorbate and were determined according to [4] using reversed-phase HPLC, where tocopherols and ascorbate were mixed with hexane and metaphosphoric acid in MagNALyser, respectively. Then, centrifugation was done (4 °C, 14,000× g, 20 min). Meanwhile, thiamine and riboflavin were separated on a reverse-phase (C18, Agilent Technologies, Santa Clara, CA, USA) column (HPLC, methanol: water) [4].

2.6. Essential oil Analysis
2.6.1. GC-MS Analysis

The tested sprouts and mature plants were air-dried, and then extraction of essential oil was done using 15 g. The dried parts were subjected to steam distillation for three hours using a Clevenger-type instrument. The essential oil contents were determined by using GC/MS according to [20]. GC–MS analysis was done by using a Thermoquest GC–MS instrument (EI mode at 70 eV), where a DB-1 fused silica capillary column (60 m * 0.25 mm i.d., film thickness 0.25 mm) was used. The temperature was increased from 40 to 250 °C at 4 °C min^{-1}, then kept constant at 250 °C for 10 min, while the temperatures of detector and injector were 300 °C and 250 °C, respectively. The carrier gas used was Helium (at a flow rate of 1.1 mL min^{-1}. Identification of the detected compounds was achieved by comparing their mass spectra with those obtained from NIST library. For the retention index (RI) of identified essential oil, please check Supplementary Table S1.

2.6.2. Determination of DAHPS

Determination of DAHPS activity was done according to [21]. Fresh samples were homogenized in 3 mL pre-cooled Tris-HCL buffer (50 mM, pH = 7.4) containing polyvinylpyrrolidone (1%), phenylmethylsulfonyl fluoride (0.1 mM), leupeptin (10 μM), and 2-mercaptoethanol (1.4 mM), The mixture was kept at 4 °C for 30 min, then centrifugation was done (12000, 20 min). The supernatant was used for the assay, whereas the mixture contained the extract (0.8 mL), Tris-HCl buffer (2.2 mL, 50 mM, pH = 7.5), phosphoenolpyruvate (0.2 mM), 0.1 mM $MnSO_4$/0.1 mM $CoCl_2$, and erythrose-4- phosphate (0.1 mM). Then, incubation was done at 30 °C for 30 min. Afterward, the reaction was started by adding the enzyme, then terminated by adding 500 μL trichloroacetic acid (25%), while the control was prepared without the enzyme. The enzyme activity was detected on the basis of the amount of enzyme used for the synthesis of 1 nmol of DAHPS per minute at 30°C. Finally, the concentration of DAHPS was detected at 549 nm.

2.6.3. Determination of PAL

For determination of PAL activity, fresh samples were homogenized in 3 mL precooled sodium borate buffer (0.1 M, pH = 8.8) containing polyvinylpyrrolidone (0.4%), EDTA (1 mM), 2-mercaptoethanol (5 mM), The mixture was kept at 4 °C for 30 min, then centrifugation was done (12000, 20 min). The supernatant was used for the assay, whereas the mixture contained the extract (0.8 mL), 2.2 mL sodium borate buffer (0.1 M, pH = 8.8) containing L-Phe (120 μM). Then, incubation was done at 25 °C for 40 min. Afterward, the reaction was terminated by adding 120 μL HCL (6 N), while the control was prepared without the enzyme. Detection of the product, trans-cinnamic acid, was done at 290 nm. The enzyme activity was detected on the basis of deamination of 1.0 nmol of L-phenylalanine to cinnamic acid per minute.

2.7. Determination of Phenolic Profile
2.7.1. Determination of Total Phenolic Content

Extraction of flavonoids and phenolic acids was done by homogenizing 100 mg of sprouts in 1 mL of 80% ethanol. Centrifugation was done for 20 min at 4 °C, and then the flavonoid content was measured by using the modified aluminum chloride colorimetric method (quercetin as a standard). Meanwhile, the total phenolic content was estimated by using a Folin–Ciocalteu assay (gallic acid as a standard) [4].

2.7.2. HPLC Analysis

The Individual flavonoids and phenolic acids were detected by homogenizing 50 mg of samples in acetone–water (4:1) for 24 h. Thereafter, phenolic compounds have been analyzed by using HPLC (SCL-10A vp, Shimadzu Corporation, Kyoto, Japan), equipped with a Lichrosorb Si-60, 7 μm, 3 × 150 mm column, diode array detector), whereas 3,5-dichloro-4-hydroxybenzoic was used as an internal standard. The mobile phase consisted of water-formic acid (90:10), and acetonitrile/water/formic acid (85:10:5). The detection of each compound's concentration was done using a calibration curve of the corresponding standard. For the retention index (RI) of identified essential oil, please check Supplementary Table S1.

2.8. Biological Activities
2.8.1. Antioxidant Activities

According to the methods outlined in [4], the antioxidant potential of the examined sprouts and plants was evaluated through different assays; i.e., diphenylpicrylhydrazyl (DPPH), ferric reducing antioxidant power (FRAP), oxygen radical absorbance capacity (ORAC), and inhibition of LDL (low-density lipoprotein) oxidation (TBARS and conjugated dienes) [22–24]. For determination of FRAP activity, FRAP reagent (180 μL) and ethanol extracts (20 μL) were added into each well in the micro-plate, then incubation was done for 30 min at 37 °C. The absorbance was detected at 593 nm with a microplate reader

(Synergy Mx, Biotek Instruments Inc., Vermont, VT, USA). Trolox was used as a standard. For determination of ORAC, 120 µL of fluorescein (112 nM) dissolved in phosphate buffer (75 mM) was added into the micro-plate. Afterward, 20 µL of the samples, 20 µL of phosphate buffer (blank) and 20 µL of trolox (standard) were added to the plate, then incubation was done for 15 min at 37 °C, and the absorbance was detected at 485/520 nm. Thereafter, 80 µL of AAPH (62 mM) (2,2′-azobis 2-methylpropionamidin dihydrochloride) was added, and then, the absorbance was detected at 485/520 nm. The difference between the two measurements was detected using a standard row. For determination of LDL, dialyzed LDL (100 µg protein/mL) was diluted in 10 mM PBS (phosphate-buffered saline containing 0.01 M phosphate-buffer and 0.15 M NaCl, pH 7.4), and incubation was done at 37 °C with or without 10 µM $CuSO_4$. Afterward, oxidation was done in the presence or absence of colostrum proteins, and then lipid peroxidation was detected. Thiobarbituric acid reactive substances (TBARS) were determined at 532 nm/600 nm, where 1,1,3,3-Tetramethoxypropane was used as a standard. Conjugated diene was detected at 232 nm of LDL solution (100 µg protein/mL) in PBS incubated with $CuSO_4$ (10 µM) in the presence or absence of different concentrations of bovine colostrums protein.

2.8.2. Hypocholesterolaemic Activity
Inhibition of Micellar Solubility of Cholesterol

In order to evaluate the effect of the tested sprouts and plants on micellar solubility of cholesterol, the protocol described in [3] was used, where the extracts were added to micellar solution [15 mM sodium phosphate, 10 mM sodium taurocholate, 2 mM cholesterol, 5 mM oleic acid, 132 mM NaCl]. Afterward, the mixture was sonicated for 2 min and incubated in a water bath at 37 °C for 24 h. Thereafter, centrifugation was done for 60 min (40,000 rpm, 20 °C). The cholesterol content was evaluated at 500 nm by using a cholesterol analysis kit (Pointe Scientific, New York, NY, USA, C7510).

Pancreatic Lipase Inhibition Assay

The method described in [3] was used to evaluate the impact of the tested extracts on pancreatic lipase by using 4-MUO as a substrate. About 0.5 mL of the extracts were taken and mixed with 0.5 mL of lipase. Centrifugation was done for 10 min at 4000 rpm, and then 2 mL of the 4-MUO solutions were added. Incubation was done at 37 °C. Aliquots of 0.2 mL were taken at different time points, and 4-MUO hydrolysis by lipase was detected at 350/450 nm. A logarithmic regression curve was created to detect IC50 values (mg/mL).

2.9. Statistical Analyses

To perform the statistical analyses, the SPSS statistical package (SPSS Inc., Chicago, IL, USA) was used. Each experiment was replicated at least two times, and for all assays, 3 to 5 replicates were used, and each replicate corresponded to a group of sprouts and mature plants harvested from a certain tray. A one-Way Analysis of Variance (ANOVA) test was applied. Tukey's test was used as the post-hoc test for separation of means ($p < 0.05$). Cluster analysis was performed by using Pearson distance metric of the MultiExperiment Viewer (MeV)™ 4 software package (version 4.5, Dana-Farber Cancer Institute, Boston, MA, USA).

3. Results and Discussion
3.1. Physical Properties and Biomass Accumulation of Anise Fruits, Sprouts, and Mature Plants as Affected by Laser Light Treatment

Laser light has been known to act as a stimulating factor for plant growth and yield, as well as the sprouting process [4,6]. The absorption of laser light at a certain wavelength could activate the photosynthesis process and increase seed internal energy, which consequently accelerates cell division and improve enzymatic activities [4,6,10]. This might also include the induction of some phytohormones, such as indole-3-acetic acid (IAA), which is important for cell division and growth in germinating seeds [8]. Such laser-provoked changes in plants could eventually lead to improved growth and biomass accumulation.

Supporting such a hypothesis, the current results have shown significant increases in length, width, thickness, pod length, and yield of laser-treated anise fruits, when compared to their respective controls (Figure 1). Meanwhile, fruit mass was not affected by laser treatment. In agreement, previous studies have demonstrated the enhancing effects of laser on growth and biomass production in many plant species, such as anise, cumin, fennel, coriander, white lupine, faba bean, lemongrass sprouts, and buckwheat sprouts [4,8–11,19]. Thus, laser light might affect plant growth, either directly through activating germination and thermodynamic parameters, or indirectly by its positive impact on the photosynthetic activity.

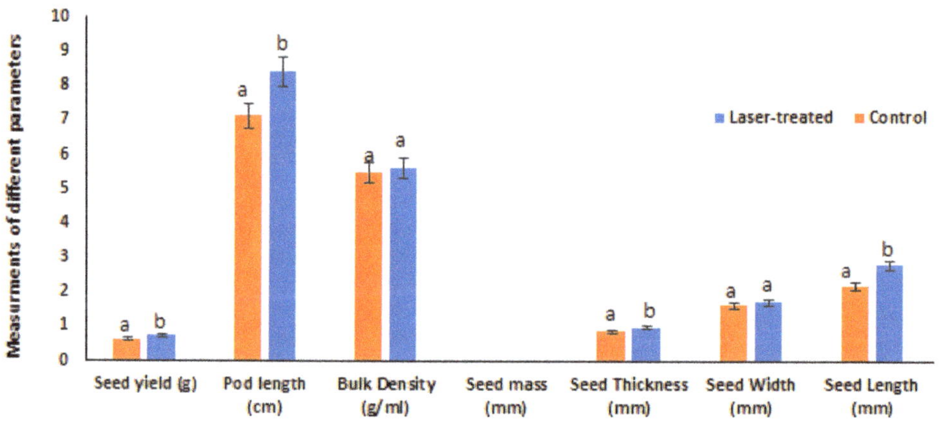

Figure 1. Physical properties in control and laser light-treated anise fruits. Data are represented by the means of at least 3 replicates ± standard error. Different small letters (a, b) above columns indicate significant differences between control and laser-treated samples at $p < 0.05$.

The present investigation has also revealed that laser light significantly induced the accumulation of photosynthetic pigments in aniseed sprouts and mature plants, whereby a significant increment was observed in pigment contents of laser-treated sprouts, particularly in the levels of chlorophyll a, b, and (a + b) which were increased by 106, 24 and 77% (Figure 2). Meanwhile, insignificant increments were detected in pigment contents of laser-treated mature plants, compared to their untreated controls. By comparing their response to laser treatment, anise sprouts seemed to better respond, thus accumulating higher photosynthetic pigments than mature plants do. In this regard, the higher chlorophyll content, under the effect of laser light, is expected to trigger the photosynthetic activity, which leads to higher sugar content, and consequently higher biomass and yield accumulation [25]. In addition, it has been reported that the photon energy of laser light could be absorbed by chlorophyll, and hence directly affects the photosynthetic activity [26]. Moreover, several studies have addressed the key role of carotenoids in the photosynthesis process, so the biosynthesis of carotenoids is expected to be influenced by light intensity and quality [27]. Consequently, the production of carotenoids might be triggered by the genes incorporated in carotenoid biosynthesis [28].

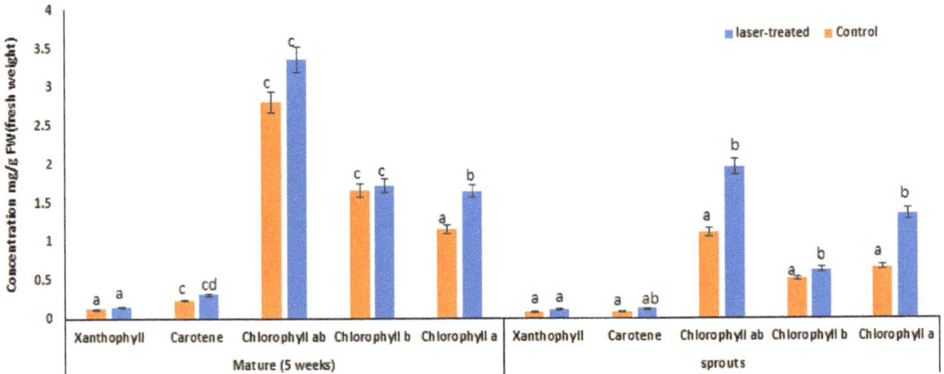

Figure 2. Pigments in sprouts and mature anise (control and laser light-treated). Data are represented by the means of at least 3 replicates ± standard error. Different small letters (a, b, c and d) above columns indicate significant differences between control and laser-treated samples at $p < 0.05$.

In accordance, several reports have dealt with the positive impact of laser light on levels of chlorophyll a and b in some plant species, such as *Isatis indogotica*, sunflower, and soybean, as well as in buckwheat sprouts [4,6,10,29]. Similarly, it has been previously shown that various light sources might exert a positive impact on chlorophyll accumulation in buckwheat sprouts [30].

3.2. Sprouting and Laser Light Induced a More Pronounced Effect on Nutritive Values of Anise

The determination of nutritive values of plants has been dependent on their bioactive contents of primary metabolites (e.g., lipid, proteins, and sugars), and secondary metabolites (e.g., phenolic compounds) [5]. The increasing of such bioactive metabolites by using promising approaches, such as laser light, could significantly enhance the plant's nutritional quality [4].

The present results have revealed that laser light treatment significantly enhanced the amount of some detected bioactive primary (lipid, proteins, and sugars) and secondary metabolites (saponins, steroids) in both sprouts and mature plants, in comparison to untreated controls (Table 1). Meanwhile, the total phenols, flavonoids, and tannins were increased only in anise sprouts. The total nutrients of fruits were almost comparable to those of control plants. So, the laser-treated anise sprouts might respond better to laser effects, hence accumulate higher content of bioactive metabolites, than the mature plants do. The increased sugar content might be associated with the enhanced photosynthetic activity, reported herein. As a result, the bioavailability of sugars could provide the carbon skeleton and energy needed for the biosynthesis of various classes of other metabolites [25,31]. This could explain the increased contents in some bioactive metabolites, reported in our study, such as phenolic compounds, saponins, and steroids. In agreement, laser light has been previously reported to increase total sugar contents in lemongrass sprouts and consequently increased their primary and secondary metabolites [11].

The laser-enhanced bioactive contents could be also attributed to the ability of laser light to activate plant metabolism, which could improve the nutrient status and plant productivity [6,32]. Moreover, sprouting has been recognized as an effective way to enhance the active primary and secondary metabolites, since seeds might be subjected to many physiological changes. For example, some bioactive metabolites, e.g., phenolic compounds, have been reported to increase gradually during sprouting in buckwheat species [33].

Sprouts have been used as rich sources of vitamins and mineral elements, whose deficiency could risk human health. Therefore, enhancing the contents of vitamins and minerals in aniseeds sprouts by using promising techniques, such as laser light, might boost their nutritional and health-promoting values. According to our results, the majority of the detected minerals have significantly increased in response to laser treatment in aniseeds sprouts (dominated by P, K, Ca, and Mg), and mature plants (mostly N), whereby anise sprouts and mature plants are likely to equally respond to laser light effects on their mineral contents (Table 1). Meanwhile, the fruits appeared to act as the control plants.

In agreement with our results, previous reports have shown laser light to enhance the levels of mineral elements in many plant species, e.g., N, P, and K in fennel and coriander [9], P and K in anise and cumin [19], K, Ca and Mg in sunflower [29], k in sugar beet [32] and K, P and Na in buckwheat sprouts [4]. The laser-enhanced effects on mineral contents could be explained by the ability of laser light to act as an inducer for more energy production from plant cell which could stimulate plant metabolism and nutrient uptake [34], or by increased root growth which consequently increases mineral uptake [35].

Table 1. Total metabolite contents, minerals, vitamins, and essential oil levels in fruits, sprouts, and mature anise (control and laser light-treated). Data are represented by the means of at least 3 replicates $\pm$ standard error. Different small letters (a, b, c, d) within a row indicate significant differences between control and laser-treated samples at $p < 0.05$.

Metabolite	Fruits	Sprouts Control	Laser-Treated	Mature Control	Laser-Treated
Total primary metabolite (mg/g FW)					
Lipid	84.7 ± 13.7a	109.1 ± 9.7a	173.2 ± 5.9b	104.4 ± 18.3b	138.2 ± 9.9c
Protein	163 ± 19.9a	171.4 ± 10.7a	239.5 ± 4.4b	185.9 ± 26a	230.3 ± 14.4b
Total Sugar	192.4 ± 26.8a	247.9 ± 8.5a	304 ± 31.6b	235.9 ± 34.1b	309.3 ± 19.3c
Ash	2.4 ± 0.16a	2.6 ± 0.2a	2 ± 0.4a	1.8 ± 0.93a	2.58 ± 0.16a
Crude fiber	2.02 ± 0.28a	2.6 ± 0.1a	3.2 ± 0.5b	2.48 ± 0.36ab	3.25 ± 0.2b
Total secondary metabolite (mg/g FW)					
Phenols	6.67 ± 0.9a	10.1 ± 0.9a	11.6 ± 0.1b	7.66 ± 1.2a	10.1 ± 0.76a
Flavonoids	0.26 ± 0.04a	0.33 ± 0a	0.5 ± 0b	0.39 ± 0.05a	0.47 ± 0.03a
Taninin	41 ± 6a	47.33 ± 0.4a	63.3 ± 6.2b	55 ± 7a	62 ± 4.5a
Saponin	4.62 ± 0.82a	7 ± 0.4a	12.9 ± 1.2b	8.51 ± 1.23b	10.15 ± 0.7c
Stroid	8.8 ± 1a	9.1 ± 0.6a	13.9 ± 0.7b	12 ± 2b	13 ± 1bc
Vitamins (mg/g FW)					
Tocopherol (Vit E)	0.49 ± 0.07a	0.63 ± 0.1a	0.91 ± 0.1b	0.61 ± 0.09a	0.79 ± 0.05ab
α-Carotene (Vit A)	0.28 ± 0.04a	0.36 ± 0.1a	0.52 ± 0.1b	0.35 ± 0.05a	0.45 ± 0.03a
β-Carotene (Vit A)	0.09 ± 0.01a	0.16 ± 0a	0.16 ± 0a	0.07 ± 0.02a	0.17 ± 0.01b
β-Cryptoxanthin (Vit A)	0.05 ± 0.01a	0.07 ± 0a	0.06 ± 0a	0.04 ± 0.01a	0.07 ± 0b
Thiamine (Vit B)	0.05 ± 0.01a	0.06 ± 0a	0.07 ± 0a	0.05 ± 0.01a	0.07 ± 0a
Phylloquinone (Vit K)	0.09 ± 0.01a	0.14 ± 0a	0.14 ± 0a	0.08 ± 0.01a	0.15 ± 0.01b
Ascorbic Acid (Vit C)	1.2 ± 0.2a	1.4 ± 0.1a	1.96 ± 0.1ab	1.5 ± 0.2a	1.7 ± 0.1ab
Minerals (mg/g DW)					
K	10.2 ± 1.4a	11.7 ± 1a	16.89 ± 1b	12.7 ± 1.7a	14.7 ± 0.9b
P	3 ± 0.5a	3.46 ± 0.4a	6.28 ± 0.5b	4.7 ± 0.6ab	5 ± 0.3ab
Ca	2.2 ± 0.3a	2.24 ± 0.3a	3.19 ± 0.5b	2.7 ± 0.3a	2.8 ± 0.2a
Mg	0.8 ± 0.17a	1.39 ± 0.5a	2.55 ± 0.4ab	1.65 ± 0.23b	2.01 ± 0.13c
Na	0.32 ± 0.04a	0.37 ± 0.1a	0.47 ± 0.1b	0.35 ± 0.05a	0.43 ± 0.03a
Fe	0.14 ± 0.02a	0.15 ± 0.02a	0.18 ± 0.02a	0.15 ± 0.02a	0.17 ± 0.01a
Zn	0.02 ± 0a	0.04 ± 0.01a	0.04 ± 0.01a	0.02 ± 0a	0.04 ± 0b
Cu	0.07 ± 0.01a	0.08 ± 0.02a	0.17 ± 0.01b	0.13 ± 0.02a	0.13 ± 0.01a
N	24.5 ± 2.9a	32.51 ± 3a	34.58 ± 3a	22.4 ± 3.6a	30.9 ± 2.2b

Table 1. *Cont.*

Metabolite	Fruits	Sprouts Control	Sprouts Laser-Treated	Mature Control	Mature Laser-Treated
Essential oils (mg/g FW)					
α-pinene	0.02 ± 0a	0.02 ± 0.01a	0.05 ± 0.02b	0.05 ± 0.01b	0.04 ± 0b
Sabinene	0.26 ± 0.03a	0.46 ± 0.22a	0.34 ± 0.07a	0.2 ± 0.04a	0.25 ± 0.03a
Myrcene	0.13 ± 0.03a	0.15 ± 0.03a	0.45 ± 0.09b	0.39 ± 0.05b	0.34 ± 0.02b
Fenchone	3.3 ± 0.43a	4.38 ± 0.67a	4.79 ± 0.75a	3.61 ± 0.55a	5.09 ± 0.32b
p-cymene	0.47 ± 0.06a	0.61 ± 0.18a	0.62 ± 0.1a	0.48 ± 0.07a	0.59 ± 0.04ab
o-isoeugenol	3.9 ± 0.36ab	2.97 ± 0.6a	4.47 ± 0.7b	3.47 ± 0.48a	4.17 ± 0.26ab
1,8-cineole	1.24 ± 0.18a	1.41 ± 0.2a	1.84 ± 0.4a	1.61 ± 0.21ab	1.72 ± 0.11ab
Cis-β-ocimene	0.26 ± 0.04a	0.33 ± 0.04a	0.51 ± 0.11b	0.4 ± 0.05a	0.47 ± 0.03a
Aα-phellandrene	0 ± 0a	0 ± 0a	0 ± 0a	00.14 ± 0a	0.012 ± 0a
Methyl chavicol	0.18 ± 0.03a	0.2 ± 0.04a	0.39 ± 0.08 b	0.16 ± 0.04a	0.23 ± 0.02ab
Endo-fenchyl acetate	0 ± 0a	0.001 ± 0a	0.002 ± 0b	0 ± 0a	0 ± 0a
p-anisaldehyde	0.07 ± 0.01ab	0.08 ± 0.04a	0.06 ± 0.01a	0.05 ± 0.01a	0.08 ± 0a
Limonene	0.12 ± 0.02a	0.18 ± 0.03a	0.35 ± 0.05b	0.23 ± 0.03ab	0.3 ± 0.02c
Stearic acid	0.03 ± 0a	0.03 ± 0.01a	0.03 ± 0.01a	0.03 ± 0a	0.03 ± 0a
2-oleoylglycerol	0.19 ± 0.03a	0.29 ± 0a	0.44 ± 0.05b	0.29 ± 0.04a	0.41 ± 0.03a
γ-himachalene	0.1 ± 0.01a	0.11 ± 0.03a	0.11 ± 0.01a	0.1 ± 0.01a	0.12 ± 0.01a
Trans-Pseudoisoeugenyl	0.13 ± 0.02a	0.15 ± 0.06a	0.16 ± 0.01a	0.14 ± 0.02a	0.17 ± 0.01b
Trans-anethole	44 ± 7a	49.99 ± 8.36a	75.17 ± 2.4b	66.4 ± 8b	70.5 ± 4bc
Amino acids					
Phenylalanine	2.0 ± 0.1a	3.2 ± 0.12b	5.8 ± 0.8d	2.8 ± 0.07b	4.2 ± 0.4c
L-phenylalanine aminolyase	25.9 ± 1.8a	33.1 ± 2.1a	57.1 ± 5.4c	28.3 ± 3.1a	49.2 ± 5.41b
DAHPS	0.1 ± 0.02a	0.44 ± 0b	0.95 ± 0.01c	0.41 ± 0b	0.8 ± 0.05c
Other related compounds					
Cinnamic acid	2.8 ± 0.1b	1.6 ± 0a	2.7 ± 0.1b	3.1 ± 0.1b	4.07 ± 0.2c
Shikimic acid	33.9 ± 1.1a	58.3 ± 6.8b	74 ± 4.0c	40.5 ± 6.2a	60.8 ± 7.4b

Regarding vitamins, the present investigation also showed that laser light treatment caused significant increases in some of the measured vitamins (mostly Vit A, E, and C in aniseeds sprouts, and Vit A and K in the mature plant) (Table 1). Laser light appears to have a more pronounced effect on the vitamin content of anise mature plant than that of the sprouts, while the fruits were significantly similar to control plants regarding their vitamin content.

Our results could be supported by recent studies by [4], who demonstrated a significant enhancement in the contents of Vit C, E, B1, and B2 in buckwheat sprouts after laser treatment. Moreover, [10] reported that laser light enhanced Vit C content in soybean. The improved vitamin content in response to laser treatment could be ascribed to activating the photosynthetic process, which consequently leads to increased carbohydrates production that might be used as a precursor for the synthesis of various classes such as vitamins [25]. In addition, the accumulation of vitamins could be also due to increasing the energy provided to the seeds after laser light exposure, which could be then converted into chemical energy and accelerated the metabolic events during the sprouting process [26].

The current investigation has also demonstrated that the chemical profile of anise essential oil contained 18 identified compounds, whereas *trans*-anethole was reported as the dominant component (Table 1). When compared to untreated control, the laser-treated anise sprouts and mature plants have exhibited significantly higher contents of most detected essential oils. Such increases were almost similar in both laser-treated sprouts and mature plants, indicating that they responded equally to laser light effects on their essential oil contents. The fruits were observed to contain almost similar amounts of essential oils to the control plants.

The enhanced levels of such essential oils could be attributed to the increased photosynthetic activity, in response to laser light, which consequently stimulated the sugar content that serves as a precursor for the synthesis of different metabolites, e.g., essential oils [11]. In agreement with our results, laser light has been demonstrated to increase the essential oil contents of lemongrass sprouts [11]. Moreover, it has been previously reported that *trans*-anethole is the main component which comprises 90% of aniseeds essential oil, being the most important constituent responsible for the distinct aroma and taste of aniseed [13,18], in addition to its variable biological properties, e.g., antibacterial, antifungal, antioxidant, and anti-migraine headache effects [36–38]. Also, other components, such as anisaldehyde, methyl chavicol, have been previously reported in high amounts in aniseed essential oil [17].

Phenylpropanoid compounds are known to be the main components of essential oils, being synthesized from phenylalanine through the cinnamic and shikimic acids pathway [21,39]. In this regard, several enzymes are involved in these biosynthetic pathways, such as 3-deoxy-D-arabinoheptulosonate-7-phosphate synthase (DAHPS) which is incorporated in the shikimic acids pathway, whereas shikimic acid is being converted into chorismite (a precursor for phenylalanine) [40]. Then, phenylalanine is converted into cinnamic acid with the aid of the phenylalanine aminolyase enzyme (PAL). Eventually, p-coumaric acid is synthesized from cinnamic acid. Both p-coumaric and cinnamic acids are required for the biosynthesis of essential oils in plants [21,41].

Thus, in order to get insight into the laser-induced changes in essential oils precursors and their related biosynthetic enzymes, the effect of laser on the levels of phenylalanine, PAL, DAHPS, cinnamic acid, and shikimic acid in anise sprouts and mature plants was investigated. The obtained results have shown that laser light treatment led to significant increments in phenylalanine, PAL, DAHPS, cinnamic acid, and shikimic acid in both sprouts and mature plants, when compared with their respective controls (Table 1). The increases in DAHPS, cinnamic acid, and shikimic acid were almost similar in both laser-treated sprouts and mature plants, while the increases in phenylalanine and PAL were more obvious in laser-treated sprouts than the mature plants, indicating that the effect of laser on essential oils precursors was more pronounced in sprouts than the mature plants. The fruits were observed to contain almost similar amounts of essential oils to the control plants. Similar to our results, eCO2 has been previously reported to exert positive effects on different stages of anise plants [42]. The obtained results could indicate the enhancement in essential oil metabolism in different stages of anise plants, exposed to laser irradiation treatment.

So, improving the essential oil content of anise sprouts and mature plants by using laser light could increase their content of anethole, thus enhancing their quality and therapeutic values, particularly antioxidant properties.

3.3. Laser Light Increased the Antioxidant Potential of Anise Sprouts and Mature Plants by Enhancing Their Phenolic Content

The antioxidant activities of plants have been mostly associated with their higher content of phenolic compounds (flavonoids and phenolic acids) [43]. Therefore, the enhancement of phenolic content of plants by application of biophysical methods, such as laser light, could effectively increase their antioxidant capacities [4]. In the present investigation, gallic and caffeic acids had the highest amount among detected phenolic acids, while quercetin and naringenin were the major flavonoids reported in both anise sprouts and mature plants (Table 2). When exposed to laser light treatment, phenolic and flavonoid profiles of anise sprouts and mature plants were positively affected, the effect that was more pronounced on sprouts than mature plants. There were also insignificant differences between the fruits and control plants regarding their phenolic contents.

Table 2. Phenolic and flavonoid profile of fruits, sprouts, and mature anise (control and laser light-treated). Data are represented by the means of at least 3 replicates ± standard error. Different small letters (a, b, c) within a row indicate significant differences between control and laser-treated samples at $p < 0.05$.

Compound	Fruits	Sprouts		Mature	
		Control	Laser-Treated	Control	Laser-Treated
Phenolic acids (µg/g DW)					
Caffeic acid	3.86 ± 0.59a	4.41 ± 0.2a	6.44 ± 0.1b	5.64 ± 0.73ab	6.08 ± 0.38b
Ferulic acid	0.03 ± 0.01a	0.04 ± 0a	0.09 ± 0c	0.06 ± 0.01b	0.07 ± 0b
Catechin	1.17 ± 0.17a	1.34 ± 0.3a	1.82 ± 0a	1.59 ± 0.21a	1.77 ± 0.11a
Galic acid	3.76 ± 0.71a	4.84 ± 0.3a	9.72 ± 0.2b	7.54 ± 0.98ab	8.16 ± 0.51b
p-Coumaric acid	1.02 ± 0.16a	1.31 ± 0.1a	2.04 ± 0a	1.58 ± 0.22a	1.88 ± 0.8a
Flavonoids (µg/g DW)					
kaempferol	0.44 ± 0.06a	0.82 ± 0a	1.05 ± 0.2a	0.57 ± 0.1a	1.5 ± 0.07a
Chlorogenic acid	0.11 ± 0.01a	0.14 ± 0a	0.16 ± 0a	0.12 ± 0.02a	0.17 ± 0.01a
Quercetin	1.54 ± 0.3a	1.98 ± 0.1a	4.28 ± 0.1b	3.32 ± 0.4ab	3.51 ± 0.2ab
Luteolin	0.04 ± 0.01a	0.07 ± 0a	0.14 ± 0b	0.08 ± 0.01a	0.12 ± 0.01ab
Apigenin	0.16 ± 0.03a	0.21 ± 0a	0.44 ± 0b	0.44 ± 0.04b	0.46 ± 0.02b
Naringenin	0.78 ± 0.14a	1.44 ± 0.1b	2.67 ± 0.1c	1.44 ± 0.23b	2.0 ± 0.14ab
Velutin	0.01 ± 0a	0.01 ± 0a	0.02 ± 0b	0.02 ± 0a	0.02 ± 0a
Tricin	0.77 ± 0.13a	1.17 ± 0.1a	2.06 ± 0ab	1.36 ± 0.2a	1.81 ± 0.11a
vitexin	0.51 ± 0.1a	0.58 ± 0a	1.21 ± 0b	1.06 ± 0.13a	1 ± 0.06a

In line with our findings, previous and recent studies confirmed that aniseeds contain high amounts of phenolic and flavonoid compounds, whereas gallic and caffeic acids were the main phenolic acids [44,45], while naringenin was the predominant flavonoid detected in anise fruits [14]. It was also reported that aniseeds contain high amounts of total phenolic and flavonoids that are the main contributors to antioxidant and other bioactivities of aniseeds [44,45]. Moreover, earlier studies recorded the enhancing effects of laser light on the total phenolic content in some plants such as sunflower, soybean, lemongrass sprouts, and buckwheat sprouts [4,10,11,29]. In a similar context, different light sources, such as light-emitting diodes (LEDs) and UV, have been previously investigated for their positive effects on increasing secondary metabolites biosynthesis, especially flavonoids in buckwheat sprouts [28,46]. The enhanced levels of phenolic and flavonoids might be ascribed to stimulated photosynthetic activity in response to laser light, which consequently increased the sugar content that acts as a precursor for the synthesis of different classes of metabolites, such as phenolic compounds [25,31]. Besides, the increases in flavonoid contents could be also attributed to the enhanced enzymatic activities of phenylalanine ammonia-lyase (PAL) and tyrosine ammonia-lyase (TAL) which are being involved in the phenylpropanoid pathway [47].

As a consequence of increasing their phenolic and essential oil contents under laser light treatment, the tested sprouts and mature plants have exhibited potent antioxidant activities as tested by different assays (DPPH, FRAP, TAC, ORAC, lipid peroxidation, TBARS and conjugated diene, and inhibition % of hemolysis) (Table 3). Both sprouts and mature plants significantly responded to laser light effects on their antioxidant capacities, while the fruits exhibited similar antioxidant capacities to those of control plants.

In this regard, aniseed has been recognized as a rich source of antioxidant constituents [14,17,36]. It was also reported that aniseeds contain high amounts of total phenolic and flavonoids compounds which are the main contributors to antioxidant and other bioactivities of aniseeds [44,45]. For instance, quercetin and naringenin, detected herein, were previously investigated for their strong antioxidant activities [14,48]. In accordance with our results, laser light has been recently found to enhance the DPPH, FRAP, and ABTS antioxidant capacities of both lemongrass and buckwheat sprouts, by increasing their phenolic and flavonoid contents [4,11]. Similarly, the application of other light sources, such as UV irradiation, enhanced the antioxidant activity of buckwheat sprouts [46].

Table 3. Antioxidant, anti-lipidemic, and anti-hemolytic activities in fruits, sprouts, and mature anise (control and laser-treated). Data are represented by the means of at least 3 replicates ± standard error. Different small letters (a, b, c) within a row indicate significant differences between control and laser-treated samples at $p < 0.05$.

Activity	Fruits	Sprouts Control	Sprouts Laser-Treated	Mature Control	Mature Laser-Treated
Antioxidant					
DPPH (%)	$36.5 \pm 0.37a$	$50.2 \pm 2.956a$	$65.61 \pm 2.6b$	$37.41 \pm 0.4a$	$54.27 \pm 0.2b$
FRAP	$6.4 \pm 2.3a$	$15.3 \pm 0.6a$	$27.6 \pm 1.9b$	$14.18 \pm 3.1b$	$24.05 \pm 1.5c$
Total antioxidant capacity (TAC) (nmol/g FW)	$9.39 \pm 1.3a$	$10.72 \pm 0.061a$	$13.23 \pm 0.2b$	$11.5 \pm 1.5a$	$13.4 \pm 0.84b$
Anti-lipid peroxidation (ORAC)	$2.5 \pm 0.37a$ $571 \pm 80a$	$3.22 \pm 0.275a$ $865.7 \pm 34.3a$	$4.39 \pm 0.07b$ $1064.6 \pm 85.5b$	$3.41 \pm 0.4b$ $821 \pm 65b$	$4.27 \pm 0.2c$ $1082 \pm 68c$
% inhibation of LDL oxidation (TBARS)	$13 \pm 2a$	$14.3 \pm 0.9a$	$28.6 \pm 3.7b$	$24 \pm 2ab$	$25 \pm 3b$
% inhibation of LDL oxidation (conjugated dienes)	$15 \pm 3a$	$15.4 \pm 0.3a$	$28.5 \pm 3.4b$	$17 \pm 3a$	$25 \pm 2b$
Anti-lipidemic					
Anti-Amylase IC_{50}(mg/mL)	$3.1 \pm 0.2a$	$2.7 \pm 0.1a$	$1.5 \pm 0.1b$	$2.96 \pm 0.22a$	$2.35 \pm 0.1b$
Anti-Lipase IC_{50}(mg/mL)	$1.32 \pm 0.1a$	$1.7 \pm 0.3a$	$0.97 \pm 0b$	$1.75 \pm 0.14a$	$1.0 \pm 0.1b$
Anti-chlostrol	$39 \pm 5.77a$	$50.2 \pm 1.9a$	$68.7 \pm 3.7b$	$43 \pm 7.5a$	$67 \pm 4.17b$
Anti-hemolytic activity					
% inhibation of hemolysis	$11 \pm 1.78a$	$14 \pm 1.1a$	$22.6 \pm 1.8b$	$18 \pm 2.38a$	$21 \pm 1.28a$

3.4. Laser Light Improved the Anti-Lipidemic Activity Particularly in Anise Sprouts

In the present study, the anti-lipidemic activity of aniseed sprouts and mature plants was evaluated by measuring their inhibitory effect on α-amylase, lipase activity, and cholesterol levels, in response to laser light. Results demonstrated superior anti-lipidemic activity after laser treatment in both treated groups (sprouts and mature plants) as indicated by decreasing α-amylase/lipase activities and cholesterol levels, compared to non-irradiated plants (Table 3). Both sprouts and mature plants significantly responded to laser light effects on their anti-lipidemic activities, while the fruits exhibited similar effects to those of control plants.

Hypolipidemic and antidiabetic activities of aniseed were previously reported and proved to be strongly correlated to its antioxidant constituents, such as phenolic compounds [15]. Similar to our results, laser light has been previously found to enhance the ability of lemongrass sprouts to decrease the levels of cholesterol, triglycerides, and low-density lipoprotein (LDL) [11]. Such hypocholesterolemic activity was reported to be associated with the availability of bioactive metabolites such as phenolic compounds, which could help in the excretion of cholesterol in the feces [49]. Moreover, the anti-lipidemic activity of many plants, such as *grewia optiva*, has been ascribed to their phenolic content, particularly quercetin, being responsible for its α-glucosidase and α-amylase activities. Quercetin has also previously exhibited better anti-lipase activity than other phytochemicals [50].

3.5. Tissue-Specific Response to Laser Light Treatment

The hierarchical clustering analysis has shown that there was an obvious tissue-specific response to laser light effects (Figure 3). Anise sprouts have been shown to be more responsive to the enhancing effect of laser light, where it had the highest contents of pigments, phenolic compounds, total proteins, sugars, vitamins, and minerals, followed by mature plants. Meanwhile, anise fruits exhibited the lowest response to laser light treatment. The data presented in Figure 3 has also shown that there were 6 clusters, whereas cluster 1 contained the highest amount of some volatile compounds (e.g., sabinene and paraanisaldehyde) which were mainly concentrated in the mature tissues under laser effects. Meanwhile, cluster 2 had the highest content of phenols, Zn, and beta carotene (mainly in the laser-irradiated sprouts). Also, cluster 3 had the highest concentration of tocopherols, ascorbic acid, sugars, crude fibers, saponins, and some minerals (e.g., K, Mg, and Fe), mostly accumulated in the laser treated-sprouts. Cluster 4 and 5 were rich in some phenolic compounds (e.g., kaempferol, chlorogenic acid, and apigenin), and Ca (accumulated in both laser treated-sprouts and mature tissues). Eventually, cluster 6 contained Cu, P, and ferulic acid (concentrated in the laser treated-sprouts). The variations among the three developmental stages might be ascribed to their diversity and ontogeny. Our results could be supported by previous studies that discussed the positive effects of laser on increasing the primary and secondary metabolites of many plants and sprouts such as buckwheat and lemongrass sprouts [4,11]. Other studies have also demonstrated that laser light could induce seed germination and thermodynamic parameters, which would be reflected in increasing photosynthesis and biochemical processes [6,7].

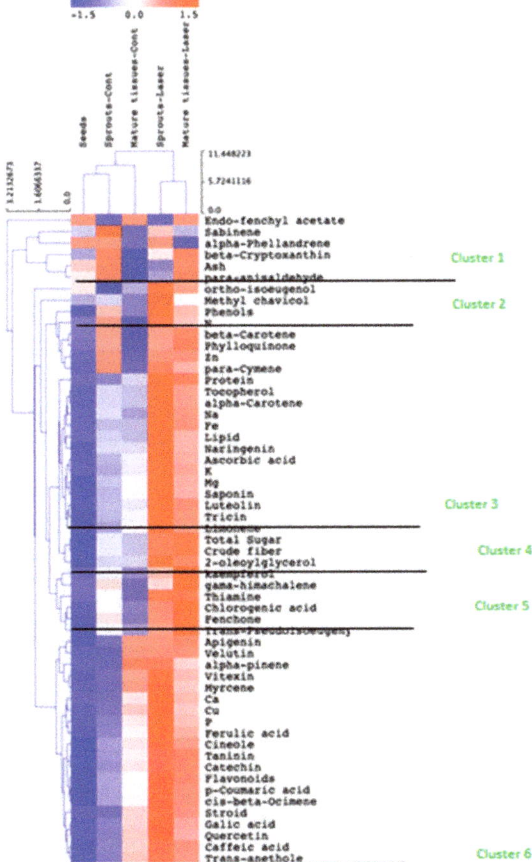

Figure 3. Stage-specific responses of anise fruits, sprouts, and mature plants to the effect of laser light treatment on the nutritional and health-promoting properties. The measured parameters are represented by contents of pigments, total nutrients, minerals, vitamins, essential oils, and phenolic compounds. Data are represented by the means of at least 3 replicates.

4. Conclusions

Based on the above results, it could be concluded that laser light could be an effective technique to enhance the growth and photosynthetic activity of anise sprouts and mature plants. Consequently, laser light increased the total bioactive metabolites as well as the levels of minerals and vitamins, with concomitant increments in antioxidant activities in both anise sprouts and mature plants. Interestingly, laser light effects might be more pronounced on anise sprouts than their mature plants. Thus, this study could support the use of laser light as an advantageous approach to increase the nutritional and health-promoting values of both anise sprouts and mature plants, preferably in the sprouting stage.

Supplementary Materials: The following are available online at https://www.mdpi.com/article/10.3390/plants10122591/s1, Table S1: Retention index essential oil measured by GC mass and retention time of phenolics and flavonoids profile measured by HPLC in fruits, sprouts and mature anise, Table S2: Retention time of phenolics and flavonoids profiles measured by HPLC in fruits, sprouts and mature anise.

Author Contributions: Conceptualization, M.K.O. and H.A.; methodology, S.A.A. and A.H.A.H.; software, S.M.A.-Q., N.A.A.-H. and H.A.; validation, M.K.O. and H.A.; formal analysis, Z.K.A. and M.A.-M.; investigation, Z.K.A., N.A.A.-H. and S.M.A.-Q.; resources, M.K.O.; data curation, W.H.A.-Q., M.H.A., S.S., A.H.A.H., H.A. and N.A.A.-H.; writing—original draft preparation, N.A.A.-H. and M.A.-M.; writing—review and editing, N.A.A.-H. and H.A. All authors have read and agreed to the published version of the manuscript.

Funding: The authors extend their appreciation to the Researchers Supporting Project number (RSP-2021/374) King Saud University, Riyadh, Saudi Arabia.

Institutional Review Board Statement: Not applicable.

Informed Consent Statement: Not applicable.

Data Availability Statement: Data presented in this study are available on reasonable request.

Acknowledgments: The authors extend their appreciation to the Researchers Supporting Project number (RSP-2021/374) King Saud University, Riyadh, Saudi Arabia.

Conflicts of Interest: The authors declare that they have no known competing financial interests or personal relationships that could have appeared to influence the work reported in this paper.

References

1. Manchali, S.; Murthy, K.N.C.; Patil, B.S. Crucial facts about health benefits of popular cruciferous vegetables. *J. Funct. Foods* **2012**, *4*, 94–106. [CrossRef]
2. Marton, M.; Mandoki, Z.S.; Csapo-Kiss, Z.S.; Csapo, J. The role of sprouts in human nutrition. A review. *Acta Univ. Sapientiae* **2010**, *3*, 81–117.
3. Almuhayawi, M.S.; Al Jaouni, S.K.; Almuhayawi, S.M.; Selim, S.; Abdel-Mawgoud, M. Elevated CO_2 improves the nutritive value, antibacterial, anti-inflammatory, antioxidant and hypocholestecolemic activities of lemongrass sprouts. *Food Chem.* **2021**, *357*, 129730. [CrossRef] [PubMed]
4. Almuhayawi, M.S.; Hassan, A.H.A.; Abdel-Mawgoud, M.; Khamis, G.; Selim, S.; Al Jaouni, S.K.; AbdElgawad, H. Laser light as a promising approach to improve the nutritional value, antioxidant capacity and anti-inflammatory activity of flavonoid-rich buckwheat sprouts. *Food Chem.* **2020**, *345*, 128788. [CrossRef] [PubMed]
5. Saleh, A.M.; Abdel-Mawgoud, M.; Hassan, A.R.; Habeeb, T.H.; Yehia, R.S.; AbdElgawad, H. Global metabolic changes induced by arbuscular mycorrhizal fungi in oregano plants grown under ambient and elevated levels of atmospheric CO_2. *Plant Physiol. Biochem.* **2020**, *152*, 255–263. [CrossRef] [PubMed]
6. Chen, Y.-P.; Yue, M.; Wang, X.-L. Influence of He–Ne laser irradiation on seeds thermodynamic parameters and seedlings growth of Isatis indogotica. *Plant Sci.* **2005**, *168*, 601–606. [CrossRef]
7. Perveen, R.; Ali, Q.; Ashraf, M.; Al-Qurainy, F.; Jamil, Y.; Raza Ahmad, M. Effects of different doses of low power continuous wave He–Ne laser radiation on some seed thermodynamic and germination parameters, and potential enzymes involved in seed germination of sunflower (*Helianthus annuus* L.). *Photochem. Photobiol.* **2010**, *86*, 1050–1055. [CrossRef]
8. Podleśny, J.; Stochmal, A.; Podleśna, A.; Misiak, L.E. Effect of laser light treatment on some biochemical and physiological processes in seeds and seedlings of white lupine and faba bean. *Plant Growth Regul.* **2012**, *67*, 227–233. [CrossRef]
9. Osman, Y.A.H.; El-Tobgy, K.M.K.; El-Sherbini, E.S.A. Effect of laser radiation treatments on growth, yield and chemical constituents of fennel and coriander plants. *J. Appl. Sci. Res.* **2009**, *5*, 244–252.

10. Asghar, T.; Jamil, Y.; Iqbal, M.; Abbas, M. Laser light and magnetic field stimulation effect on biochemical, enzymes activities and chlorophyll contents in soybean seeds and seedlings during early growth stages. *J. Photochem. Photobiol. B Biol.* **2016**, *165*, 283–290. [CrossRef]

11. Okla, M.K.; El-Tayeb, M.A.; Qahtan, A.A.; Abdel-Maksoud, M.A.; Elbadawi, Y.B.; Alaskary, M.K.; Balkhyour, M.A.; Hassan, A.H.A.; AbdElgawad, H. Laser Light Treatment of Seeds for Improving the Biomass Photosynthesis, Chemical Composition and Biological Activities of Lemongrass Sprouts. *Agronomy* **2021**, *11*, 478. [CrossRef]

12. Tepe, A.S.; Tepe, B. Traditional use, biological activity potential and toxicity of Pimpinella species. *Ind. Crops Prod.* **2015**, *69*, 153–166. [CrossRef]

13. Ullah, H.; Honermeier, B. Fruit yield, essential oil concentration and composition of three anise cultivars (*Pimpinella anisum* L.) in relation to sowing date, sowing rate and locations. *Ind. Crops Prod.* **2013**, *42*, 489–499. [CrossRef]

14. Rebey, I.B.; Wannes, W.A.; Kaab, S.B.; Bourgou, S.; Tounsi, M.S.; Ksouri, R.; Fauconnier, M.L. Bioactive compounds and antioxidant activity of Pimpinella anisum L. accessions at different ripening stages. *Sci. Hortic.* **2019**, *246*, 453–461. [CrossRef]

15. Rajeshwari, U.; Shobha, I.; Andallu, B. Comparison of aniseeds and coriander seeds for antidiabetic, hypolipidemic and antioxidant activities. *Spat. DD* **2011**, *1*, 9–16. [CrossRef]

16. Rajeshwari, C.U.; Abirami, M.; Andallu, B. In vitro and in vivo antioxidant potential of aniseeds (*Pimpinella anisum*). *Asian J. Exp. Biol. Sci.* **2011**, *2*, 80–89.

17. Gülçin, İ.; Oktay, M.; Kıreçcı, E.; Küfrevıoğlu, Ö.İ. Screening of antioxidant and antimicrobial activities of anise (*Pimpinella anisum* L.) seed extracts. *Food Chem.* **2003**, *83*, 371–382. [CrossRef]

18. Tabanca, N.; Demirci, B.; Ozek, T.; Kirimer, N.; Baser, K.H.C.; Bedir, E.; Khan, I.A.; Wedge, D.E. Gas chromatographic–mass spectrometric analysis of essential oils from Pimpinella species gathered from Central and Northern Turkey. *J. Chromatogr. A* **2006**, *1117*, 194–205. [CrossRef]

19. El Tobgy, K.M.K.; Osman, Y.A.H.; El Sherbini, E.A. Effect of laser radiation on growth, yield and chemical constituents of anise and cumin plants. *J. Appl. Sci. Res.* **2009**, *5*, 522–528.

20. Habeeb, T.H.; Abdel-Mawgoud, M.; Yehia, R.S.; Khalil, A.M.A.; Saleh, A.M.; AbdElgawad, H. Interactive Impact of Arbuscular Mycorrhizal Fungi and Elevated CO_2 on Growth and Functional Food Value of Thymus vulgare. *J. Fungi* **2020**, *6*, 168. [CrossRef]

21. Wang, Z.; Xiao, S.; Wang, Y.; Liu, J.; Ma, H.; Wang, Y.; Tian, Y.; Hou, W. Effects of light irradiation on essential oil biosynthesis in the medicinal plant Asarum heterotropoides Fr. Schmidt var. mandshuricum (Maxim) Kitag. *PLoS ONE* **2020**, *15*, e0237952. [CrossRef] [PubMed]

22. Naudts, K.; Van den Berge, J.; Farfan Vignolo, E.R.; Rose, P.; Abdelgawad, H.; Ceulemans, R.; Janssens, I.A.; Asard, H.; Nijs, I. Future climate alleviates stress impact on grassland productivity through altered antioxidant capacity. *Environ. Exp. Bot.* **2013**, *99*, 150–158. [CrossRef]

23. Casasole, G.; Raap, T.; Costantini, D.; AbdElgawad, H.; Asard, H.; Pinxten, R.; Eens, M. Neither artificial light at night, anthropogenic noise nor distance from roads are associated with oxidative status of nestlings in an urban population of songbirds. *Comp. Biochem. Physiol. Part A Mol. Integr. Physiol.* **2017**, *210*, 14–21. [CrossRef]

24. Huang, D.; Ou, B.; Hampsch-Woodill, M.; Flanagan, J.A.; Prior, R.L. High-throughput assay of oxygen radical absorbance capacity (ORAC) using a multichannel liquid handling system coupled with a microplate fluorescence reader in 96-well format. *J. Agric. Food Chem.* **2002**, *50*, 4437–4444. [CrossRef] [PubMed]

25. Li, P.; Li, H.; Zong, Y.; Li, F.Y.; Han, Y.; Hao, X. Photosynthesis and metabolite responses of Isatis indigotica Fortune to elevated [CO_2]. *Crop. J.* **2017**, *5*, 345–353. [CrossRef]

26. Aladjadjiyan, A. The use of physical methods for plant growing stimulation in Bulgaria. *J. Cent. Eur. Agric.* **2007**, *8*, 369–380.

27. Wu, J.; Gao, X.; Zhang, S. Effect of laser pretreatment on germination and membrane lipid peroxidation of Chinese pine seeds under drought stress. *Front. Biol. China* **2007**, *2*, 314–317. [CrossRef]

28. Tuan, P.A.; Thwe, A.A.; Kim, Y.B.; Kim, J.K.; Kim, S.-J.; Lee, S.; Chung, S.-O.; Park, S.U. Effects of white, blue, and red light-emitting diodes on carotenoid biosynthetic gene expression levels and carotenoid accumulation in sprouts of tartary buckwheat (*Fagopyrum tataricum* Gaertn.). *J. Agric. Food Chem.* **2013**, *61*, 12356–12361. [CrossRef]

29. Perveen, R.; Jamil, Y.; Ashraf, M.; Ali, Q.; Iqbal, M.; Ahmad, M.R. He-Ne Laser-Induced Improvement in Biochemical, Physiological, Growth and Yield Characteristics in Sunflower (*Helianthus annuus* L.). *Photochem. Photobiol.* **2011**, *87*, 1453–1463. [CrossRef]

30. Nam, T.-G.; Lim, Y.J.; Eom, S.H. Flavonoid accumulation in common buckwheat (Fagopyrum esculentum) sprout tissues in response to light. *Hortic. Environ. Biotechnol.* **2018**, *59*, 19–27. [CrossRef]

31. Al Jaouni, S.; Saleh, A.M.; Wadaan, M.A.M.; Hozzein, W.N.; Selim, S.; AbdElgawad, H. Elevated CO_2 induces a global metabolic change in basil (*Ocimum basilicum* L.) and peppermint (*Mentha piperita* L.) and improves their biological activity. *J. Plant. Physiol.* **2018**, *224*, 121–131. [CrossRef] [PubMed]

32. Sacala, E.; Demczuk, A.; Grzys, E.; Prosba-Bialczyk, U.; Szajsner, H. Impact of presowing laser irradiation of seeds on sugar beet properties. *Int. Agrophysics* **2012**, *26*, 295–300. [CrossRef]

33. Rauf, M.; Yoon, H.; Lee, S.; Lee, M.-C.; Oh, S.; Choi, Y.-M. Evaluation of sprout growth traits and flavonoid content in common and tartary buckwheat germplasms. *Plant. Breed. Biotechnol.* **2019**, *7*, 375–385. [CrossRef]

34. Dinani, H.J.; Eslami, P.; Mortazaeinezhad, F.; Ghahrizjani, R.T. Effects of laser radiation on the growth indicators of Kelussia odoratissima Mozaff. Medical plant. In Proceedings of the 2019 Photonics North (PN), Quebec City, QC, Canada, 21–23 May 2019; pp. 1–8.

35. Shafique, H.; Jamil, Y.; ul Haq, Z.; Mujahid, T.; Khan, A.U.; Iqbal, M.; Abbas, M. Low power continuous wave-laser seed irradiation effect on Moringa oleifera germination, seedling growth and biochemical attributes. *J. Photochem. Photobiol. B Biol.* **2017**, *170*, 314–323.

36. Bettaieb Rebey, I.; Bourgou, S.; Aidi Wannes, W.; Hamrouni Selami, I.; Saidani Tounsi, M.; Marzouk, B.; Fauconnier, M.-L.; Ksouri, R. Comparative assessment of phytochemical profiles and antioxidant properties of Tunisian and Egyptian anise (*Pimpinella anisum* L.) seeds. *Plant. Biosyst. Int. J. Deal. All Asp. Plant. Biol.* **2018**, *152*, 971–978. [CrossRef]

37. Ibrahim, M.K.; Mattar, Z.A.; Abdel-Khalek, H.H.; Azzam, Y.M. Evaluation of antibacterial efficacy of anise wastes against some multidrug resistant bacterial isolates. *J. Radiat. Res. Appl. Sci.* **2017**, *10*, 34–43. [CrossRef]

38. Mosavat, S.H.; Jaberi, A.R.; Sobhani, Z.; Mosaffa-Jahromi, M.; Iraji, A.; Moayedfard, A. Efficacy of Anise (*Pimpinella anisum* L.) oil for migraine headache: A pilot randomized placebo-controlled clinical trial. *J. Ethnopharmacol.* **2019**, *236*, 155–160. [CrossRef]

39. Maeda, H.; Dudareva, N. The shikimate pathway and aromatic amino acid biosynthesis in plants. *Annu. Rev. Plant. Biol.* **2012**, *63*, 73–105. [CrossRef]

40. Sangwan, N.S.; Farooqi, A.H.A.; Shabih, F.; Sangwan, R.S. Regulation of essential oil production in plants. *Plant. Growth Regul.* **2001**, *34*, 3–21. [CrossRef]

41. Vialart, G.; Hehn, A.; Olry, A.; Ito, K.; Krieger, C.; Larbat, R.; Paris, C.; Shimizu, B.; Sugimoto, Y.; Mizutani, M. A 2-oxoglutarate-dependent dioxygenase from Ruta graveolens L. exhibits p-coumaroyl CoA 2′-hydroxylase activity (C2′ H): A missing step in the synthesis of umbelliferone in plants. *Plant. J.* **2012**, *70*, 460–470. [CrossRef]

42. Balkhyour, M.A.; Hassan, A.H.A.; Halawani, R.F.; Summan, A.S.; AbdElgawad, H. Effect of Elevated CO_2 on Seed Yield, Essential Oil Metabolism, Nutritive Value, and Biological Activity of *Pimpinella anisum* L. Accessions at Different Seed Maturity Stages. *Biology* **2021**, *10*, 979. [CrossRef]

43. Masella, R.; Di Benedetto, R.; Varì, R.; Filesi, C.; Giovannini, C. Novel mechanisms of natural antioxidant compounds in biological systems: Involvement of glutathione and glutathione-related enzymes. *J. Nutr. Biochem.* **2005**, *16*, 577–586. [CrossRef]

44. Christova-Bagdassarian, V.L.; Bagdassarian, K.S.; Atanassova, M.S. Phenolic compounds and antioxidant capacity in Bulgarian plans (dry seeds). *Int. J. Adv. Res.* **2013**, *1*, 186–197.

45. Sakr, A.A.E.; Taha, K.M.; Abozid, M.M. Comparative Study Between Anise Seeds and Mint Leaves (Chemical Composition, Phenolic Compounds and Flavonoids). *Menoufia J. Agric. Biotechnol.* **2019**, *4*, 53–60. [CrossRef]

46. Tsurunaga, Y.; Takahashi, T.; Katsube, T.; Kudo, A.; Kuramitsu, O.; Ishiwata, M.; Matsumoto, S. Effects of UV-B irradiation on the levels of anthocyanin, rutin and radical scavenging activity of buckwheat sprouts. *Food Chem.* **2013**, *141*, 552–556. [CrossRef] [PubMed]

47. Lim, J.-H.; Park, K.-J.; Kim, B.-K.; Jeong, J.-W.; Kim, H.-J. Effect of salinity stress on phenolic compounds and carotenoids in buckwheat (*Fagopyrum esculentum* M.) sprout. *Food Chem.* **2012**, *135*, 1065–1070. [CrossRef] [PubMed]

48. Zhang, M.; Swarts, S.G.; Yin, L.; Liu, C.; Tian, Y.; Cao, Y.; Swarts, M.; Yang, S.; Zhang, S.B.; Zhang, K. Antioxidant properties of quercetin. In *Oxygen Transport to Tissue XXXII*; LaManna, J., Puchowicz, M., Xu, K., Harrison, D., Bruley, D., Eds.; Springer: Boston, MA, USA, 2011; Volume 701, pp. 283–289.

49. Oboh, G.; Ademosun, A.O.; Akinleye, M.; Omojokun, O.S.; Boligon, A.A.; Athayde, M.L. Starch composition, glycemic indices, phenolic constituents, and antioxidative and antidiabetic properties of some common tropical fruits. *J. Ethn. Foods* **2015**, *2*, 64–73. [CrossRef]

50. Zahoor, M.; Bari, W.U.; Zeb, A.; Khan, I. Toxicological, anticholinesterase, antilipidemic, antidiabetic and antioxidant potentials of Grewia optiva Drummond ex Burret extracts. *J. Basic Clin. Physiol. Pharmacol.* **2020**, *31*, 1–16. [CrossRef]

Article

Bacterial Endophytes as a Promising Approach to Enhance the Growth and Accumulation of Bioactive Metabolites of Three Species of *Chenopodium* Sprouts

Mohammed S. Almuhayawi [1,*], Mohamed Abdel-Mawgoud [2,*], Soad K. Al Jaouni [3], Saad M. Almuhayawi [4], Mohammed H. Alruhaili [1], Samy Selim [5] and Hamada AbdElgawad [6]

[1] Department of Medical Microbiology and Parasitology, Faculty of Medicine, King Abdulaziz University, Jeddah 21589, Saudi Arabia; malruhaili@kau.edu.sa
[2] Department of Medicinal and Aromatic Plants, Desert Research Centre, Cairo 11753, Egypt
[3] Hematology/Pediatric Oncology, Yousef Abdulatif Jameel Scientific Chair of Prophetic Medicine Application, Faculty of Medicine, King Abdulaziz University, Jeddah 21589, Saudi Arabia; saljaouni@kau.edu.sa
[4] Department of Otolaryngology-Head and Neck Surgery, Faculty of Medicine, King Abdulaziz University, Jeddah 21589, Saudi Arabia; salmehawi@kau.edu.sa
[5] Department of Clinical Laboratory Sciences, College of Applied Medical Sciences, Jouf University, Sakaka 72388, Saudi Arabia; sabdulsalam@ju.edu.sa
[6] Department of Botany and Microbiology, Faculty of Science, Beni-Suef University, Beni-Suef 62521, Egypt; hamada.abdElgawad@uantwerpen.be
* Correspondence: msalmuhayawi@kau.edu.sa (M.S.A.); mohamed_drc@yahoo.com (M.A.-M.)

Citation: Almuhayawi, M.S.; Abdel-Mawgoud, M.; Al Jaouni, S.K.; Almuhayawi, S.M.; Alruhaili, M.H.; Selim, S.; AbdElgawad, H. Bacterial Endophytes as a Promising Approach to Enhance the Growth and Accumulation of Bioactive Metabolites of Three Species of *Chenopodium* Sprouts. *Plants* **2021**, *10*, 2745. https://doi.org/10.3390/plants10122745

Academic Editors: Beatrice Falcinelli and Angelica Galieni

Received: 2 November 2021
Accepted: 9 December 2021
Published: 13 December 2021

Publisher's Note: MDPI stays neutral with regard to jurisdictional claims in published maps and institutional affiliations.

Abstract: Sprouts are regarded as an untapped source of bioactive components that display various biological properties. Endophytic bacterium inoculation can enhance plant chemical composition and improve its nutritional quality. Herein, six endophytes (Endo 1 to Endo 6) were isolated from *Chenopodium* plants and morphologically and biochemically identified. Then, the most active isolate Endo 2 (strain JSA11) was employed to enhance the growth and nutritive value of the sprouts of three *Chenopodium* species, i.e., *C. ambrosoides*, *C. ficifolium*, and *C. botrys*. Endo 2 (strain JSA11) induced photosynthesis and the mineral uptake, which can explain the high biomass accumulation. Endo 2 (strain JSA11) improved the nutritive values of the treated sprouts through bioactive metabolite (antioxidants, vitamins, unsaturated fatty acid, and essential amino acids) accumulation. These increases were correlated with increased amino acid levels and phenolic metabolism. Consequently, the antioxidant activity of the Endo 2 (strain JSA11)-treated *Chenopodium* sprouts was enhanced. Moreover, Endo 2 (strain JSA11) increased the antibacterial activity against several pathogenic bacteria and the anti-inflammatory activities as evidenced by the reduced activity of cyclooxygenase and lipoxygenase. Overall, the Endo 2 (strain JSA11) treatment is a successful technique to enhance the bioactive contents and biological properties of *Chenopodium* sprouts.

Keywords: bacterial endophytes; *Chenopodium* sp.; sprouts; photosynthesis; amino acid metabolism; phenolics metabolism; antioxidant; anti-inflammatory

1. Introduction

Sprouts have been recognized as excellent sources of bioactive phytochemicals, such as proteins, vitamins, minerals, and phenolic compounds, which exhibit a variety of nutritive and health promoting characteristics, e.g., anticancer and antioxidative activities [1,2]. The reduced levels of antinutritional factors also give the sprouts additional advantages over seeds and mature plants, thus, sprouts could be qualified as natural healthy foods [2]. Therefore, the application of various techniques for improving the nutraceutical and functional food values of sprouts and their contents of bioactive metabolites has recently been a research hotspot [3,4]

Bacterial endophytes are a large group of microorganisms colonizing the internal plant tissues, mainly in the roots. Such a symbiotic relationship between plants and endophytic bacteria is supposed to lead to a set of biochemical and physiological changes, which eventually offer many benefits to the host plant, either directly or indirectly [5]. The direct effect of endophytes includes the production of growth regulators, phosphate solubilization, and nitrogen fixation, while the indirect effect occurs in response to pathogen infection to enhance the plant's resistance against diseases [5–7]. The production of siderophore by endophytic bacteria could also protect the plant against phytopathogens through reducing the availability of iron for some pathogens [8]. Bacterial endophytes have also been known to be rich sources of bioactive secondary metabolites [9]. Therefore, bacterial endophytes could play a significant role in plant growth, by induction of such growth promoting effects on plants, which recommend them as excellent candidates for improving plant yield, as well as the accumulation of bioactive compounds [10,11]. For instance, a positive effect of bacterial strains on the bioactive content of sesquiterpenoid of *Atractylodes lancea* was previously reported [11]. In addition, bacterial endophytes have been utilized to improve the quality and nutritive value of *Lyophyllum decastes* by enhancing its bioactive components [12]. However, few attempts have been made to explore the bacterial endophyte-induced changes in the plant metabolome. Therefore, detailed metabolic studies are important in order to understand (i) the changes associated with the plant–endophyte interaction and (ii) how the plant's bioactive metabolites respond to such an interaction to improve a plant's nutritional and pharmaceutical values.

Chenopodium species belong to the *Amaranthaceae* family. Their seeds and sprouts have been known for their high nutritive values. *Chenopodium* sprouts are considered as sources for proteins, vitamins, antioixdants, and carbohydrates, e.g., *Chenopodium quinoa* [13,14] and *Chenopodium formosanum* Koidz. [15]. Sprouting has been reported to induce modifications in storage proteins in quinoa, with a concomitant increase in some metals, such as Cu and Zn [13]. For instance, *Chenopodium ambrosioides* has been traditionally used for the treatment of bronchitis, tuberculosis, vomiting, and skin ulcerations. It has also anthelmintic, anti-inflammatory, antipyretic, and analgesic properties [16]. In addition, *C. ambrosioides* has been considered as a rich source of carotenoids, proteins, and fats [17]. Similarly, *C. botrys* has been used for treating coughs and abdominal pain. It has also some medicinal properties, such as antidiuretic, antispasmodic, anticonvulsant, anti-inflammatory, antidiabetic, antibacterial, antifungal, and antiviral [17]. *C. botrys* also contains a variety of bioactive compounds, mainly flavonoids, anthraquinones, saponins, and tannins [18].Therefore, enhancing the phytochemical content of such medicinal plants and their sprouts by using ecofriendly approaches, such as bacterial endophytes, could increase their health promoting and nutritive values. For instance, calcium as an essential nutrient for plants was applied to improve growth, antioxidant levels, and nutrient availability in *Chenopodium formosanum* sprouts [15]. Similarly, high CO_2 was applied to increase the growth of *Chenopodium album* sprouts [19]. This ecofriendly approach could also support their traditional uses as well as their application in pharmaceutical industries and also to ensure safety to the environment. To our knowledge, the influence of endophytic bacteria on the growth and bioactive metabolites of the selected sprouts (*Chenopodium ambrosoides*, *Chenopodium ficifolium*, and *Chenopodium botrys*) has not been previously investigated. In addition, endophytic bacteria have not been intensively studied when compared with endophytic fungi [9,20]. Thus, the present study was conducted with the aim to isolate and identify a bioactive endophyte to improve *Chenopodium* growth and tissue quality. For this purpose, we investigated the effects induced by the endophytic bacteria *Streptomyces* on growth, physiology, levels of minerals, and some primary and secondary metabolites in the tested *Chenopodium* sprouts. Additionally, the concomitant biological activities, i.e., antioxidant, antimicrobial, and anti-inflammatory activities were measured. We hypothesize that the use of a bioenhancer, such as endophytic bacteria, could elevate the health promoting and nutritive values of the tested sprouts.

2. Results and Discussion

2.1. Characterization of the Isolated Endophytic Bacteria

In the current investigation, a total of six endophytic bacterial strains were isolated and characterized (Endo 1 to Endo 6) on the basis of morphological and biochemical traits. Most of the tested endophytic bacteria showed variability in substrate color and aerial mycelia, besides their ability to produce diffusible pigments (Table 1). The aerial hyphae of the examined endophytes were also observed with long spiral spore chains, long rectiflexible spore chains, or verticillate spores, such as those previously reported for other bacterial strains [21]. The morphological examinations of the six isolates indicated that these isolates belong to the genus *Streptomyces*, where its morphologically related genera have extensively branched mycelia [22].

Table 1. The morphological and biochemical characterization of bacterial isolates, whereas the signs + and − indicate presence or absence, respectively. Data are represented by the means of at least 3 replicates, and error bars represent standard deviations.

	Isolate	Endo 1	Endo 2	Endo 3	Endo 4	Endo 5	Endo 6
Colony	Aerial mycelium	+	−	+	+	−	+
	Pigmentation	+	+	+	+	+	+
Spore chain	Spiral	+	−	−	+	+	−
	Rectiflexibles	−	+	+	−	−	−
	Verticillate	−	−	−	−	−	+
Spore color	Yellow	+	−	−	−	−	−
	Orange	−	+	+	+	−	+
	Red	−	−	−	−	+	−
N source utilization	L-Cysteine	+	+	+	+	+	−
	L-Phenylalanine	−	−	+	−	−	
	L-Histidine	−	−	+	+	+	−
	L-Lysine	+	+	+	−	−	+
	L-Asparagine	+	-	+	+	+	+
	L-Arginine	+		−	−	−	−
	L-proline	+	+	+	+	−	
	L-Valine	−	−	+	−	−	+
	Tyrosine	+	+	−	+		+
C source utilization	D-fructose	−	−	+	+	+	−
	D-glucose	+	+	−	+	−	−
	Sucrose	+	+	+	-	+	+
	Maltose	−	−	+	−	+	
	Raffinose	+	+	−	+	+	+
	Lactose	−	−		+	+	−
	Galactose	+	+	−	+	+	−
	Meso-Inositol	+	+	+	−	+	+
	Cellulose	−	−	+	+	+	+
	Xylose	+	+	+	−	+	−
	Dextran	+	+	−	−	−	−

Table 1. *Cont.*

	Isolate	Endo 1	Endo 2	Endo 3	Endo 4	Endo 5	Endo 6
Enzyme activity	Catalase	+	−	+	−	+	−
	Peroxidase	+	−	+	+	+	+
	Starch hydrolysis	+		+	−	−	+
	Gelatin liquefication	+	+	−	+	−	+
	Casein hydrolysis	−	−	+	+	+	+
	Lipolysis	+	+	+	+	−	+
	Citrate utilization	+	+	+	+	+	+
	H_2S Production	−	+	−	+	+	+
	Nitrate reduction	+	+	+	−	−	−
	Urease	+	+	+		−	−
	L-Asparaginase	−	+	+	+	+	−
	L-Glutaminase	+	+	+	+	+	+
Biological activity	Antioxidant Activity (FRAP)	42.4 ± 3.1	66.6 ± 4.0	21.8 ± 1.6	33.1 ± 1.5	24 ± 1.4	32.4 ± 2.3
	Antioxidant Activity DPPH (%)	68.7 ± 2.8	73.7 ± 3.8	34.2 ± 2.1	20.8 ± 1.5	27.5 ± 1.4	48.2 ± 2.9
	Antioxidant Activity (ABTS%)	30.5 ± 1.0	59.6 ± 1.9	39.7 ± 1.3	34.8 ± 1.1	23.2 ± 0.8	53.6 ± 1.7
	Phosphate Solubilization (mg/mL)	5.7 ± 0.8	7.2 ± 0.2	4.5 ± 0.1	3.8 ± 1.5	2.7 ± 1.4	7.3 ± 1.0
Bioactive compounds production	Total flavonoids (mg/100 g bacteria)	5 ± 1.19	8 ± 1.53	5.8 ± 0.84	5.3 ± 1.36	6.4 ± 0.86	7.5 ± 1.18
	Total Phenols (mg/100 g bacteria)	36.8 ± 1.06	45.4 ± 1.31	23.7 ± 0.6	42.2 ± 1.2	23.7 ± 0.6	33.8 ± 0.9
	Tocopherols (mg/g bacteria)	0.3 ± 0.01	0.5 ± 0.01	0.2 ± 0.01	0.5 ± 0.01	0.3 ± 0.01	0.2 ± 0.01
	Flavonoids (mg/100 g bacteria)						
	Quercetin		1.63 ± 0.1				
	Quercetrin		1.41 ± 0.5				
	Luteolin		0.70 ± 0.1				
	Apigenin		4.11 ± 0.7				
	Isoquercetrin		10.2 ± 1.6				
	Rutin		1.27 ± 0.4				
	Ellagic acid		0.71 ± 0.1				
	Velutin		0.30 ± 0.0				
	Naringenin		1.12 ± 0.3				
	Genistein		0.95 ± 0.2				
	Daidzein		1.08 ± 0.1				
	Fisetin		0.74 ± 0.1				
	O-hydroxydaidzein		1.13 ± 0.1				
	IAA-Me		1.15 ± 0.21				
	ABA		0.29 ± 0.1				
	GA		0.16 ± 0.07				
	Sidephore Catechol		7.3 ± 0.3				
	Sidephore Salicylate		9.23 ± 0.42				

Regarding the biochemical and physiological attributes, most isolated endophytes were capable of utilizing different carbon sources (e.g., galactose and sucrose) as well as nitrogen sources (e.g., cysteine and tyrosine) in addition to their ability to produce antioxidant metabolites (phenolics and tocopherols) and several bioactive enzymes, such as the antioxidant enzymes (catalase and peroxidase), N metabolic enzymes (e.g., L-asparaginase, glutaminase, urease and nitrate reduction), and C metabolic enzymes (e.g., starch and casein hydrolysis and lipolysis). Moreover, endophytic isolates showed a high antioxidant capacity (FRAP, DPPH, and ABTS%) and solubilization of tricalcium phosphate (Table 1). The production of such enzymes could enable the tested bacteria to survive during unfavorable conditions [23]. In this regard, it was found that the utilization of carbon and nitrogen sources by some bacterial strains might affect their production of bioactive secondary metabolites [24].

2.2. Selection of the Most Active Endophytic Bacteria

Bacterial endophytes have been considered as rich sources of bioactive secondary metabolites, such as flavonoids, phenolics, alkaloids, and terpenoids [9]. The assumption has also been that the presence of such bioactive compounds in bacterial endophytes might be responsible for their enhanced biological activities [20]. This can significantly contribute to improving plant growth and biological activity, as indicated by previous reports that described the significant impact of bacterial endophytes on enhancing the nutritive value of tartary buckwheat sprouts by increasing their total and individual flavonoid contents, e.g., rutin and quercetin [25]. Interestingly, a positive correlation was observed between the phenolic content and antioxidant activities of all the tested endophytic bacteria. The most active isolate endophyte 2 (Endo 2) was selected for the availability of its carbon and N sources, and their utilization and metabolism (e.g., starch hydrolysis, nitrate reduction, urease, and L-asparaginase) and also for the highest production of phenolic and tocopherol content, antioxidant activities, and tricalcium phosphate solubilization. In this regard, bacterial endophytes that are rich in phenolics and flavonoids showed highly different biological activities that directly or indirectly support plant growth. Moreover, phenolics production supports the endophytes to establish exclusive symbiotic relationships with plants [26].

This strain was further analyzed for its phenolic profile, whereas 13 phenolic acids and flavonoids were quantified, being dominated by isoquercetrin and apigenin. In this regard, bacterial endophytes are known to contain high levels of phenolic compounds, which might account for their biological properties, particularly the antimicrobial activity [27]. In addition, the production of phytohormones, such as gibberellins (GA), abscisic acid (ABA), and indole acetic acid (IAA), as well as siderophore was also confirmed in the selected endophytic bacterium (Endo 2) as an indication for the growth promoting potential of such an endophyte. In this regard, bacterial endophytes have been known to produce GA and IAA, which may exert growth promoting effects on plants [20]. IAA might also induce a protective effect against environmental stress conditions by enhancing various cellular defense mechanisms [28]. More interesting, the production of siderophores by bacterial endophytes could enable them to overcome some adverse environmental conditions and also reduce the availability of iron for some plant pathogens [8]. Some endophytic bacteria, such as *Stenotrophomonas maltophilia*, have been shown to synthesis catechol-type siderophores [29]. Thus, (Endo 2) was selected for improving the nutritive values of the tested sprouts, i.e., *C. ambrosoides*, *C. ficifolium*, and *C. botrys*.

2.3. Molecular Characterization of the Most Active Isolate

According to the 16S rRNA gene sequence analysis, the Endo 2 strain is affiliated within the genus *Streptomyces* (strain JSA11). With a similarity of >97%, it is closely related to unidentified species within the same genus as shown in Figure 1. The 16S rRNA gene data of the actinobacterial strain reported in this study have been deposited in the NCBI and GenBank nucleotide sequence databases under the accession number (MZ489115).

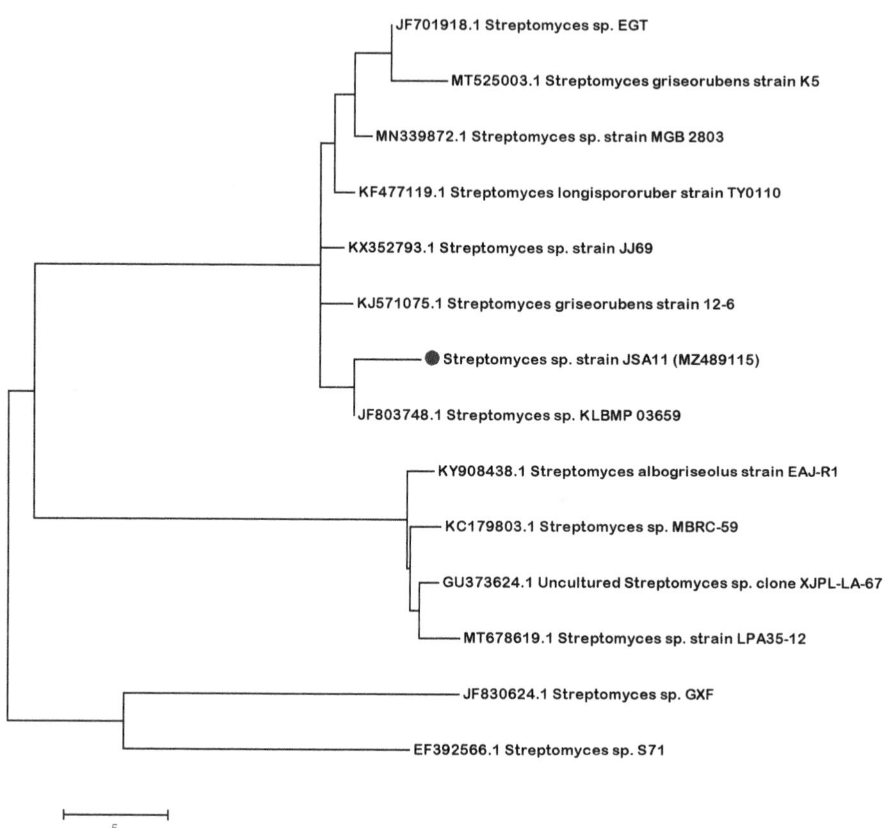

Figure 1. Neighbor joining tree (partial sequences ~ 950 bp) showing the phylogenetic relationships of actinobacterial 16S rRNA gene sequence of potential strains to closely related (S ≥ 97%) sequences from the GenBank database.

2.4. Bacterial Endophytes Promoted Photosynthesis and Biomass Production of Chenopodium Sprouts

Endophytic bacteria have been assumed to promote plant photosynthetic activity by enhancing the chlorophyll content [30]. Supporting such a hypothesis, the present investigation has clearly revealed that the inoculation of the tested sprouts with *Streptomyces* (strain JSA11) has resulted in significant increases in photosynthetic activity and respiration rate, when compared with the control sprouts (Table 2). In this regard, endophytes have been known to have positive impacts on photosynthesis, which consequently could enhance photosynthetic light and carbon reactions [31]. Similar to our results, sugar beet plants inoculated with endophytes have been shown to have a higher dark respiration rate (i.e., higher light saturation point, light compensation point, and photochemical efficiency) than noninoculated plants [30]. The increased photosynthetic capacity might be due to the ability of enodophytic bacteria to produce some compounds which enhance the electron transport system, which in turn could provide the NADPH and ATP needed for carbon assimilation [30].

The results obtained and shown in (Table 2) have also demonstrated the positive impact of *Streptomyces* (strain JSA11) on chlorophyll and pigment contents of the tested sprouts, whereby significant increases in chlorophyll a and b were observed for *C. ficifolium* and *C. botrys* but not for *C. ambrosoides* under endophytic bacterial stimulation. In addition, all the examined species showed a significant enhancement in β-carotene; however, chlorophyll a and b were enhanced only in *C. botrys* when grown under endophytic bacterial inoculation. The increased chlorophyll contents in response to bacterial inoculation could be due to enhancing the chloroplast metabolism [30]. Such a result is in accordance with those previously obtained by [20], who demonstrated that the inoculation of tomato plants with bacterial endophytes led to increments in chlorophyll contents. Moreover, the biosynthesis of IAA by endophytic bacteria has been demonstrated to enhance the production of different pigments and metabolites [32]. Previous findings have dealt with the positive role of inoculating sunflower plants with growth promoting bacteria, which led to enhanced levels of chlorophyll a and b and carotenoids [33]. Increased levels of chlorophyll and photosynthetic pigments were also reported in mustard plants in response to growth promoting bacteria, which could effectively enhance the enzymes responsible for chlorophyll biosynthesis [34]. Thus, endophytic bacteria could lead to a better growth performance of plants through their positive impact on photosynthesis and chlorophyll content.

Table 2. Photosynthetic-related parameters and pigment contents of control and endophytic bacterial-treated *Chenopodium ambrosoides*, *Chenopodium ficifolium*, and *Chenopodium botrys* sprouts. Data are represented by the means of at least 3 replicates ± standard deviations. Different small letter superscripts (a, b) within a row indicate significant differences between control and endophytic bacterial samples.

	C. ambrosoides		C. ficifolium		C. botrys	
	Control	Endo	Control	Endo	Control	Endo
Photosynthetic Related Parameters (μmol CO_2 m^{-2} s^{-1})						
Photosynthesis	10.1 ± 0.8 [a]	11.5 ± 0.8 [b]	11.3 ± 1.1 [a]	13.1 ± 0.7 [b]	9.7 ± 1 [a]	12.8 ± 1.2 [b]
Respiration	1.4 ± 5.4 [a]	1.9 ± 0.0 [ab]	1.2 ± 0.06 [a]	2.0 ± 0.1 [b]	1.1 ± 0.0 [a]	1.9 ± 0.1 [b]
Pigments (mg/gFW)						
Chl [a]	2.07 ± 0.4 [a]	2.42 ± 0.38 [a]	1.65 ± 0.2 [a]	2.78 ± 0.4 [b]	1.76 ± 0.3 [a]	2.55 ± 0.4 [b]
Chl [b]	1.04 ± 0.1 [a]	1.46 ± 0.2 [a]	1.02 ± 0.1 [a]	1.65 ± 0.3 [ab]	1 ± 0.05 [a]	1.89 ± 0.41 [b]
Chl [a+b]	3.11 ± 0.5 [a]	3.8 ± 0.3 [a]	2.68 ± 0.4 [a]	4.4 ± 0.41 [a]	2.7 ± 0.3 [a]	4.44 ± 0.8 [b]
Beta-carotene	0.06 ± 0.01 [a]	0.11 ± 0.02 [b]	0.06 ± 0.0 [a]	0.12 ± 0.02 [b]	0.07 ± 0.01 [a]	0.11 ± 0.01 [ab]

As a consequence of improved photosynthesis, the biomass production (expressed as FW) of all the *Streptomyces* (strain JSA11)-inoculated sprouts was significantly enhanced when compared with the untreated control plants, whereas *C. ficifolium* appeared to have the highest biomass accumulation under both the control and bacterial inoculation conditions (Figure 2). In addition, the water content of all the tested sprouts was not affected by the bacterial treatments (Figure 2). This may indicate that endophyte treatment increased more dry weight accumulation (e.g., increased primary metabolites such as sugars, proteins, and fatty acids). This was also consistent with the observed increase in photosynthesis.

In accordance, the higher yield of runner bean plants inoculated with rhizobacteria has been found to be associated with higher photosynthetic activities [35]. Consequently, the higher the photosynthetic activity, the more sugar synthesis by plants, which could lead to a higher growth rate. For instance, sugar beet plants inoculated with endophytes have been found to accumulate higher total carbohydrates than the control plants [30]. However, other reports have demonstrated that additional carbohydrates, synthesized by host plants, might be diverted to feed endophytes at the expense of increasing biomass accumulation [31,36]. This could explain the mutualistic association between endophytes and the host plants, whereas the host plants could benefit from an increased growth rate. In return, endophytes could gain more nutrients from the host plant [30].

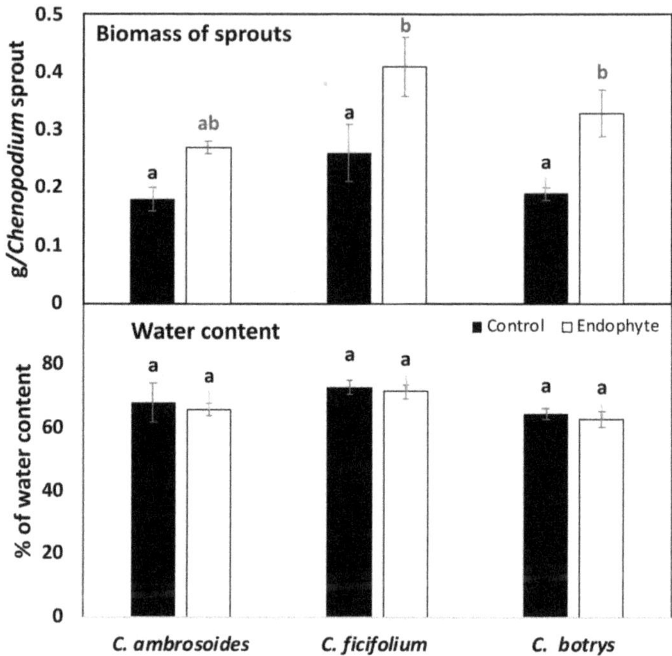

Figure 2. Biomass and water content of control and endophytic bacterial-treated *Chenopodium ambrosoides*, *Chenopodium ficifolium,* and *Chenopodium botrys* sprouts. Data are represented by the means of at least 3 replicates, and error bars represent standard deviations. Different small letters above bars indicate significant differences between means at $p < 0.05$.

On the other hand, the growth promoting effects of endophytic bacteria on plants might be ascribed to their ability to produce growth regulators, such as GA and IAA, which consequently stimulate cell elongation and differentiation in plants [20,32]. Moreover, the production of IAA by endophytic bacteria has been shown to improve photosynthesis, which would be positively reflected on the better growth performance of the target plants [32]. For example, bacterial endophytes have been previously reported to induce positive effects on growth and biomass production i.e., higher fresh weight (FW) and dry weight (DW) in buckwheat sprouts [37], as well as many plant species, such as tomato plants [20] and sugarcane plants [30]. Furthermore, other mechanisms triggered by endophytic bacteria might be involved in plant growth promotion, possibly through enhanced nutrient uptake and antagonistic effects against phytopathogens, nitrogen fixation, or the production of phytohormones, siderophores, and secondary metabolites [5,11]. The endophytes-produced organic acids were also assumed to play a role in plant growth promoting traits and the protection against pathogens [5].

2.5. Improved Minerals and Vitamins Contents by Endophytic Bacterial Treatment Contribute to Enhancing the Nutritive Value of Chenopodium Sprouts

Endophytic bacteria have been supposed to enhance the nutrient absorption by plants, particularly N, P, K, Ca, and Mg, leading to better nutrient use efficiency [38]. In the present study, the mineral profile of the investigated sprouts was analyzed, whereas nine mineral elements were determined, i.e., K, Na, Ca, Cu, Fe, P, Zn, Mn, and Mg (Table 3). When inoculated with *Streptomyces* (strain JSA11), the tested plants showed marked increases in almost all the detected elements. The percentage of increase in K content was recoded in both the control and inoculated plants. It increased by about 5%, 110%, and 100% in *C. ambrosoides*, *C. ficifolium*, and *C. botrys*, respectively, when compared with the control. The enhanced nutrient uptake by plants under endophytic bacterial treatment could be explained by the ability of bacterial endophytes to decrease the endogenous ethylene, hence allowing more nutrient absorption by increased root growth, in addition to the effective role of IAA, produced by endophytic bacteria, in stimulating the uptake of nutrients by plants [39]. Several reports could support our results and explain the endophytic bacterial-induced increments in minerals. For instance, the contribution of endophytic bacteria to phosphate solubilization processes has been well documented. On the one hand, the release of organic acids converts insoluble phosphates into a soluble form, thus improving P utilization efficiency and making it available for plants and, on the other hand, it also increases soil fertility [40]. The increases in K contents in plants could be explained by the ability of endophytic bacteria to solubilize K salts into more soluble forms, which consequently could be taken up by the plant roots, thus enhancing plant growth [10]. Moreover, the Cu content has been previously reported to increase in *Brassica napus* as a result of bacterial inoculation [41]. Regarding the mineral content of the *Chenopodium* species, it has been previously found that the most abundant macroelement in *C. botrys* was K, followed by Ca, P, and Mg, while the predominant microelement in *C. botrys* was Fe, followed by Na, Mn, Zn, and Cu [42].

Table 3. Proximate composition, vitamins, minerals of control, and endophytic bacterial-treated *Chenopodium ambrosoides, Chenopodium ficifolium,* and *Chenopodium botrys* sprouts. Data are represented by the means of at least 3 replicates ± standard deviations. Different small letter superscripts (a, b) within a row indicate significant differences between control and endophytic bacterial samples.

	C. ambrosoides		C. ficifolium		C. botrys	
	Control	**Endo**	**Control**	**Endo**	**Control**	**Endo**
Minerals (mg/gDW)						
K	13.9 ± 0.4 [a]	15.97 ± 1.2 [ab]	11.4 ± 0.12 [a]	23.2 ± 1 [b]	15.5 ± 1 [a]	29.8 ± 1.3 [b]
Na	1.8 ± 0.04 [a]	2.1 ± 0.3 [b]	1.5 ± 0.01 [a]	3.2 ± 0.1 [b]	1.9 ± 0.1 [a]	4.1 ± 0.1 [b]
Ca	1.46 ± 0.3 [a]	2.1 ± 0.12 [b]	1.4 ± 0.17 [a]	2.28 ± 0.1 [b]	1.5 ± 0.2 [a]	1.95 ± 0.02 [a]
Cu	0.006 ± 0 [a]	0.008 ± 0 [ab]	0.007 ± 0 [a]	0.011 ± 0 [b]	0.01 ± 0 [a]	0.013 ± 0 [a]
Fe	0.13 ± 0.02 [a]	0.19 ± 0 [b]	0.16 ± 0.01 [a]	0.2 ± 0.02 [a]	0.2 ± 0.01 [a]	0.33 ± 0.04 [ab]
P	0.96 ± 0.1 [a]	1.4 ± 0.05 [ab]	1.13 ± 0.07 [a]	1.62 ± 0.05 [b]	1.06 ± 0.0 [a]	1.64 ± 0.2 [a]
Zn	0.09 ± 0 [a]	0.12 ± 0.01 [ab]	0.11 ± 0 [a]	0.14 ± 0.01 [a]	0.12 ± 0.0 [a]	0.16 ± 0.01 [a]
Mn	0.028 ± 0 [a]	0.024 ± 0 [a]	0.027 ± 0 [a]	0.03 ± 0 [a]	0.02 ± 0 [a]	0.031 ± 0 [ab]
Mg	2.8 ± 0.3 [a]	2.3 ± 0.2 [a]	3.18 ± 0.39 [a]	3.9 ± 0.2 [a]	3.1 ± 0.5 [a]	5.18 ± 0.1 [b]
Vitamins (mg/gFW)						
Vitamin C	2.09 ± 0.07 [a]	2.34 ± 0.42 [b]	1.92 ± 0.06 [a]	3.04 ± 0.61 [b]	1.7 ± 0.72 [a]	2.7 ± 0.8 [b]
Vitamin E	4.71 ± 0.2 [a]	6.69 ± 0.23 [b]	4.25 ± 0.42 [a]	6.54 ± 0.58 [b]	3.46 ± 0.2 [a]	6.4 ± 0.6 [b]
Thiamin	0.49 ± 0.1 [a]	0.53 ± 0.14 [a]	0.42 ± 0.05 [a]	0.58 ± 0.13 [a]	0.35 ± 0.07 [a]	0.64 ± 0.1 [b]
Riboflavin	0.15 ± 0.03 [a]	0.18 ± 0.05 [a]	0.13 ± 0.03 [a]	0.18 ± 0.02 [ab]	0.09 ± 0.01 [a]	0.2 ± 0.04 [b]
Proximate composition (mg/gFW)						
Total proteins	9.1 ± 1 [a]	6.5 ± 4.8 [b]	11.2 ± 1.8 [a]	16 ± 3.7 [b]	9.1 ± 1 [a]	13.1 ± 3.2 [b]
Fat	123 ± 5.4 [a]	125 ± 15 [a]	151.3 ± 6 [a]	140. ± 6 [b]	123 ± 5.4 [a]	150.1 ± 16 [b]
Crude Fiber	6.6 ± 0.6 [a]	6.8 ± 1.8 [a]	8.1 ± 0.7 [a]	12.3 ± 1 [b]	6.6 ± 0.6 [a]	10.1 ± 0.8 [b]
Ash	3.8 ± 0.5 [a]	3.6 ± 0.4 [a]	3.9 ± 0.5 [a]	5.3 ± 0.8 [b]	3.4 ± 0.5 [a]	4.6 ± 0.7 [b]
Carbohydrate	6.3 ± 0.3 [a]	5.9 ± 0.2 [a]	7.7 ± 0.4 [a]	11.1 ± 0.8 [b]	6.3 ± 0.3 [a]	10.1 ± 2 [b]

The results of the current investigation also indicate that the vitamin C and E contents were significantly enhanced in the three species in response to inoculation with *Streptomyces* (strain JSA11) when compared with the control plants (Table 3). In addition, riboflavin and thiamin only increased in *C. botrys*. Meanwhile, vitamin E had the highest concentration among the detected vitamins. Our results could be supported by previous reports of [43], who found that vitamin E was among the main components of *C. ambrosioides*.

The proximate composition of the examined sprouts showed that the total protein, crude fiber, ash, and carbohydrate contents were remarkably enhanced in both *C. ficifolium* and *C. botrys* inoculated with *Streptomyces* (strain JSA11), while such parameters were not significantly changed in *C. ambrosoides*, except for the total proteins. (Table 3). At the same time, bacterial endophytes caused a significant increment in fat content in both *C. ficifolium* and *C. botrys*. In line with our results, the total sugar and protein contents of mushrooms were found to be enhanced in response to bacterial endophytes, resulting in a higher plant nutritive value [12]. Bacterial treatment has also caused elevations in sugar levels in mustard plants [34]. Further, the growth promoting rhizobacteria have been reported to be potent inducers of the plant nutritive value through increasing total protein and carbohydrate content in *Phaseolus coccineus* [35]. In this regard, the endophytic bacterial-induced increase in plant protein content might be attributed to the higher N fixation by the endophytic bacteria [35].

In the same context, the proximate composition of both *C. ambrosioides* and *C. botrys* (i.e., moisture content, ash content, crude protein, crude fiber, crude fat, and carbohydrates) has been previously analyzed and proven to fall within the standard values for some vegetable drugs, which could support the use of this plant in pharmaceutical preparations [44]. *C. ambrosioides* has been reported to be a rich source of carotenoids, as well as proteins and fats [17]. Overall, the enhanced levels of minerals, vitamins, sugars, and proteins in *Chenopodium* sprouts in response to bacterial endophytes treatments could support the plant nutritive value

2.6. Endophytic Bacterial Treatment Improved the Functional Food Value of Chenopodium Sprouts through Enhancing Their Bioactive Primary Metabolites Levels

The enhanced photosynthetic capacity as a result of endophytic bacterial inoculation is supposed to increase the plant's sugar content, which in turn, could be further utilized to meet the energy required for the production of different types of bioactive metabolites [11].

In the current study, the three tested species showed a qualitatively similar, but quantitatively different amino acid profile (Table 4). Apparently, most of the detected amino acids in all the investigated plants were significantly promoted when inoculated with *Streptomyces* (strain JSA11) when compared with control plants. However, some amino acids remained unaffected when treated with endophytic bacteria, e.g., arginine, alanine, histidine, and valine in the case of *C. ficifolium* and *C. botrys* and alanine and isoleucine in the case of *C. ambrosoides*. Notably, the highest concentrations were reached for leucine, lysine, arginine, glutamic acid, and glutamine in all the tested sprouts. In this regard, the uptake of amino acids by plants was greatly affected by the N fixation by microbial communities [45]. Similar to our findings, bacterial treatment has been previously reported to induce higher concentrations of asparagine, therionine, serine, glutamine, alanine, valine, methionine, leucine, tyrosine, phenyl alanine, lysine, histidine, arginine, and proline in mustard plants [34]. Various amino acids (e.g., histidine, leucine, methionine, and tyrosine) were also significantly increased in mushrooms treated with bacterial endophytes [12]. Moreover, the endophytic bacterial association with sugarcane plants led to an accumulation of some amino acids, such as alanine and glycine, while others (e.g., proline and aspartate) were reduced [38]. Such variability might be dependent on the plant species or growth conditions. The majority of the detected amino acids, such as valine, alanine, leucine, proline, isoleucine, serine, threonine, phenyl alanine, asparagine, and tyrosine were previously reported in *C. ambrosioides* [43].

Table 4. Amino acids and amino acids-related enzymes, organic acids, and fatty acids of control and endophytic bacterial -treated *Chenopodium ambrosoides, Chenopodium ficifolium,* and *Chenopodium botrys* sprouts. Data are represented by the means of at least 3 replicates ± standard deviations. Different small letter superscripts (a, b) within a row indicate significant differences between control and endophytic bacterial samples.

	C. ambrosoides		*C. ficifolium*		*C. botrys*	
	Control	Endo	Control	Endo	Control	Endo
Amino Acid Metabolism (µg/gFW)						
Glutamic acid	14.6 ± 1.1 [a]	19.2 ± 1.8 [b]	14.98 ± 1 [a]	18.4 ± 2.7 [b]	14.5 ± 1.4 [a]	19.9 ± 1.3 [b]
Glutamine	12.8 ± 0.9 [a]	16.4 ± 1.1 [b]	12.5 ± 0.9 [a]	16.0 ± 1.2 [b]	12.6 ± 1.3 [a]	17.4 ± 1.3 [b]
Serine	6.39 ± 0.3 [a]	8.61 ± 1.1 [b]	6.7 ± 0.5 [a]	8.8 ± 0.8 [b]	9.8 ± 2.4 [a]	11.8 ± 1.9 [c]
Glycine	8.3 ± 0.13 [a]	9.6 ± 0.0 [b]	7.6 ± 0.12 [a]	9.5 ± 0.7 [b]	7.3 ± 0.66 [a]	9.7 ± 0.18 [b]
Arg	13.6 ± 0.5 [a]	16 ± 1.5 [ab]	20.9 ± 3.5 [a]	23.3 ± 2.3 [b]	21.6 ± 3.8 [a]	24.2 ± 1.8 [b]
Alanine	2.4 ± 0.3 [a]	2.48 ± 0.8 [a]	3.5 ± 0.16 [a]	2.7 ± 0.5 [b]	3.4 ± 0.8 [a]	2.85 ± 0.5 [a]
Histidine	5.2 ± 0.48 [a]	5.7 ± 0.2 [ab]	4.98 ± 0.8 [a]	5.2 ± 0.29 [a]	6.1 ± 0.42 [a]	4.6 ± 0.7 [b]
Valine	5.3 ± 0.7 [a]	6.95 ± 1 [b]	5.05 ± 1.3 [a]	5.72 ± 0.9 [a]	5.9 ± 0.8 [a]	6.0 ± 1.04 [a]
Methionine	2.2 ± 0.19 [a]	2.9 ± 0.3 [ab]	2.01 ± 0.2 [a]	3.5 ± 0.4 [b]	1.65 ± 0.1 [a]	3.31 ± 0.0 [b]
Cystine	1.4 ± 0.03 [a]	2.1 ± 0.18 [b]	1.7 ± 0.06 [a]	2.3 ± 0.2 [b]	1.4 ± 0.12 [a]	1.9 ± 0.1 [ab]
Isoleucine	4.2 ± 0.2 [a]	5.48 ± 0.3 [a]	7.15 ± 1 [a]	7.3 ± 1.5 [a]	5.2 ± 0.79 [a]	7.8 ± 0.9 [b]
Leucine	13.8 ± 0.4 [a]	14 ± 0.9 [ab]	11.5 ± 1.4 [a]	12.0 ± 1.1 [a]	11.1 ± 0.7 [a]	13.0 ± 1 [ab]
Tyrosine	5.6 ± 0.7 [a]	6.3 ± 0.5 [ab]	5.47 ± 0.4 [a]	7.2 ± 0.2 [b]	5.3 ± 0.1 [a]	6.4 ± 0.6 [ab]
Lysine	13.5 ± 0.3 [a]	21.1 ± 1.6 [b]	12.8 ± 0.6 [a]	21.3 ± 1.3 [b]	13.3 ± 0.7 [a]	25.8 ± 2.5 [b]
Threonine	4.6 ± 0.3 [a]	5.3 ± 0.6 [ab]	4.4 ± 0.4 [a]	6.3 ± 0.2 [b]	4.9 ± 0.1 [a]	5.9 ± 0.2 [ab]
Trep	0.4 ± 0.04 [a]	0.61 ± 0.01	0.4 ± 0.04 [a]	0.6 ± 0.03	0.4 ± 0.04 [a]	0.67 ± 0.04
Amino Acid Biosynthesis Enzymes						
GS (nmol γ-glutamyl hydroxamate/mg protein min^{-1})	5.4 ± 0.4 [a]	7.14 ± 0.5 [b]	5.51 ± 0.4 [a]	6.8 ± 0.7 [ab]	5.44 ± 0.5 [a]	7.4 ± 0.7 [b]
DHDPS (nmol o-ABA and L-2,3- dihydrodipicolinate adduct/mg protein min^{-1})	2.09 ± 0.3 [a]	2.4 ± 0.2 [a]	2.09 ± 0.3 [a]	3.5 ± 0.2 [b]	2.5 ± 0.4 [a]	3.1 ± 0.3 [b]
CGS (nmol L-cystathionine/mg protein min^{-1})	0.012 ± 0 [a]	0.033 ± 0 [b]	0.013 ± 0 [a]	0.039 ± 0 [b]	0.014 ± 0 [a]	0.37 ± 0 [b]
Organic Acid (mg/g FW)						
Oxalic	9.08 ± 0.8 [a]	9.5 ± 0.4 [a]	3.5 ± 0.5 [a]	4.73 ± 4.7 [b]	1.92 ± 0.3 [a]	4.73 ± 0.5 [b]
Malic	2.3 ± 0.31 [a]	3.58 ± 0.2 [ab]	2.92 ± 0.3 [a]	2.67 ± 0.6 [a]	2.1 ± 0.22 [a]	4.27 ± 0.9 [b]
Succinic	2.17 ± 0 [a]	2.03 ± 0 [a]	1.75 ± 0.3 [a]	3.5 ± 0.3 [a]	3.2 ± 1.2 [a]	5.9 ± 0.9 [b]
Citric	1.3 ± 0.04 [a]	1.9 ± 0.09 [a]	2.96 ± 0.29 [a]	2.9 ± 0.2 [a]	2.88 ± 0.5 [a]	2.5 ± 0.5 [ab]
Lactic	0.2 ± 0.1 [a]	0.24 ± 0.1 [a]	0.32 ± 0.09 [a]	0.26 ± 0.07 [a]	0.33 ± 0.05 [a]	0.49 ± 0.05 [ab]
Fatty Acids (mg/g FW)						
Tetradecanoic (C14:0)	0.7 ± 0.1 [a]	0.8 ± 0.1 [ab]	0.8 ± 0.1 [a]	0.98 ± 0.09 [b]	0.66 ± 0.06 [a]	1.02 ± 0.1 [b]
Pentadecanoic (C16:0)	12 ± 2 [a]	15.8 ± 1 [ab]	11.6 ± 0.6 [a]	15.5 ± 0.7 [ab]	18 ± 1.6 [a]	18.1 ± 1.3 [a]
Eicosanoic (C20:0)	0.8 ± 0.03 [a]	0.96 ± 0.1 [ab]	1.1 ± 0.2 [a]	1.02 ± 0.1 [a]	0.88 ± 0.09 [a]	0.96 ± 0.1 [a]
Docosanoic (C22:0)	0.8 ± 0.1 [a]	1.45 ± 0.2 [b]	0.91 ± 0.1 [a]	1.29 ± 0.1 [ab]	0.91 ± 0.1 [a]	1.2 ± 0.1 [ab]
Octadecanoic (C18:0)	6.9 ± 0.7 [a]	10.2 ± 1.8 [b]	7.2 ± 0.8 [a]	8.9 ± 0.9 [ab]	7.2 ± 1.2 [a]	9.5 ± 1.2 [b]
Pentacosanoic (C24:0)	0.14 ± 0.0 [a]	0.13 ± 0.02 [a]	0.12 ± 0.01 [a]	0.12 ± 0.01 [a]	0.13 ± 0.01 [a]	0.15 ± 0.1 [a]
Pentadecanoic (C16:1)	1.42 ± 0.1 [a]	1.3 ± 0.09 [a]	1.4 ± 0.23 [a]	2.3 ± 0.2 [b]	1.55 ± 0.3 [a]	2.5 ± 0.1 [b]
Pentadecanoic (C16:1)	0.89 ± 0.0 [a]	0.74 ± 0.09 [a]	0.69 ± 0.03 [a]	0.9 ± 0.14 [b]	0.44 ± 0.1 [a]	0.43 ± 0.06 [a]
Pentadecanoic (C16:3)	1.0 ± 0.5 [a]	0.59 ± 0.2 [a]	0.9 ± 0.4 [a]	0.52 ± 0.2 [a]	0.73 ± 0.3 [a]	0.6 ± 0.2 [a]

Table 4. *Cont.*

	C. ambrosoides		*C. ficifolium*		*C. botrys*	
	Control	**Endo**	**Control**	**Endo**	**Control**	**Endo**
Octadecanoic (C18:1)	3.73 ± 0. [a]	3.96 ± 0.8 [a]	2.97 ± 0.4 [a]	4.45 ± 0.6 [b]	3.46 ± 0.6 [a]	4.48 ± 0.7 [b]
Octadecanoic (C18:2)	21.5 ± 0.9 [a]	24.2 ± 1.4 [a]	25.1 ± 1.9 [a]	22.5 ± 0.8 [a]	24.2 ± 1 [a]	20.8 ± 0.5 [a]
Heptadecanoic (C18:3)	12.1 ± 3. [a]	25.64 ± 2.3 [b]	23.3 ± 3.3 [a]	29.68 ± 6 [b]	25.81 ± 4.1 [a]	26.52 ± 0.4 [a]
Heptadecanoic (C18:4)	1.0 ± 0.06 [a]	1.27 ± 0.14 [a]	1.04 ± 0.3 [a]	1.2 ± 0.08 [a]	1.13 ± 0.2 [a]	1.16 ± 0.09 [a]
Tetracosanoic (C20:3)	0.11 ± 0.01 [a]	0.26 ± 0.0 [b]	0.22 ± 0.07 [a]	0.23 ± 0.04 [a]	0.2 ± 0.06 [a]	0.19 ± 0.04 [a]

In order to obtain more insight into the amino acid-related changes under endophytic bacterial treatments, the glutamine synthases (GS), dihydrodipicolinate synthase (DHDPS), and cystathionine γ-synthase (CGS) were evaluated as being the key enzymes involved in N metabolism and the synthesis of glutamine, lysine, and methionine [4]. The results obtained show remarkable increases in such enzymes in the *Streptomyces* (strain JSA11)-treated sprouts when compared with controls (Table 4). Our results could be supported by previous studies that described the increased enzymatic activity of glutathione synthase (GS) in sugarcane plants associated with bacterial endophytes [38]. However, it has been shown that endophytic bacteria decreased the tyrosine aminotransferase activity in the hairy roots of *Salvia miltiorrhiza* [46]. Such a difference might depend on the interacting species or growth conditions.

From the present results, it is also clear that the tested species shared the same fatty acid profile qualitatively but not quantitatively, but pentadecanoic (C16:0) and octadecanoic (C18:2) showed the highest concentrations among the saturated and unsaturated fatty acids, respectively, in all the examined plants (Table 4). Some of the detected fatty acids (e.g., tetradecanoic, docosanoic, and octadecanoic) were significantly enhanced, while others (e.g., eicosanoic, pentacosanoic, and octadecanoic) were not affected in all the *Streptomyces* (strain JSA11)-treated sprouts when compared with the untreated control plants. Supporting our results, it has been previously shown that *C. ambrosioides* had higher levels of unsaturated fatty acids (dominated by linolenic and linoleic) than saturated fatty acids [47]. In addition, several fatty acids have been characterized from *C. ambrosioides*, e.g., palmitic, octadecadienoic, stearic, palmitic, octadecadienoic, and stearic acids [43].

Regarding their content of organic acids, the examined *Chenopodium* sprouts contained five organic acids, i.e., oxalic, malic, succinic, citric, and lactic acids (Table 4). Similarly, *C. ambrosoides* has been previously demonstrated to contain oxalic, ascorbic, quinic, fumaric, malic, lactic, succinic, and citric acids with oxalic acid being the most dominant one [43,47]. The application of *Streptomyces* (strain JSA11) endophytic bacterial treatment did not induce significant elevations in most of the detected organic acids, except for malic acid (in *C. botrys*), oxalic acid (in *C. ficifolium* and *C. botrys*), and lactic and succinic acids (in *C. botrys*). Consistent with our results, the organic acid content of mushrooms (e.g., oxalic, tartaric, formic, malic, and fumaric acids) could be significantly altered by bacterial endophytes to improve the plant's nutritive value [12].

2.7. Endophytic Bacteria Stimulated the Antioxidant Activities of Chenopodium Sprouts through Enhancing Their Phenolic Content

The functional food value of plants has been assumed to be associated with their levels of secondary metabolites [48] that could be enhanced under endophytic bacterial stimulations through their ability to produce IAA, which in turn is likely to be utilized by plants to synthesize sugars. Thus, the increased amount of sugar might be further employed for the production of bioactive secondary metabolites [11]. Additionally, the endophytic bacterial-secreted metabolites were hypothesized to play a role in the production of some plant secondary metabolites [11].

The results obtained herein show that *C. ambrosoides* had higher amounts of total phenolic and flavonoids contents than *C. ficifolium* and *C. botrys* under control conditions

(Figure 3). Meanwhile, *Streptomyces* (strain JSA11) bacterial endophytes caused increases in the total flavonoids in all the tested sprouts, as well as increases in total phenolics in only *C. ficifolium* and *C. botrys*. Likewise, high phenolic and flavonoid contents were previously observed for tartary buckwheat sprouts inoculated with growth promoting bacteria [25]. Moreover, *C. ambrosioides* and *C. botrys* were shown to possess high total flavonoids and phenolic contents [49,50].

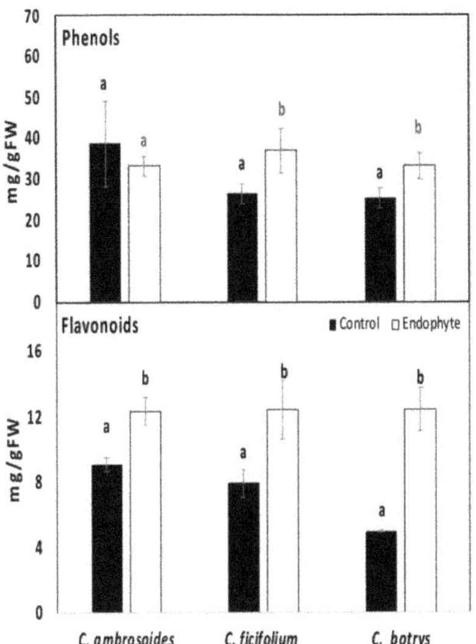

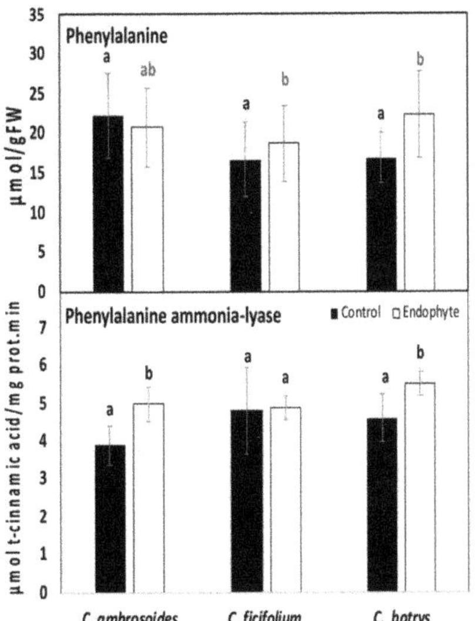

Figure 3. Phenolic, flavonoids, and phenolic biosynthesis enzymes of control and endophytic bacterial-treated *Chenopodium ambrosoides*, *Chenopodium ficifolium*, and *Chenopodium botrys* sprouts. Data are represented by the means of at least 3 replicates, and error bars represent standard deviations. Different small letters above bars indicate significant differences between means at $p < 0.05$.

In order to follow the metabolic pathway responsible for such increases in polyphenols, the enzymatic activity of phenylalanine ammonia lyase (PAL), a key enzyme in polyphenol production, was evaluated [4]. PAL catalyzes the nonoxidative elimination of ammonia from its substrate (phenylalanine) to give trans-cinnamic acid. Bacterial endophytes were proposed to play a significant role in the synthesis of plant secondary metabolites via their regulatory effect on enzyme activities [46]. Our results indicate that the *Streptomyces* (strain JSA11) endophytic bacterial treatment led to significant increases in PAL enzyme activity in both *C. ambrosoides* and *C. botrys*, while significant increments were reported for t substrate phenylalanine in *C. ficifolium* and *C. botrys*, in comparison to the untreated control plants (Figure 3). Similarly, previous studies have investigated the increments in such enzyme in response to eCO$_2$ [4]. However, previous reports have shown endophytic bacteria to decrease phenylalanine ammonia lyase activity in the hairy roots of *Salvia miltiorrhiza* [46]. Such variability might be dependent on the plant species or growth conditions. Thus, the regulation of plant secondary metabolites by endophytic bacteria might be governed by their impact on enzymatic activity.

The increments in the phenolic content in the studied species in response to the *Streptomyces* (strain JSA11) inoculation were also observed to be concomitant with dramatic increases in antioxidant activities (evaluated by FRAP, ABTS, and DPPH assays) (Table 5). Meanwhile, no significant changes were observed for only *C. ambrosoides* regarding DPPH activity. In accordance with our results, high antioxidant capacities were obtained for some plants, such as tomato plants, when treated with growth promoting bacteria [51]. The endophytes-provoked antioxidant capacities might be exerted as a result of an accumulation of reactive oxygen species by the endophytes or plants. Furthermore, *C. ambrosioides* extracts have previously shown high DPPH scavenging activity and TBARS inhibition [47], while its oil displayed high scavenging antioxidant activity, tested by different assays, including ABTS, DPPH, β-carotene-linoleic acid, and reducing power assays [16,52]. Such activity might be attributed to its rich content of phenolic compounds [47], as well as monoterpene hydrocarbons, particularly α-terpinene [52]. Previous studies have also shown *C. botrys* extracts to display high total antioxidant capacities (evaluated by DPPH, FRAP, and ABTS assays), superoxide anion (O_2^-), nitric oxide (NO), and hydroxyl (HO) radicals [50].

Table 5. Antioxidant activity of control and endophytic bacterial-treated *Chenopodium ambrosoides*, *Chenopodium ficifolium*, and *Chenopodium botrys* sprouts. Data are represented by the means of at least 3 replicates ± standard deviations. Different small letter superscripts (a, b, and c) within a row indicate significant differences between control and endophytic bacterial samples.

	C. ambrosoides		*C. ficifolium*		*C. botrys*		Extraction Solvent *	Cipro-Floxacin
	Control	Endophyte	Control	Endophyte	Control	Endophyte		
Antioxidant Activities								
FRAP (mg gallic acid/gFW)	16.2 ± 1.2 [a]	37.5 ± 1 [c]	18.3 ± 0.6 [b]	33.06 ± 3 [c]	11.5 ± 0.1 [a]	21.4 ± 6 [c]	0.5 ± 0	-
ABTS (µmol trolox/gFW)	2.0 ± 0.1 [a]	1.9 ± 0.2 [b]	1.37 ± 0.1 [a]	1.9 ± 0.01 [b]	0.93 ± 0.1 [a]	2.2 ± 0.3 [b]	0.7 ± 0	-
DPPH%	18.8 ± 0.7 [a]	13.7 ± 1 [a]	12.2 ± 1 [a]	16.37 ± 3 [b]	12.7 ± 1 [a]	17.06 ± 3 [b]	0.1 ± 0	-

* Ex. Solvent is the extraction solvent (80% of ethanol).

2.8. Endophytic Bacteria Enhanced the Chenopodium Biological Activities

2.8.1. Antibacterial Activities

The *Chenopodium* species have been known for their promising antimicrobial activities [17]. According to our results, the extracts of all tested species were shown to induce antimicrobial activities against several bacterial and fungal species, whereas the most potent effect was clearly shown by all plants against *Aspergillus flavus*, based on the inhibition zone diameter (Table 6). Interestingly, such potential activities were significantly enhanced in the examined species in response to the *Streptomyces* (strain JSA11) inoculation, while no changes were obtained for some sprouts regarding their activity against certain organisms (e.g., *Enterococcus faecalis* in the case of *C. ficifolium* and *C. botrys*). In accordance, several reports have investigated the potential antimicrobial effects of the *Chenopodium* species. For instance, *C. ambrosioides* oil has previously exhibited potent fungi toxic activity against *Aspergillus flavus* [16]. Meanwhile, *C. ambrosioides* had also antimicrobial activity against *Staphylococcus aureus*, *Micrococcus luteus*, *Bacillus cereus*, *Escherichia coli*, *Pseudomonas aeruginosa* [52], *Bacillus subtilis*, *Aspergillus oryzae*, and *Aspergillus niger* [49]. Moreover, the antifungal activities of *C. ambrosioides* oil on *Candida albicans*, *C. glabrata*, *C. krusei*, and *C. parapsilosis* have been previously investigated [52], the activity that might be attributed to the presence of monoterpene hydrocarbons [52], which could affect the bacterial membranes. Similar to our results, *C. ambrosioides* extracts have been previously revealed to exert an antibacterial effect against *Staphylococcus aureus* and *Enterococcus faecalis*. Moreover, *C. ambrosioides* had antimicrobial potency against *Paenibacillus apiarus*, *Paenibacillus thiaminolyticus*, *Mycobacterium tuberculosis*, *M. smegmatis*, and *M. avium*. Such activity could be ascribed to the presence of phenolic compounds [53]. On the other hand, *C. botrys* extracts have been shown to possess promising antimicrobial potency against *Staphylococcus aureus* and *Klebsiella pneumonia* [18].

Table 6. Antibacterial activity of control and endophytic bacterial-treated *Chenopodium ambrosoides*, *Chenopodium ficifolium*, and *Chenopodium botrys* sprouts. Data are represented by the means of at least 3 replicates ± standard deviations. Different small letter superscripts (a, b) within a row indicate significant differences between control and endophytic bacterial samples.

	C. ambrosoides		*C. ficifolium*		*C. botrys*			
	Control	Endo	Control	Endo	Control	Endo	Ex. Solvent	Cipro-Floxacin
	Antibacterial Activities (Zone Inhibition, mm)							
S. saprophyticus ATCC 19701	14.3 ± 0.4 [a]	25.5 ± 1 [b]	16.5 ± 0.3 [a]	20 ± 2.3 [a]	19 ± 0.1 [a]	22.5 ± 0.6 [b]	1.5 ± 0.1	33.4 ± 2.3
S. epidermidis ATCC 12228	10.9 ± 0.3 [a]	22.0 ± 1 [b]	14.5 ± 0.9 [a]	17.6 ± 2 [ab]	16. ± 0.3 [b]	19.92 ± 1 [ab]	0.9 ± 0	23.6 ± 5
E. faecalis ATCC 10541	15.0 ± 0.3 [a]	20 ± 0.9 [b]	17.8 ± 0.4 [a]	15 ± 1.7 [ab]	16 ± 0.1 [a]	19.1 ± 1 [ab]	1.1 ± 0	21.5 ± 4.0
S. salivarius ATCC25975	13.2 ± 0.8 [a]	17 ± 0.4 [ab]	16.2 ± 0.3 [a]	20 ± 0.4 [a]	13 ± 0.4 [a]	17.09 ± 1 [ab]	0.89 ± 0	30.6 ± 2.4
E. coli ATCC 29998	14.52 ± 1 [a]	18.3 ± 1 [ab]	19.5 ± 0.6 [a]	21 ± 0.7 [a]	16 ± 1.0 [a]	19.29 ± 1 [ab]	1.2 ± 0	25.9 ± 0.7
S.typhimurium ATCC14028	16.2 ± 0.3 [a]	20.6 ± 1 [ab]	19.1 ± 0 [a]	23 ± 0.6 [b]	18 ± 0.2 [a]	20.5 ± 1 [ab]	0.5 ± 0	22.1 ± 0.6
P. aeruginosa ATCC10145	16.9 ± 0.5 [a]	33 ± 1.6 [b]	20.2 ± 0.7 [a]	22 ± 1 [a]	24 ± 0.2 [a]	29 ± 1 [b]	1 ± 0	22.8 ± 1 [a]
P. vulgaris ATCC8427	16. ± 0.4 [a]	26.3 ± 1 [b]	18.7 ± 0.4 [a]	17 ± 0.5 [a]	20 ± 0.1 [a]	23.9 ± 0.8 [a]	1.9 ± 0.1	27.1 ± 1.5
E.r aerogenes ATCC 13048	15.4 ± 0.3 [a]	28.6 ± 1 [b]	17.8 ± 0.4 [a]	22 ± 0.4 [b]	21 ± 0.2 [a]	25 ± 0.7 [a]	0.7 ± 0.0	24 ± 0.7
S. marcescens ATCC99006	1.9 ± 0.06 [a]	7.3 ± 0.6 [b]	2.2 ± 0.05 [a]	5.9 ± 0.5 [b]	4 ± 0.05 [a]	5.5 ± 0.3 [a]	0.9 ± 0.1	8.1 ± 0.5
S. typhimurium ATCC14028	16 ± 0.5 [a]	10 ± 0.5 [ab]	19.2 ± 0.4 [a]	17 ± 0.4 [a]	12 ± 0.1 [a]	13.89 ± 1 [a]	1.1 ± 0	22 ± 0.2
C. albicans ATCC90028	6.9 ± 0.2 [a]	9.8 ± 0.5 [b]	8.0 ± 0.19 [a]	7.9 ± 0.9 [a]	8.1 ± 0.1 [a]	9.3 ± 0.4 [a]	0.07 ± 0	8.9 ± 0.4
C. glabrata ATCC90030	2.3 ± 0.07 [a]	4.8 ± 0.2 [b]	2.9 ± 0.14 [a]	3.1 ± 0.1 [a]	3 ± 0.03 [a]	4.18 ± 0.1 [a]	0.2 ± 0	4.0 ± 0.01
A. flavus ATCC9170	22 ± 0.5 [a]	43 ± 1.1 [b]	25.7 ± 0.6 [a]	33.2 ± 5 [b]	29 ± 0.3 [a]	36 ± 1 [b]	20.5 ± 1	33.29 ± 5 [b]

2.8.2. Anti-Inflammatory Activities of Chenopodium Sprouts

In the present investigation, the *Streptomyces* (strain JSA11) bacterial endophytes treatments have resulted in remarkable reductions in COX-2 and LOX activities of both *C. ficifolium* and *C. botrys*, while no significant effect was induced on *C. ambrosoides* when compared with their respective controls (Table 7). Unlike our results, *C. ambrosioides* extracts have been previously shown to exert anti-inflammatory effects against some inflammatory mediators (e.g., K, NO, TNF-α, and PGE2), an effect that might be due to the presence of bioactive compounds, such as phenolics and flavonoids [54]. On the other hand, it has been recently shown that the application of other elicitors, such as eCO_2 or laser light, has effectively enhanced the anticancer and anti-inflammatory activities of both buckwheat and lemongrass sprouts [3,4].

Table 7. Anti-inflammatory activity of control and endophytic bacterial-treated *Chenopodium ambrosoides*, *Chenopodium ficifolium*, and *Chenopodium botrys* sprouts. Data are represented by the means of at least 3 replicates ± standard deviations. Different small letter superscripts (a, b, and c) within a row indicate significant differences between control and endophytic bacterial samples.

Anti-Inflammatory	*C. ambrosoides*		*C. ficifolium*		*C. botrys*	
	Control	Endo	Control	Endo	Control	Endo
Cyclooxygenase-2 (µg/mL)	1.1 ± 0.2 [a]	0.73 ± 0.0 [ab]	1.2 ± 0.1 [a]	0.6 ± 0.04 [b]	1.3 ± 1 [a]	0.45 ± 0 [c]
Lipoxygenase (µg/mL)	7.2 ± 0.1 [a]	5.7 ± 0.5 [ab]	6.7 ± 4 [a]	4.6 ± 0.6 [b]	6.8 ± 0.7 [a]	3.1 ± 0.6 [c]

2.9. Species-Specific Response to Endophytic Bacterial Treatment

The hierarchical clustering data represented in (Figure 4) show that there was a clear sprout species-specific response to the effect of the *Streptomyces* (strain JSA11) inoculation. *C. botrys* showed a more prominent response to the enhancing effect of *Streptomyces* bacterial endophytes, where it had the highest content of pigments, vitamins, amino acids, fatty acids, and minerals. Consequently, *C. botrys* showed the highest anti-inflammatory activity, followed by *C. ficifolium*. Meanwhile, *C. ambrosioides* exhibited the highest response to bacterial endophytes regarding the antimicrobial activity. The variations among the three species might be attributed to species diversity and ontogeny [3].

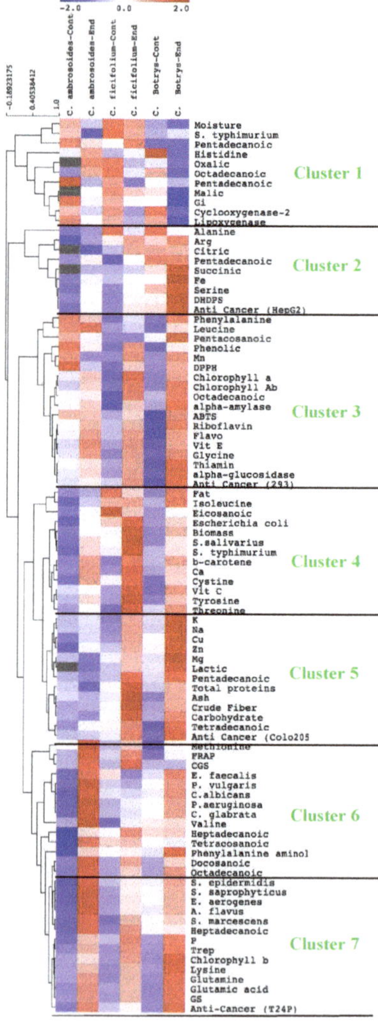

Figure 4. Species-specific responses of *Chenopodium ambrosoides*, *Chenopodium ficifolium*, and *Chenopodium botrys* sprouts to the effect of endophytic bacterial treatment on the nutritional and health-promoting properties. The measured parameters are represented by amino, fatty, and organic acids; antioxidant capacity; the contents of pigments, minerals, and vitamins; and antibacterial, antidiabetic, anti-inflammatory, and anticancer activities. Data are represented by the means of at least 3 replicates.

3. Material and Methods

3.1. Experimental Setup, Plant Materials, and Growth Conditions

3.1.1. Isolation and Characterization of Endophyte Isolates

Endophytic bacteria were isolated from *Chenopodium* leaves and stems, and the isolates were prepared only from *C. ambrosoides*. The dissection of *Chenopodium* leaves was completed, and then they were washed with water and cut into small pieces (1 cm). The dissected pieces were rinsed first with Tween 20 (0.1%) for 30 s), then with Na-hypochlorite (1%) for 5 min, and finally with distilled H_2O for 5 min. Afterwards, the dissected pieces were surface sterilized in ethanol (70%) for 5 min, and then they were air-dried in a flow chamber. Plant tissue pieces were transferred to plates of a glycerol–yeast–agar medium supplemented with nystatin (50 $\mu g\ L^{-1}$). About 1 g of sterilized plant tissue was added to 10 mL of an aqueous saline solution (0.9% NaCl), and then the solution was well mixed and heated at 50 °C for 30 min. Serial dilutions were prepared with the pour plate method. After incubation of 14 days at 27 °C, the selected colonies were purified on the used medium (glycerol–yeast–agar) for 8 days at 27 °C. Further, the purified colonies were maintained on starch casein agar as agar-slants at 4 °C and as suspensions at −20 °C in 20% glycerol until use [55]. The morphological identification was performed by examining cover slips of the isolates with a light microscope. Cultural characteristics in different media were observed after incubation at 27 °C (7–14 days). The bioactive isolates were morphologically characterized using Bergey's manual key. Biochemical features including IAA and siderophore production were measured [56,57] to investigate the growth promoting potential of endophytic bacteria. In addition, the carbon and nitrogen utilization; hydrolysis of starch, casein and lipid; nitrate reduction, gelatin liquefication, and the ability to produce H_2S, urease, L-asparaginase, L-glutaminase, and peroxidase were carried out following the methods of [58]. The ability to utilize nitrogen sources was determined in a basal medium containing glucose (2 g), $MgSO_4.7H_2O$ (0.125 g), $FeSO_4.7H_2O$ (0.002 g), K_2HPO_4 (0.25 g), NaCI (0.5 g), and agar (3.0 g) in 200 mL of water.

For antioxidant activities measurement, the purified endophyte strain was cultured on glycerol–yeast extract–agar (glycerol 5 mL, yeast extract 2 g, K_2HPO_4 1 g, agar 15 g, and distilled water 1000 mL) plates at 28 °C for 7 days. The *Streptomyces* strain was inculcated into glycerol–yeast extract broth in Erlenmeyer flasks for propagation. The culture was incubated on a rotary shaker at 180 rpm, at 28 °C, and for 7 days. Then, *Streptomyces* cells were collected by centrifugation for 10 min at 8000× *g* at 4 °C. The precipitated cells were washed in a sterile saline solution, and the cell suspension was freeze-dried and repeatedly extracted with ethanol. After removing the ethanol solvent using a rotary vacuum evaporator at 37 °C, the final extract (30 mg) was obtained and suspended in 10 mL of 80% ethanol (3 mg/mL). Total antioxidant activity, namely, ferric reducing antioxidant power (FRAP), 2-diphenyl-1-picrylhydrazyl (DPPH), and 2,2′-azino-bis (3-ethylbenzothiazoline-6-sulfonic acid) (ABTS) were measured by using an 80% ethanol extract (2 mg/mL). More details about antioxidant measurements are in Section 3.5.1.

The detection of phenolic and flavonoid compounds in the bacterial endophytes was performed via the HPLC method after extraction of 50mg of the cell biomass in methanol according to [3,21]. The compounds were identified by using the standards and their relative retention times. About 25 standards were used and bought from Sigma-Aldrich. The used phenolic and flavonoid standards are commonly present in plants. The peak area of each standard could be used as an indication for the concentrations of each compound. For the detection of the target compounds, about 50 mg of ground sprout samples were mixed with acetone/water (4:1). The HPLC system (SCL-10 AVP, Japan) was provided with a Lichrosorb Si-60, 7 μm, 3 × 150 mm column, and DOD detector. The mobile phase was a mixture of water/formic acid (90:10), as well as acetonitrile/water/formic acid (85:10:5) at a flow rate of 0.8 mL/min. Meanwhile, the internal standard was 3,5-dichloro-4-hydroxybenzoic.

3.1.2. Extraction of DNA

To molecularly identify the bioactive isolate, DNA from the isolates was extracted according to [59]. High quality extracted DNA was used for amplification of the 16S rRNA gene (primers were 1498R (5-ACGGCTACCTTGTTACGACTT-3) and 27F (5-GAGTTTGATC CTGGCTCA-3)) [60,61] Invitrogen (Waltham, MA, USA). The samples were prepared with 50 ng/µL of a PCR product then delivered to the company MacroGen in Korea (http://www.dna.macrogen.com, accessed on 12 May 2021) for nucleotide sequencing.

The evolutionary history was inferred by using the maximum likelihood method and the Tamura 3-parameter model [62]. We generated the tree with the highest log likelihood (-13039.1216). After obtaining the initial tree for the heuristic search (neighbor joining and BioNJ algorithms), a matrix of pairwise distances was estimated (maximum composite likelihood approach), and finally the topology with superior log likelihood value was selected. There was a total of 683 positions in the final dataset. Evolutionary analysis was conducted in MEGA6 [63]. BLAST (http://www.ncbi.nlm.nih.gov/BLAST, accessed on 28 September 2021) was applied for sequence analysis, and then MEGAX software was used for cluster analysis.

3.1.3. Seed Inoculation with Bacterial Endophyte and Growth of Chenopodium Sprouts

The *Chenopodium* seeds (*C. ambrosoides*, *C. ficifolium*, and *C. botrys*) were brought from the Agricultural Research Centre, Giza, Egypt. They were surface sterilized (2.5% sodium hypochlorite, 5 min) and rinsed thoroughly in sterile distilled water. Afterwards, sterilized *Chenopodium* seeds were soaked in a liquid suspension of inoculants at 25% concentrations (2.5×10^{-7} cfu.mL^{-1}) for 6 h at room temperature, and the control was soaked in distilled water. The control and treated *Chenopodium* seeds were transferred to the sterilized trays filled with vermiculite and irrigated every day. The size of each tray was (8 cm × 12 cm × 4 cm), where sprouts of each treatment were grown in different trays. The *Chenopodium* sprouts were cultured in growth chambers at 24 °C, photosynthetically active radiation (PAR) of 200 µmol m^{-2} s^{-1}, 16 h light/8 h dark cycle, and 60% relative humidity. After eight days of cultivation, 30 sprouts from each tray of 8 × 12 × 4 cm in size (a biological replicate) were harvested and weighed to determine the fresh sprout weight. Finally, the sprouts were frozen in liquid nitrogen at −196 °C and stored for further analyses. At least four biological replicates were used for evaluation of chemical and biological attributes. The percentage of water content of the sprouts was estimated by drying fresh sprouts in an oven at 70 °C for 3 days. After determination of the dried weight of the sprouts, the water content was measured as a percentage. The fresh sprout weight (FW) was expressed as biomass (g/sprout).

3.2. Determination of Photosynthetic Rate

An EGM-4 infrared gas analyzer (PP Systems, Hitchin, UK) was used for determination of photosynthetic rate, from 180 s measurements of net CO_2 exchange (NE).

3.3. Pigment Analysis

The homogenization of sprouts (about 200 mg) was carried out in acetone (7000 rpm, 1 min) by using a MagNALyser (Roche, Vilvoorde, Belgium). Then they were centrifuged (14,000× *g*, 4 °C, 20 min). The supernatant was filtered and analyzed by using HPLC (Shimadzu SIL10-ADvp, reversed-phase, at 4 °C) [3]. Carotenoids were separated on a silica-based C18 column, and two solvents were used: acetonitrile/methanol/water (81:9:10) and methanol/ethyl acetate (68:32). In addition, chlorophyll a and b and β-carotene were extracted and detected at 4 wavelengths (420, 440, 462, and 660 nm) by using a diode-array detector (Shimadzu SPDM10Avp).

3.4. Determination of the Nutritional Value

In order to give insight into the functional food value of *Chenopodium* sprouts, the levels of amino, organic, and fatty acids, as well as minerals and vitamins profiles were evaluated as described below.

3.4.1. Proximate Composition Analysis

Carbohydrate levels were evaluated following the method of [64] from each *Chenopodium* sprout treated and nontreated with bacterial endophytes. The protein concentration was also measured for each sprout sample (0.2 g FW) [65]. Total lipid concentrations were assessed, where the sprout samples were homogenized in a 2:1 mixture of chloroform/methanol (*v/v*) [66]. Then sprouts were centrifuged for 15 min at 3000× *g*. The pellets were redissolved in a 4:1 mixture of toluene/ethanol (*v/v*). After concentration, the total lipid content was calculated. The extracted lipids were determined by gravimetric analysis and expressed as weight (g) per fresh weight (g) of sprout. Crude fibers were extracted, where sprouts were gelatinized (heat-stable alpha-amylase, pH 6, 100 °C for 25 min), and then enzymatically digested (protease: pH 7.5, 60 °C, 25 min and amyloglucosidase: pH 6, 0 °C, 30 min) to remove undesirable protein and starch [67]. Fibers were precipitated in ethanol for washing, and then after washing, the residues were weighed.

3.4.2. Elemental Analysis

The detection of mineral elements was performed according to [68], whereas 200 mg from bacterial endophytic-treated and control sprouts were digested in 5:1 (*v:v*) HNO_3/H_2O solution. Thereafter, macro and microelements were evaluated (inductively coupled plasma mass spectrometry, ICP-MS, Finnigan Element XR, and Scientific, Bremen, Germany). Nitric acid (1%) was used as blank.

3.4.3. Amino Acids Levels and Metabolism

The method described in [69] was used, whereas 100 mg of each plant was homogenized in 5 mL of 80% ethanol at 5000 rpm for 1 min. After centrifugation (14,000× *g* for 25 min), the supernatant was resuspended in 5 mL chloroform. Thereafter, 1 mL of H_2O was used for the residue extraction. The supernatant and pellet were resuspended in chloroform and centrifuged (8000× *g*, 10 min). Finally, the amino acids were quantified (Waters Acquity UPLC TQD device coupled to a BEH amide column), the elution (A, 84% ammonium formate, 6% formic acid, and 10% acetonitrile, *v/v*), and (B, acetonitrile and 2% formic acid, *v/v*).

Following the protocol of [4], glutamine synthase (GS) activity was determined, and the extraction was performed in (100 mg mL^{-1} Tris-HCl (50 mM), pH 7.4, 2% polyvinylpyrrolidone, 4 mM DTT, $MgC1_2$ (10 mM), EDTA (1 mM), 10% glycerol, and 2 mM PMSF). Then, the GS activity was evaluated in a Tris-acetate reaction buffer (Tris-acetate, 200 mM, pH 6.4), as evidenced by the production of γ-glutamyl hydroxamate. Dihydrodipicolinate synthase (DHDPS) activity was performed according to [70]. Tested sprouts without L-aspartate-b-semialdehyde were used as negative controls. The reaction was performed at 36.5 °C to allow the adduct formation between the reaction product and o-ABA. The reaction was stopped by addition of 12% of trichloroacetic acid (TCA). After dark, incubation for 60 min, samples were measured at 550 nm. Cystathionine γ- synthase (CGS) was extracted in 20 mM MOPS for 15 min at 4 °C, and the supernatants were mixed with a reaction buffer containing L-cysteine (2 mM), PLP (100 μM), AVG (200 μM), and O-phospho-homoserine (5 mM). L-cystathionine formation was separated on a phenomenex Hyperclone C18 BDS column (Dionex HPLC system) [71].

3.4.4. Organic Acid Analysis

The detection of organic acids was conducted in sprouts samples (200 mg) by using HPLC (0.001 N sulfuric acid, at 210 nm, 0.6 mL min^{-1}) [72]. A liquid chromatographer (Dionex, Sunnyvale, CA, USA), and a LED model detector (Ultimate 3000) were used for detection. In cooperation, an EWPS-3000SI autosampler, a TCC-3000SD column thermostat, and an LPG-3400A pump were included into the system. Chromeleon v.6.8 computer software was also applied. Then, separation was performed on an Aminex HPH-87 H (300 × 7.8 mm) column with IG Cation H (30 × 4.6) precolumn of Bio-Red firm at a temperature of 65 °C.

3.4.5. Fatty Acids Analysis

The detection of fatty acids using GC/MS (Hewlett Packard, Palo Alto, CA, USA) equipped with an HP-5 MS column (30 m × 0.25 mm × 0.25 mm), where 200 mg were taken from the sprout samples for extraction [73]. The database NIST 05 and Golm Metabolome were applied (http://gmd.mpimp-golm.mpg.de, accessed on 8 December 2021).

3.4.6. Vitamin Level Analysis

The contents of thiamine and riboflavin were determined in sprouts (about 200 mg fresh samples) by using UV and/or fluorescence detectors [3]. Separation was performed on a reverse-phase (C18) column (HPLC, methanol/water). Vitamin C (ascorbate) was determined by HPLC analysis (Shimadzu, Hertogenbosch, The Netherlands). Sprouts tissues were extracted in 1 mL of ice-cold 6% (w/v) meta-phosphoric acid, and antioxidants were separated on a reversed phase HPLC column [74]. Vitamin E (tocopherols) was extracted with hexane. The dried extract was resuspended in hexane, and tocopherols were separated and quantified by HPLC (Shimadzu, Hertogenbosch, The Netherlands) (normal phase conditions, Particil Pac 5 μm column material, length 250 mm, i.d. 4.6 mm) [74].

3.4.7. Determination of Phenolics Levels and Their Biosynthetic Enzyme Activity

The total phenolic and flavonoid contents were evaluated by homogenizing (120 mg of sprouts in 80% ethanol). Centrifugation was performed at (4 °C, 20 min), and then the phenolic content was determined by using a Folin–Ciocalteu assay, where gallic acid was used as a standard. The flavonoid content was evaluated following the modified aluminum chloride colorimetric method, where quercetin was applied as a standard [3].

Phenylalanine ammonia-lyase (PAL) was extracted from 0.25 g (FW) frozen plant material in 1 mL sodium borate buffer (200 mM, pH 8.8) and assayed in a Tris-HCl (100 mM, pH 8.8) reaction buffer containing L-phenylalanine (40 mM) by measuring the absorbance of the produced transcinnamic acid at 290 nm [68]. Samples sprouts were replaced with water to serve as a negative control.

3.4.8. Phosphate Solubilization

The phosphate (PO_4) solubilization was measured by incubating the samples into a medium (100 mL) containing tri-Ca-PO_4 for a week at 30 °C. The pH was recorded (pH meter, Germany). The produced soluble PO_4 was spectrophotometrically measured [75].

3.5. Biological Activities

3.5.1. Antioxidant Activities

The antioxidant capacities of the sprouts were evaluated by using different assays [3]. For determination of ferric reducing antioxidant power (FRAP), about 0.1 g was extracted in 80% ethanol. Then, centrifugation took place (14,000 rpm, 20 min). Afterwards, about 0.1 mL extract was mixed with FRAP reagent (20 mM $FeCl_3$ in 0.25 M acetate buffer). For determination of the 2,2′-azino-bis(3-ethylbenzothiazoline-6-sulfonic acid) (ABTS), the ABTS was mixed with 2.4 mM potassium persulphate. The absorbance was detected at 734 nm, while detection of DPPH activity was performed by using 0.1 mL of the extract and 0.25 mL of the DPPH reagent. The absorbance was detected at 517 nm.

3.5.2. Antibacterial Activity

The standard dilution method was used for measuring the antibacterial activity of the sprouts [4]. The sprout extract (100 mg) was mixed with 1 mL dimethylosulfoxide

(DMSO). Then, 0.1 mL of liquid culture of standard strain (*Staphylococcus aureus* ATCC 6538 P) was diluted 1:10.000 in the same medium (number of inoculums contained 104–105 bacterial cells in 1 mL) and was added to the media. Afterwards, the samples were incubated at 37 °C for 18 h. The MIC (Minimal Inhibitory Concentration) of the tested sprouts was detected, and the antibacterial activity was tested against *Pseudomonas aeruginosa* (ATCC10145), *Candida glabrata* (ATCC90030), *Proteus vulgaris* (ATCC8427), *Enterobacter aerogenes* (ATCC 13048), *Staphylococcus saprophyticus* (ATCC 19701), *E. coli* (ATCC 29998), *Salmonella typhimurium* (ATCC14028), *S. epidermidis* (ATCC 12228), *Candida albicans* (ATCC90028), *Salmonella typhimurium* (ATCC14028), *Enterococcus faecalis* (ATCC 10541), *Streptococcus salivarius* (ATCC25975), *Aspergillus flavus* (ATCC9170), and *Serratia marcescens* (ATCC99006). Antibiotics, such as ciprofloxacin at 25 mg/mL and 100% DMSO, were used as positive and negative controls, respectively.

3.5.3. Determination of Lipoxygenase (LOX) and Cyclooxygenase (COX) Activities

Sprouts tissue (1.5 g) from each of the *Chenopodium* species investigated was extracted with 80% ethanol (10 mL). After shaking for 60 min, samples were centrifuged at 2500 rpm for 15 min, and the supernatants were filtered (Whatman No.1). After drying, the extract was reconstituted in 100% dimethyl sulphoxide (DMSO) (Merck Schuchardt OHG) at 10 mg/mL and tested in the assays.

Lipoxygenase activity was evaluated by using linoleic acid as a substrate and LOX as an enzyme [4]. About 10 µL of the sprout extract was mixed with 90 µL LOX (400 U/mL). Then, the mixture was incubated at 25 °C for 5 min. Afterwards, 100 µL of linoleic acid (0.4 mM) was added, and then incubation was performed again for 20 min at 25 °C. Thereafter, about 100 µL of ferrous orange xylenol reagent in 10 µM $FeSO_4$, 30 mM H_2SO_4, 90% methanol, and 100 µM xylenol orange were added. The absorbance was detected at 560 nm, and the percentage of inhibition was evaluated.

Cyclooxygenase-2 activity inhibition was detected by applying the manufacturer's instructions for the COX assay kit (No. 560131; Cayman Chemical, Ann Arbor, MI, USA). Incubation was performed for 90 min at 25 °C. The absorbance was detected at 420 nm, and the percentage of inhibition was measured [4].

3.6. Statistical Analyses

Statistical analyses were determined by using the SPSS package (SPSS Inc., Chicago, IL, USA). A one-way analysis of variance (ANOVA) was used for all data per species. Tukey's multiple range test ($p < 0.05$) was performed as the post hoc test for mean separations. Each experiment was performed in three replicates ($n = 3$). All parameters were subjected to Pearson's distance metric cluster analysis by using the MultiExperiment Viewer (MeV) TM4 software package (Dana-Farber Cancer Institute, Boston, MA, USA).

4. Conclusions

Based on our obtained results, a selected bioactive endophyte *Streptomyces* (strain JSA11) could represent a promising approach to enhance the growth and yield of the tested *Chenopodium* sprouts and to increase their contents of bioactive metabolites, such as amino acids, organic acids, vitamins, and minerals. At the ecological level, endophytic bacteria could represent an ecofriendly approach and possibly an alternative to chemical fertilizers, with efficient plant growth promoting activity. Meanwhile, at the economic level, endophytic bacterial treatment and also sprouting are considered low-cost techniques, particularly when combined together to ensure higher plant productivity as well as increased levels of phytochemicals and biological activities.

Author Contributions: M.S.A. and S.S. planned and designed the research; M.S.A., M.H.A., S.M.A. and M.A.-M. performed the experiments; M.S.A., S.K.A.J., H.A. and M.A.-M. analyzed the data; M.S.A., M.H.A. and S.S. contributed to the reagents /chemicals. M.S.A. and M.A.-M. provided a draft version of the manuscript, and H.A. revised and finalized the manuscript. All authors have read and agreed to the published version of the manuscript.

Funding: The authors extend their appreciation to the Deputyship for Research & Innovation, Ministry of Education in Saudi Arabia for funding this research work through the project number IFPHI-211-140-2020 and King Abdulaziz University, DSR, Jeddah, Saudi Arabia.

Institutional Review Board Statement: Not applicable.

Informed Consent Statement: Not applicable.

Data Availability Statement: Data presented in this study are available on reasonable request.

Conflicts of Interest: The authors declare that they have no known competing financial interests or personal relationships that could have appeared to influence the work reported in this paper.

References

1. Manchali, S.; Murthy, K.N.C.; Patil, B.S. Crucial facts about health benefits of popular cruciferous vegetables. *J. Funct. Foods* **2012**, *4*, 94–106.
2. Marton, M.; Mandoki, Z.S.; Csapo-Kiss, Z.S.; Csapo, J. The role of sprouts in human nutrition. A review. *Acta Univ. Sapientiae* **2010**, *3*, 81–117.
3. Almuhayawi, M.S.; Hassan, A.H.A.; Abdel-Mawgoud, M.; Khamis, G.; Selim, S.; Al Jaouni, S.K.; AbdElgawad, H. Laser light as a promising approach to improve the nutritional value, antioxidant capacity and anti-inflammatory activity of flavonoid-rich buckwheat sprouts. *Food Chem.* **2020**, *345*, 128788. [CrossRef]
4. Almuhayawi, M.S.; Al Jaouni, S.K.; Almuhayawi, S.M.; Selim, S.; Abdel-Mawgoud, M. Elevated CO_2 improves the nutritive value, antibacterial, anti-inflammatory, antioxidant and hypocholestecolemic activities of lemongrass sprouts. *Food Chem.* **2021**, *357*, 129730. [CrossRef]
5. Santoyo, G.; Moreno-Hagelsieb, G.; Del Carmen Orozco-Mosqueda, M.; Glick, B.R. Plant growth-promoting bacterial endophytes. *Microbiol. Res.* **2016**, *183*, 92–99. [CrossRef] [PubMed]
6. Lodewyckx, C.; Vangronsveld, J.; Porteous, F.; Moore, E.R.B.; Taghavi, S.; Mezgeay, M.; Der Lelie, D.V. Endophytic bacteria and their potential applications. *CRC Crit. Rev. Plant Sci.* **2002**, *21*, 583–606. [CrossRef]
7. Chaturvedi, H.; Singh, V.; Gupta, G. Potential of bacterial endophytes as plant growth promoting factors. *J. Plant Pathol. Microbiol.* **2016**, *7*, 9. [CrossRef]
8. Jasim, B.; Joseph, A.A.; John, C.J.; Mathew, J.; Radhakrishnan, E.K. Isolation and characterization of plant growth promoting endophytic bacteria from the rhizome of Zingiber officinale. *3 Biotech* **2014**, *4*, 197–204. [CrossRef] [PubMed]
9. Brader, G.; Compant, S.; Mitter, B.; Trognitz, F.; Sessitsch, A. Metabolic potential of endophytic bacteria. *Curr. Opin. Biotechnol.* **2014**, *27*, 30–37. [CrossRef] [PubMed]
10. Aeron, A.; Maheshwari, D.K.; Meena, V.S. Endophytic bacteria promote growth of the medicinal legume *Clitoria ternatea* L. by chemotactic activity. *Arch. Microbiol.* **2020**, *202*, 1049–1058. [CrossRef] [PubMed]
11. Zhou, J.; Sun, K.; Chen, F.; Yuan, J.; Li, X.; Dai, C. Endophytic Pseudomonas induces metabolic flux changes that enhance medicinal sesquiterpenoid accumulation in Atractylodes lancea. *Plant Physiol. Biochem.* **2018**, *130*, 473–481. [CrossRef]
12. Xu, M.; Bai, H.-Y.; Fu, W.-Q.; Sun, K.; Wang, H.-W.; Xu, D.-L.; Dai, C.-C.; Jia, Y. Endophytic bacteria promote the quality of Lyophyllum decastes by improving non-volatile taste components of mycelia. *Food Chem.* **2021**, *336*, 127672. [CrossRef] [PubMed]
13. Suárez-Estrella, D.; Bresciani, A.; Iametti, S.; Marengo, M.; Pagani, M.A.; Marti, A. Effect of sprouting on proteins and starch in quinoa (*Chenopodium quinoa* Willd.). *Plant Foods Hum. Nutr.* **2020**, *75*, 635–641. [CrossRef]
14. Paśko, P.; Sajewicz, M.; Gorinstein, S.; Zachwieja, Z. Analysis of selected phenolic acids and flavonoids in Amaranthus cruentus and Chenopodium quinoa seeds and sprouts by HPLC. *Acta Chromatogr.* **2008**, *20*, 661–672. [CrossRef]
15. Chao, Y.-Y.; Wang, W.-J.; Liu, Y.-T. Effect of Calcium on the Growth of Djulis (*Chenopodium formosanum* Koidz.) Sprouts. *Agronomy* **2021**, *11*, 82. [CrossRef]
16. Kumar, R.; Mishra, A.K.; Dubey, N.K.; Tripathi, Y.B. Evaluation of *Chenopodium ambrosioides* oil as a potential source of antifungal, antiaflatoxigenic and antioxidant activity. *Int. J. Food Microbiol.* **2007**, *115*, 159–164. [CrossRef]
17. Kokanova-Nedialkova, Z.; Nedialkov, P.; Nikolov, S. The genus Chenopodium: Phytochemistry, ethnopharmacology and pharmacology. *Pharmacogn. Rev.* **2009**, *3*, 280.
18. Ullah, F.; Iqbal, N.; Ayaz, M.; Sadiq, A.; Ullah, I.; Ahmad, S.; Imran, M. DPPH, ABTS free radical scavenging, antibacterial and phytochemical evaluation of crude methanolic extract and subsequent fractions of Chenopodium botrys aerial parts. *Pak. J. Pharm. Sci.* **2017**, *30*, 761–766.
19. Sera, B.; Stranák, V.; Serý, M.; Tichý, M.; Spatenka, P. Germination of Chenopodium album in response to microwave plasma treatment. *Plasma Sci. Technol.* **2008**, *10*, 506. [CrossRef]
20. Khan, A.L.; Waqas, M.; Kang, S.-M.; Al-Harrasi, A.; Hussain, J.; Al-Rawahi, A.; Al-Khiziri, S.; Ullah, I.; Ali, L.; Jung, H.-Y. Bacterial endophyte Sphingomonas sp. LK11 produces gibberellins and IAA and promotes tomato plant growth. *J. Microbiol.* **2014**, *52*, 689–695. [CrossRef] [PubMed]
21. Almuhayawi, M.S.; Mohamed, M.S.M.; Abdel-Mawgoud, M.; Selim, S.; Al Jaouni, S.K.; AbdElgawad, H. Bioactive Potential of Several Actinobacteria Isolated from Microbiologically Barely Explored Desert Habitat, Saudi Arabia. *Biology* **2021**, *10*, 235. [CrossRef]

22. Manfio, G.P. Towards minimal standards for the description of Streptomyces species. *Biotekhnologiya* **1995**, *8*, 228–237.

23. Boroujeni, M.E.; Arijit, D.; Prashanthi, K.; Sandeep, S.; Sourav, B. Enzymatic screening and random amplified polymorphic DNA fingerprinting of soil streptomycetes isolated from Wayanad district in Kerala, India. *J. Biol. Sci.* **2012**, *12*, 43–50.

24. Ser, H.-L.; Law, J.W.-F.; Chaiyakunapruk, N.; Jacob, S.A.; Palanisamy, U.D.; Chan, K.-G.; Goh, B.-H.; Lee, L.-H. Fermentation conditions that affect clavulanic acid production in Streptomyces clavuligerus: A systematic review. *Front. Microbiol.* **2016**, *7*, 522. [CrossRef]

25. Briatia, X.; Azad, M.O.K.; Khanongnuch, C.; Woo, S.H.; Park, C.H. Effect of Endophytic Bacterium Inoculation on Total Polyphenol and Flavonoid Contents of Tartary Buckwheat Sprouts. *Korean J. Crop Sci.* **2018**, *63*, 57–63.

26. Mandal, S.M.; Chakraborty, D.; Dey, S. Phenolic acids act as signaling molecules in plant-microbe symbioses. *Plant Signal. Behav.* **2010**, *5*, 359–368. [CrossRef]

27. Abdelmohsen, U.R.; Grkovic, T.; Balasubramanian, S.; Kamel, M.S.; Quinn, R.J.; Hentschel, U. Elicitation of secondary metabolism in actinomycetes. *Biotechnol. Adv.* **2015**, *33*, 798–811. [CrossRef]

28. Bianco, C.; Imperlini, E.; Calogero, R.; Senatore, B.; Pucci, P.; Defez, R. Indole-3-acetic acid regulates the central metabolic pathways in Escherichia coli. *Microbiology* **2006**, *152*, 2421–2431. [CrossRef]

29. Ryan, R.P.; Monchy, S.; Cardinale, M.; Taghavi, S.; Crossman, L.; Avison, M.B.; Berg, G.; Van Der Lelie, D.; Dow, J.M. The versatility and adaptation of bacteria from the genus Stenotrophomonas. *Nat. Rev. Microbiol.* **2009**, *7*, 514–525. [CrossRef] [PubMed]

30. Shi, Y.; Lou, K.; Li, C. Growth and photosynthetic efficiency promotion of sugar beet (*Beta vulgaris* L.) by endophytic bacteria. *Photosynth. Res.* **2010**, *105*, 5–13. [CrossRef]

31. Rho, H.; Kim, S.-H. Endophyte effects on photosynthesis and water use of plant hosts: A meta-analysis. In *Functional Importance of the Plant Microbiome*; Springer: Cham, Switzerland, 2017; pp. 43–69.

32. Duca, D.; Lorv, J.; Patten, C.L.; Rose, D.; Glick, B.R. Indole-3-acetic acid in plant–microbe interactions. *Antonie Van Leeuwenhoek* **2014**, *106*, 85–125. [CrossRef]

33. Saleem, M.; Asghar, H.N.; Zahir, Z.A.; Shahid, M. Impact of lead tolerant plant growth promoting rhizobacteria on growth, physiology, antioxidant activities, yield and lead content in sunflower in lead contaminated soil. *Chemosphere* **2018**, *195*, 606–614. [CrossRef]

34. Kang, S.-M.; Radhakrishnan, R.; You, Y.-H.; Joo, G.-J.; Lee, I.-J.; Lee, K.-E.; Kim, J.-H. Phosphate solubilizing Bacillus megaterium mj1212 regulates endogenous plant carbohydrates and amino acids contents to promote mustard plant growth. *Indian J. Microbiol.* **2014**, *54*, 427–433. [CrossRef]

35. Stefan, M.; Munteanu, N.; Stoleru, V.; Mihasan, M. Effects of inoculation with plant growth promoting rhizobacteria on photosynthesis, antioxidant status and yield of runner bean. *Rom. Biotechnol. Lett.* **2013**, *18*, 8132–8143.

36. Lemoine, R.; La Camera, S.; Atanassova, R.; Dédaldéchamp, F.; Allario, T.; Pourtau, N.; Bonnemain, J.-L.; Laloi, M.; Coutos-Thévenot, P.; Maurousset, L. Source-to-sink transport of sugar and regulation by environmental factors. *Front. Plant Sci.* **2013**, *4*, 272. [CrossRef] [PubMed]

37. Briatia, X.; Jomduang, S.; Park, C.H.; Lumyong, S.; Kanpiengjai, A.; Khanongnuch, C. Enhancing growth of buckwheat sprouts and microgreens by endophytic bacterium inoculation. *Int. J. Agric. Biol.* **2017**, *19*, 374–380. [CrossRef]

38. Cipriano, M.A.P.; Freitas-Iório, R.D.P.; Dimitrov, M.R.; De Andrade, S.A.L.; Kuramae, E.E.; Silveira, A.P.D.D. Plant-Growth Endophytic Bacteria Improve Nutrient Use Efficiency and Modulate Foliar N-Metabolites in Sugarcane Seedling. *Microorganisms* **2021**, *9*, 479. [CrossRef] [PubMed]

39. Ahemad, M.; Kibret, M. Mechanisms and applications of plant growth promoting rhizobacteria: Current perspective. *J. King Saud Univ.* **2014**, *26*, 1–20. [CrossRef]

40. Taurian, T.; Anzuay, M.S.; Angelini, J.G.; Tonelli, M.L.; Ludueña, L.; Pena, D.; Ibáñez, F.; Fabra, A. Phosphate-solubilizing peanut associated bacteria: Screening for plant growth-promoting activities. *Plant Soil* **2010**, *329*, 421–431. [CrossRef]

41. Ren, X.-M.; Guo, S.-J.; Tian, W.; Chen, Y.; Han, H.; Chen, E.; Li, B.-L.; Li, Y.-Y.; Chen, Z.-J. Effects of plant growth-promoting bacteria (PGPB) inoculation on the growth, antioxidant activity, Cu uptake, and bacterial community structure of rape (*Brassica napus* L.) grown in Cu-contaminated agricultural soil. *Front. Microbiol.* **2019**, *10*, 1455. [CrossRef]

42. Andov, L.A.; Karapandzova, M.; Stefkov, G.; Cvetkovikj, I.; Baceva, K.; Stafilov, T.; Kulevanova, S. The content of some biogenic elements in *Chenopodium album* L. and *Chenopodium botrys* L.(Amaranthaceae) from Macedonian flora. *Maced. Pharm. Bull.* **2016**, *62*, 499–500.

43. Reyes-Becerril, M.; Angulo, C.; Sanchez, V.; Vázquez-Martínez, J.; López, M.G. Antioxidant, intestinal immune status and anti-inflammatory potential of *Chenopodium ambrosioides* L. in fish: In vitro and in vivo studies. *Fish Shellfish Immunol.* **2019**, *86*, 420–428. [CrossRef]

44. Aborisade, A.B.; Adetutu, A.; Owoade, A.O. Phytochemical and proximate analysis of some medicinal leaves. *Clin. Med. Res.* **2017**, *6*, 209. [CrossRef]

45. Whiteside, M.D.; Garcia, M.O.; Treseder, K.K. Amino acid uptake in arbuscular mycorrhizal plants. *PLoS ONE* **2012**, *7*, e47643. [CrossRef]

46. Yan, Y.; Zhang, S.; Zhang, J.; Ma, P.; Duan, J.; Liang, Z. Effect and mechanism of endophytic bacteria on growth and secondary metabolite synthesis in Salvia miltiorrhiza hairy roots. *Acta Physiol. Plant.* **2014**, *36*, 1095–1105. [CrossRef]

47. Barros, L.; Pereira, E.; Calhelha, R.C.; Dueñas, M.; Carvalho, A.M.; Santos-Buelga, C.; Ferreira, I.C.F.R. Bioactivity and chemical characterization in hydrophilic and lipophilic compounds of *Chenopodium ambrosioides* L. *J. Funct. Foods* **2013**, *5*, 1732–1740. [CrossRef]

48. Saleh, A.M.; Abdel-Mawgoud, M.; Hassan, A.R.; Habeeb, T.H.; Yehia, R.S.; AbdElgawad, H. Global metabolic changes induced by arbuscular mycorrhizal fungi in oregano plants grown under ambient and elevated levels of atmospheric CO_2. *Plant Physiol. Biochem.* **2020**, *151*, 255–263. [CrossRef] [PubMed]

49. Ajaib, M.; Hussain, T.; Farooq, S.; Ashiq, M. Analysis of antimicrobial and antioxidant activities of *Chenopodium ambrosioides*: An ethnomedicinal plant. *J. Chem.* **2016**, *2016*, 4827157. [CrossRef]

50. Ozer, M.S.; Sarikurkcu, C.; Tepe, B. Phenolic composition, antioxidant and enzyme inhibitory activities of ethanol and water extracts of *Chenopodium botrys*. *RSC Adv.* **2016**, *6*, 64986–64992. [CrossRef]

51. Chen, L.; Xu, M.; Zheng, Y.; Men, Y.; Sheng, J.; Shen, L. Growth promotion and induction of antioxidant system of tomato seedlings (*Solanum lycopersicum* L.) by endophyte TPs-04 under low night temperature. *Sci. Hortic.* **2014**, *176*, 143–150. [CrossRef]

52. Brahim, M.A.S.; Fadli, M.; Hassani, L.; Boulay, B.; Markouk, M.; Bekkouche, K.; Abbad, A.; Ali, M.A.; Larhsini, M. *Chenopodium ambrosioides* var. ambrosioides used in Moroccan traditional medicine can enhance the antimicrobial activity of conventional antibiotics. *Ind. Crops Prod.* **2015**, *71*, 37–43. [CrossRef]

53. Jesus, R.S.; Piana, M.; Freitas, R.B.; Brum, T.F.; Alves, C.F.S.; Belke, B.V.; Mossmann, N.J.; Cruz, R.C.; Santos, R.C.V.; Dalmolin, T.V. In vitro antimicrobial and antimycobacterial activity and HPLC–DAD screening of phenolics from *Chenopodium ambrosioides* L. *Braz. J. Microbiol.* **2018**, *49*, 296–302. [CrossRef]

54. TrivellatoGrassi, L.; Malheiros, A.; Meyre-Silva, C.; Da Silva Buss, Z.; Monguilhott, E.D.; Fröde, T.S.; Da Silva, K.A.B.S.; De Souza, M.M. From popular use to pharmacological validation: A study of the anti-inflammatory, anti-nociceptive and healing effects of *Chenopodium ambrosioides* extract. *J. Ethnopharmacol.* **2013**, *145*, 127–138. [CrossRef] [PubMed]

55. Taechowisan, T.; Peberdy, J.F.; Lumyong, S. Isolation of endophytic actinomycetes from selected plants and their antifungal activity. *World J. Microbiol. Biotechnol.* **2003**, *19*, 381–385. [CrossRef]

56. Gordon, S.A.; Weber, R.P. Colorimetric estimation of indoleacetic acid. *Plant Physiol.* **1951**, *26*, 192. [CrossRef]

57. Schwyn, B.; Neilands, J.B. Universal chemical assay for the detection and determination of siderophores. *Anal. Biochem.* **1987**, *160*, 47–56. [CrossRef]

58. Williams, S.T. Genus Streptomyces waksman and henrici 1943. *Bwergey's Man. Syst. Bacteriol.* **1989**, *4*, 2452–2492.

59. Hong, K.; Gao, A.-H.; Xie, Q.-Y.; Gao, H.G.; Zhuang, L.; Lin, H.-P.; Yu, H.-P.; Li, J.; Yao, X.-S.; Goodfellow, M. Actinomycetes for marine drug discovery isolated from mangrove soils and plants in China. *Mar. Drugs* **2009**, *7*, 24–44. [CrossRef] [PubMed]

60. Stackebrandt, E.; Goodfellow, M. *Nucleic Acid Techniques in Bacterial Systematics*; Wiley: Hoboken, NJ, USA, 1991; ISBN 0471929069.

61. Reddy, G.S.; Aggarwal, R.K.; Matsumoto, G.I.; Shivaji, S. Arthrobacter flavus sp. nov., a psychrophilic bacterium isolated from a pond in McMurdo Dry Valley, Antarctica. *Int. J. Syst. Evol. Microbiol.* **2000**, *50*, 1553–1561. [CrossRef]

62. Tamura, K. Estimation of the number of nucleotide substitutions when there are strong transition-transversion and G+ C−content biases. *Mol. Biol. Evol.* **1992**, *9*, 678–687. [PubMed]

63. Tamura, K.; Stecher, G.; Peterson, D.; Filipski, A.; Kumar, S. MEGA6: Molecular evolutionary genetics analysis version 6.0. *Mol. Biol. Evol.* **2013**, *30*, 2725–2729. [CrossRef]

64. Wong, K.H.; Cheung, P.C.K. Nutritional evaluation of some subtropical red and green seaweeds: Part I—proximate composition, amino acid profiles and some physico-chemical properties. *Food Chem.* **2000**, *71*, 475–482. [CrossRef]

65. Lowry, O.H.; Rosebrough, N.J.; Farr, A.L.; Randall, R.J. Protein measurement with the Folin phenol reagent. *J. Biol. Chem.* **1951**, *193*, 265–275. [CrossRef]

66. Shiva, S.; Enninful, R.; Roth, M.R.; Tamura, P.; Jagadish, K.; Welti, R. An efficient modified method for plant leaf lipid extraction results in improved recovery of phosphatidic acid. *Plant Methods* **2018**, *14*, 14. [CrossRef] [PubMed]

67. Lee, S.C.; Prosky, L.; Vries, J.W.D. Determination of total, soluble, and insoluble dietary fiber in foods—Enzymatic-gravimetric method, MES-TRIS buffer: Collaborative study. *J. AOAC Int.* **1992**, *75*, 395–416. [CrossRef]

68. AbdElgawad, H.; Peshev, D.; Zinta, G.; Van den Ende, W.; Janssens, I.A.; Asard, H. Climate extreme effects on the chemical composition of temperate grassland species under ambient and elevated CO_2: A comparison of fructan and non-fructan accumulators. *PLoS ONE* **2014**, *9*, e92044.

69. Sinha, A.K.; Giblen, T.; AbdElgawad, H.; De Rop, M.; Asard, H.; Blust, R.; De Boeck, G. Regulation of amino acid metabolism as a defensive strategy in the brain of three freshwater teleosts in response to high environmental ammonia exposure. *Aquat. Toxicol.* **2013**, *130–131*, 86–96. [CrossRef] [PubMed]

70. Kumpaisal, R.; Hashimoto, T.; Yamada, Y. Purification and characterization of dihydrodipicolinate synthase from wheat suspension cultures. *Plant Physiol.* **1987**, *85*, 145–151. [CrossRef]

71. Ravanel, S.; Gakière, B.; Job, D.; Douce, R. Cystathionine γ-synthase from *Arabidopsis thaliana*: Purification and biochemical characterization of the recombinant enzyme overexpressed in Escherichia coli. *Biochem. J.* **1998**, *331*, 639–648. [CrossRef] [PubMed]

72. Hamad, I.; Abdelgawad, H.; Al Jaouni, S.; Zinta, G.; Asard, H.; Hassan, S.; Hegab, M.; Hagagy, N.; Selim, S. Metabolic analysis of various date palm fruit (*Phoenix dactylifera* L.) cultivars from Saudi Arabia to assess their nutritional quality. *Molecules* **2015**, *20*, 13620–13641. [CrossRef]

73. Habeeb, T.H.; Abdel-Mawgoud, M.; Yehia, R.S.; Khalil, A.M.A.; Saleh, A.M.; AbdElgawad, H. Interactive Impact of Arbuscular Mycorrhizal Fungi and Elevated CO_2 on Growth and Functional Food Value of Thymus vulgare. *J. Fungi* **2020**, *6*, 168. [CrossRef] [PubMed]
74. Farfan-Vignolo, E.R.; Asard, H. Effect of elevated CO_2 and temperature on the oxidative stress response to drought in *Lolium perenne* L. and *Medicago sativa* L. *Plant Physiol. Biochem.* **2012**, *59*, 55–62. [CrossRef] [PubMed]
75. King, E.J. The colorimetric determination of phosphorus. *Biochem. J.* **1932**, *26*, 292–297. [CrossRef] [PubMed]

Article

Impact of Sprouting under Potassium Nitrate Priming on Nitrogen Assimilation and Bioactivity of Three *Medicago* Species

Ahlem Zrig [1,*], **Ahmed Saleh** [2], **Foued Hamouda** [3], **Mohammad K. Okla** [4], **Wahidah H. Al-Qahtani** [5], **Yasmeen A. Alwasel** [4], **Abdulrahman Al-Hashimi** [4], **Momtaz Y. Hegab** [6], **Abdelrahim H. A. Hassan** [7] and **Hamada AbdElgawad** [8]

1   Faculty of Sciences of Gabès-City Erriadh, Zrig, Gabes 6072, Tunisia
2   Department of Botany, Faculty of Science, Cairo University, Giza 12613, Egypt; asaleh@sci.edu.eg
3   Research Unit in Enterprise and Decisions, Higher Institute of Management, Road Jilani Habib, Gabes 6002, Tunisia; foha2001@gmail.com
4   Botany and Microbiology Department, College of Science, King Saud University, Riyadh 11451, Saudi Arabia; okla103@yahoo.com (M.K.O.); yasmeen@ksu.edu.sa (Y.A.A.); aalhashimi@ksu.edu.sa (A.A.-H.)
5   Department of Food Sciences and Nutrition, College of Food and Agriculture Sciences, King Saud University, Riyadh 11451, Saudi Arabia; wahida@ksu.edu.sa
6   Resarch Institute of Medicinal and Aromatic Plants, Beni-Suef University, Beni-Suef 62511, Egypt; momtaz.hegab@science.bsu.edu.eg
7   Department of Food Safety and Technology, Faculty of Veterinary Medicine, Beni-Suef University, Beni-Suef 62511, Egypt; abdelrahim@vet.bsu.edu.eg
8   Integrated Molecular Plant Physiology Research, Department of Biology, University of Antwerp, 2020 Antwerp, Belgium; hamada.abdelgawad@uantwerpen.be
*   Correspondence: ahlem18zrig@yahoo.fr; Tel.: +216-97-901-249

Citation: Zrig, A.; Saleh, A.; Hamouda, F.; Okla, M.K.; Al-Qahtani, W.H.; Alwasel, Y.A.; Al-Hashimi, A.; Hegab, M.Y.; Hassan, A.H.A.; AbdElgawad, H. Impact of Sprouting under Potassium Nitrate Priming on Nitrogen Assimilation and Bioactivity of Three *Medicago* Species. *Plants* **2022**, *11*, 71. https://doi.org/10.3390/plants11010071

Academic Editors: Beatrice Falcinelli and Angelica Galieni

Received: 28 November 2021
Accepted: 23 December 2021
Published: 27 December 2021

Publisher's Note: MDPI stays neutral with regard to jurisdictional claims in published maps and institutional affiliations.

Abstract: Edible sprouts are rich in flavonoids and other polyphenols, as well as proteins, minerals, and vitamins. Increasing sprout consumption necessitates improving their quality, palatability, and bioactivity. The purpose of this study was to test how KNO_3 priming affects the sprouting process species on three *Medicago* species (*Medicago indicus*, *Medicago interexta*, and *Medicago polymorpha*) and their nutritional values. Targeted species of *Medicago* were primed with KNO_3, and the levels of different primary and secondary metabolites were determined. KNO_3 induced biomass accumulation in the sprouts of the three species, accompanied by an increased content of total mineral nutrients, pigments, vitamins, and essential amino acids. Besides, our results showed that KNO_3 enhanced the activity of nitrate reductase (NR), glutamate dehydrogenase (GDH), and glutamine synthetase (GS) enzymes, which are involved in the nitrogen metabolism and GOGAT cycle, which, in turn, increase the nitrogen and protein production. KNO_3 treatment improved the bioactive compound activities of *Medicago* sprouts by increasing total phenolic and flavonoid contents and enhancing the antioxidant and antidiabetic activities. Furthermore, species-specific responses toward KNO_3 priming were noticeable, where *Medicago interexta* showed the highest antioxidant and antidiabetic activities, followed by *Medicago polymorpha*. Overall, this study sheds the light on the physiological and biochemical bases of growth, metabolism, and tissue quality improvement impact of KNO_3 on *Medicago* sprouts.

Keywords: sprouts; *Medicago* species; priming; KNO_3; nitrogen assimilation; bioactivity

1. Introduction

One of the natural processing methods for increasing the nutritional value and health qualities of foods is the sprouting of seeds. This approach has been employed in Eastern countries for a long time [1]. Since the sprouts are eaten so early in their growth cycle, their nutrient content remains very high [2]. Sprouts are excellent sources of protein, vitamins, and minerals, as well as essential nutrients for promoting health, such as glucosinolates and phenolic components [3]. Besides, the phytochemicals, enzymes, and amino acid contents in sprouts are very beneficial for human health [2]. On the other hand, some public concerns stem from the danger of bacterial contamination (e.g., *E. coli, Salmonella enterica*, and *Vibrio cholerae*) of sprouts because they are typically prepared at home and served as salad ingredients without any thermal or other sanitary treatment [4]. Different seeds may be sprouted for human consumption, including legume seeds (bean, pea, lentil, soybean), grains (rye, wheat, barley, oats), and, more recently, seeds of certain vegetables (alfalfa, radish). Due to consumer demand for minimally processed, additive-free, more sustainable, nutritional, and balanced foods, sprouting of seeds is gaining popularity in all countries [4].

Medicago genus is one of the first plants to be cultivated. It literally means "Father of All Foods." It is also known as "the queen of forages" since it has been employed in the food business as a cheap source of protein, particularly as a fodder plant. *Medicago* sprouts contain high amounts of vitamins A and C, coumestrol, liquiritigenin, isoliquiritigenin, loliolide, and saponins [4]. Furthermore, the high content of protein, and bioactive compounds found in both the aerial and root sections of the lucerne plant, have sparked a lot of interest in its cultivation [5]. Because of their high level of bioactive phytochemicals, such as phenolic components, saponins (hederagenin and soyasapogenol), and essential amino acids, lucerne formulations have antifungal, antibacterial, insecticidal, and nematicidal characteristics (valine, leucine, threonine, and lysine). In addition, *Medicago* species can be used as an effective functional ingredient in the dietary prevention and treatment of several metabolic conditions, particularly the metabolic syndrome, due to its high content of proteins, minerals, isoflavones, and other substances with estrogenic activity, anti-inflammatory properties, and antioxidant activity [6]. In this regard, *Medicago* sprout seed sprouting is one of the processing ways for increasing the nutritional content of this leguminous. Human consumption of lucerne is generally modest; however, there has been a growing interest in using this plant as green salad sprouts, pills, or drinks for their influence on blood cholesterol in various nations [7].

Potassium (K) is an essential plant nutrient that regulates a variety of metabolic activities, including protein synthesis and glucose metabolism [8]. Exogenous application of K fertilizers, such as potassium nitrate (KNO_3), monopotassium phosphate (KH_2PO_4), and potassium sulfate (K_2SO_4), has been shown to improve nutrient uptake, plant growth, and photosynthesis, as well as mitigate the abiotic stress [9]. The same as K, nitrogen (N) is an essential element that has a direct impact on plant development and physiological processes [10]. N assimilation is an important physiological process that influences plant productivity and quality [10]. Nitrate and ammonium are the most important sources of nitrogen for plant growth and development, with nitrate being more important [11]. Nitrate reductase (NR) and nitrite reductase (NIR) are enzymes that catalyze the conversion of nitrate to nitrite and then nitrite to ammonium [12]. In a cyclic manner, glutamine synthetase (GS, EC 6.3.1.2) and glutamate synthase (EC 1.4.7.1) or glutamine 2-oxoglutarate aminotransferase (GOGAT, EC 1.4.7.1) absorb ammonium to create distinct amino acids [13]. Amino acids, non-pharmacological and non-toxic universal nutrients, are the primary building blocks for protein synthesis in cells [14]. Finally, proteins are classified, modified, transported, and stored to become constituent parts of plant life [15].

Seed priming is a pre-sowing technique for controlling seedling growth by modulating pre-germination metabolic activities prior to radicle emergence, which improves germination rate and plant output in general [16]. Hydro-priming (soaking in water), osmo-priming (soaking in osmotic solutions, such as polyethylene glycol, sodium, and potassium salts), solid matrix priming, biopriming (coating with bacteria, such as *Pseudomonas aureofaciens*, AB254), and treatment with plant growth regulators (PGRs), combined with priming medium, are all examples of priming methods [16]. Priming with KNO_3 improves seed germination, mineral composition, proline, amylase, and protein pattern [17]. It improves seedling establishment and vigor and has a remarkable role in the pre-sowing accomplishment of germination phases [18]. However, the detailed metabolic events induced by KNO_3

during the sprouting process are not quite understood. Therefore, the purpose of this study was to assess the impact of KNO_3 priming on growth, N metabolism, tissue quality, and biological values in the sprouts of three species of *Medicago*.

2. Materials and Methods

2.1. Plant Material and Growth Conditions

Seeds of *Medicago* (*Medicago indicus*, *Medicago interexta*, and *Medicago polymorpha*) were collected from the Agricultural Research Center (Giza, Egypt), where they were collected during filed trips to different locations in Egypt (Giza and Ismailia). Before being stored in distilled water overnight, the seeds were rinsed in distilled water and submerged in $5\ g\ L^{-1}$ sodium hypochlorite for 1 h. Two hundred seeds were clustered into a group soaked in distilled water (control), and the second group was soaked in a KNO_3 solution of 25 mM, for 16 h. The applied growth promoting concentration of KNO_3 was selected according to a pilot experiment, where five concentrations (0 and 10, 15, 25, 50 mM) were tested. Each species seeds were evenly placed on 10 vermiculite-lined trays, which were irrigated every two days with Milli-Q water, and each tray received 150 mL of aquaponic water. The experiment was conducted in a growth cabinet under controlled conditions (25 °C, 16 h light/8 h dark cycle, and PAR of 400 $\mu mol\ m^{-2}\ s^{-1}$ and relative humidity of 60% per day). After ten days, the fresh mass of each sprout was measured and then stored at -80 °C for further biochemical analyses. Experiments were repeated 4 times, and 20 plants that were pooled from each tray and treated (biological replicate) were used for each measurement.

2.2. Analysis of Mineral Contents

Two hundred milligrams of each KNO_3-primed *Medicago* sprout was processed in an oven with an HNO_3/H_2O solution (5:1 v/v) to determine macro and micro-elements. The concentrations of macro-minerals and trace elements at 25 °C were measured using inductively coupled plasma mass spectrometry (ICP-MS, Finnigan Element XR, and Scientific, Bremen, Germany), with nitric acid in one percent employed as standards [19]. Tandard solution of multielement was used for calibration solution preparation at concentrations form 0 to 50 ppm in 2% HNO_3. Nitric acid was used as blank. *Medicago* sprout samples were vaporized, atomized, and ionized inside the chamber of the plasma. The limits of detection (LODs) and quantification (LOQs) values ranged from 0.0002 to 0.01 $ug\ kg^{-1}$ and 0.004 to 0.3 $\mu g\ kg^{-1}$, respectively.

2.3. Determination of Leaf Pigments

The frozen *Medicago* sprout samples (0.5 g) were homogenized in acetone for 1 min at 7000 rpm using a MagNALyser (Roche, Vilvoorde, Belgium), and then centrifuged for 20 min at $14,000 \times g$ at 4 °C [20]. Acrodisc GHP filter (0.45 μm/13 mm) was used to filter the supernatant. The solution was then evaluated by HPLC (Shimadzu, SPDM10Avp, Japan, Tokyo) at reversed-phase and at 4 °C [19]. Pigment and carotenoid separation was carried out on a silica-based C18 column (Waters, Spherisorb, 5 m ODS1, 4.6 250 mm) with two different solvents: (A) acetonitrile: methanol: water in the ratio of 81:9:10, and (B) methanol: ethyl acetate in the ratio of 68:32. A diode-array detector (Shimadzu SPDM10Avp) was used to analyze the extraction of chlorophyll a and b, beta-carotene, and xanthophylls at four distinct wavelengths (420, 440, 462, and 660 nm), respectively.

2.4. Determination of Amino Acids

Medicago sprouts (300 mg fresh weight (FW)) samples were extracted in methanol [19]. GC/MS (Hewlett Packard, Palo Alto, CA, USA) analysis was carried out, and samples were separated on a HP-5 MS column. Two hundred milligrams of FW sprout leaves were homogenized in 80% aqueous ethanol and centrifuged at $22,000 \times g$ for 25 min to measure the amino acids. The supernatant was evaporated, and the precipitates were resuspended in chloroform. The pellet was re-dissolved in chloroform and was filtered (0.2-μm Millipore microfilters). Amino acids levels were measured by Waters Acquity UPLC-tqd system at 37 °C, low pressure, and mobile phase acetonitrile/water ration 60/40, with a measurement at 254 nm. The result was expressed in mg/g dry weight of the sample.

2.5. Determination of Polyphenols and Flavonoid Contents

One hundred milligrams of frozen sprouts were homogenized in 1 mL of 80% ethanol (*v*/*v*) to extract polyphenols and flavonoids [21]. The supernatant was utilized to determine the total phenolic and flavonoid contents after centrifugation at 4 °C for 20 min. A Folin–Ciocalteu test, with gallic acid as a standard, was used to assess phenolic content. The modified aluminum chloride colorimetric method was used to quantify flavonoid concentration, utilizing quercetin as a standard [22].

2.6. Determination of Vitamin Contents

Using UV and/or fluorescence detectors, the amounts of ascorbate, tocopherols, thiamine, and riboflavin in sprouts were measured [19]. For thiamine, and riboflavin extraction, 250 mg of sample were extracted in 0.1 N HCl for 30 min [23]. Samples were separate through a 5 μm C18 Luna Phenomenex stainless steel column (250 × 4.6 mm i.d.). The mobile phase (methanol:sodium acetate (40:60 *v*/*v*) was used, and the fluorometric detection was performed at 453 nm and 580 nm, for riboflavin, and 366 nm and 453 nm, for thiamine. At 4 °C, ascorbate (vitamin C) was extracted in 1 mL of 6% (*w*/*v*) meta-phosphoric acid and separated using reversed-phase HPLC with a UV detector (100 mm 4.6 mm Polaris C18-A, 3 lm particle size; 40 °C, isocratic flow rate: 1 mL min1, elution buffer: 2 mM KCl, pH 2.5 with O-phosphoric acid). Tocopherol (vitamin E) was separated on a Particil Pac 5 m column (length 250 mm, i.d. 4.6 mm) and measured using HPLC (Shimadzu's Hertogenbosch, normal phase conditions) and a fluorometric detector (excitation at 290 nm and emission at 330 nm). On a reverse-phase (C18) column, riboflavin and thiamine were separated (HPLC, methanol:water as mobile phase, and fluorescence as a detector).

2.7. Determination of Total Proteins

The protein was measured according to the Folin-Lowry. Two-tenths of a gram of frozen *Medicago* sprouts were homogenized in chloroform/methanol (2:1, *v*/*v*) solution and centrifuged for 15 min at 3000× *g* to measure the total proteins content [24].

2.8. Determination of N, Ammonium, and Nitrate Contents

The amount of nitrate in the water was determined using Cataldo et al.'s method [25]. Here, 0.1 mL filtrate and 0.4 mL 5% salicylic acid in concentrated H_2SO_4 made up the reaction mixture. After cooling at ambient temperature for 15 min, 9.5 mL 2 M NaOH was progressively added to elevate the pH above 12. The absorbance was measured at 410 nm when the solution was cooled to room temperature. Nitrate concentration was determined using a KNO_3 calibration curve and represented in mg NO_3-g^{-1} FW. The amount of ammonium in the sample was determined using the indophenol blue colorimetry method at 630 nm. The standard was ammonium chloride. Total N was measured using fine ground leaf dry samples (0.2 g) digested with H_2SO_4–H_2O_2 at 260 °C.

2.9. Determination of Antioxidant and Antidiabetic Activities

2.9.1. Antioxidant Activity

Each *Medicago* sprout sample (0.1 g) was extracted in 80 percent ethanol and centrifuged for 20 min at 14,000 rpm [26]. The experiment was carried out in vitro using ferric reducing antioxidant power (FRAP) to determine antioxidant capabilities. In this case, we used 0.25 mL of FRAP reagent, mixing $FeCl_3$ (20 mM) in acetate buffer (0.25 M, pH 3.6) at room temperature with 0.1 mL of diluted extract. For concentration calculation, calibration curve was performed using the standard Trolox (0.05–1 mM) as standard.

2.9.2. Antidiabetic Activity

α-Amylase Inhibition Assay

The inhibition of pancreatic α-amylase inhibition was measured using *Medicago* sprout extract mixed with reaction solution starch (1 g/L) and phosphate buffer (pH 6.9). Then, 3 U/mL amylase enzyme was added to start the process [27]. After 10 min of incubation, 0.5 mL dinitro salicylic (DNS) reagent was added to terminate the reaction. The reaction mixture was heated to 100 °C for 10 min. Finally, the mixes received 0.5 mL of a 40 percent potassium sodium tartrate solution. At 540 nm, the absorbance was measured.

α-Glucosidase Inhibition Assay

The sprout hydroethanolic extract was combined with -glucosidase (2 U/mL) and incubated at 37 °C for 5 min to determine the inhibition of α-glucosidase [27]. After adding 1 mM para-nitrophenyl glucopyranoside dissolved in 50 mM phosphate buffer, the reaction buffer was incubated for 20 min at 37 °C (pH 6.8). A solution of sodium carbonate (1 M) was added to stop the process. The amount of para-nitrophenolate produced by para-nitrophenyl glucopyranoside was measured at 405 nm, and the inhibitory activity of -glucosidase was computed. The α-glucosi- dase inhibitory activity was expressed as percent inhibition and determined as follows: %inhibition = [(average A 405 control − average A 405 extractÞ/average A 405 control × 100].

2.9.3. Glycemic Index GI

The GI was determined using an in vitro starch hydrolysis method [27]. The process begins with the incubation of *Medicago* sprouts in a reaction buffer of HCl-KCl buffer (pH 1.5) for one hour at 40 °C under shaking conditions with pepsin (100 mg/mL). The mixture was then diluted in phosphate buffer (pH 6.9) before being incubated at 37 °C with α-amylase. 1 mL of aliquots were collected every 30 min and boiled for 20 min to cease the activity of the amylase enzyme. To convert the remaining starch to glucose, 0.4 M sodium acetate buffer (pH 4.75) and 60 L amyloglucosidase were added. For 50 min, the reaction mixture was incubated at 60 °C. Approximately 0.6 mL aliquots were collected and incubated at 37 °C for 35 min with 1.2 mL glucose oxidase/peroxidase. The mixture's absorbance was measured at 500 nm. The proportion of hydrolyzed starch at different times (0, 30, 60, 90, 120, and 180 min) was used to calculate the starch digestion rate. The area under the hydrolysis curve (AUC, 0–180 min) and hydrolysis were computed. The hydrolysis index was then computed by multiplying the difference between the AUC for a sample and the AUC for a standard by 100.

2.10. Statistical Analyses

The R statistics package was used to conduct the statistical analysis (Gplot, Agricola). All data were subjected to a one-way analysis of variance (ANOVA). As a post-hoc test for mean separations, Tukey's Test ($p = 0.05$) was used. Each experiment was repeated at least three times ($n = 3$). The R software created hierarchical clustering using Heatmap (Pearson correlation). The COrplot package was used to do correlation analysis on all of the data.

3. Results

3.1. Growth and Photosynthetic Pigments of Medicago Sprouts

The fresh weight of sprouts showed a significant difference as a consequence of KNO_3 priming. As shown in Figure 1, the sprout primed with KNO_3 exhibited a significant increase in fresh weight (FW) in the three species, as compared to the control. Among the three species, the highest and most significant increase in FW was measured from *Medicago interexta* seeds priming with KNO_3 by 53%, while the FW of both *Medicago indicus* and *Medicago polymorpha* primed with KNO_3 increased only by 34% and 35%, respectively, as compared to control. In terms of the impact of KNO_3 priming on photosynthetic pigments, there was a clear rising trend in *Medicago interexta* sprout leaf pigments, which showed increases in chlorophyll a (Chla), chlorophyll b (Chlb), and total chlorophyll of 53%, 60%, and 56%, respectively, as compared to control (Figure 1). Additionally, significant increases in Chla, Chlb, and total chlorophyll by 46%, 75%, and 56%, respectively, were recorded in *Medicago polymorpha*, as compared to control. Meanwhile, *Medicago indicus* sprouts showed an increase in chla by 38% and in total chlorophyll by 27%, while a slight increase was recorded in chlb. Furthermore, a significant amount of variation in carotenoid compounds in response to KNO_3 priming was associated with differences among *Medicago* species. Indeed, the highest increase of β-carotene, lutein, and Neoxanthin by 54% 58%, and 70% was recorded in *Medicago interexta* sprouts (Figure 1). However, violaxanthin pigment was significantly increased in *Medicago polymorpha* and *Medicago indicus* by 36% and 35%, respectively, as compared to control.

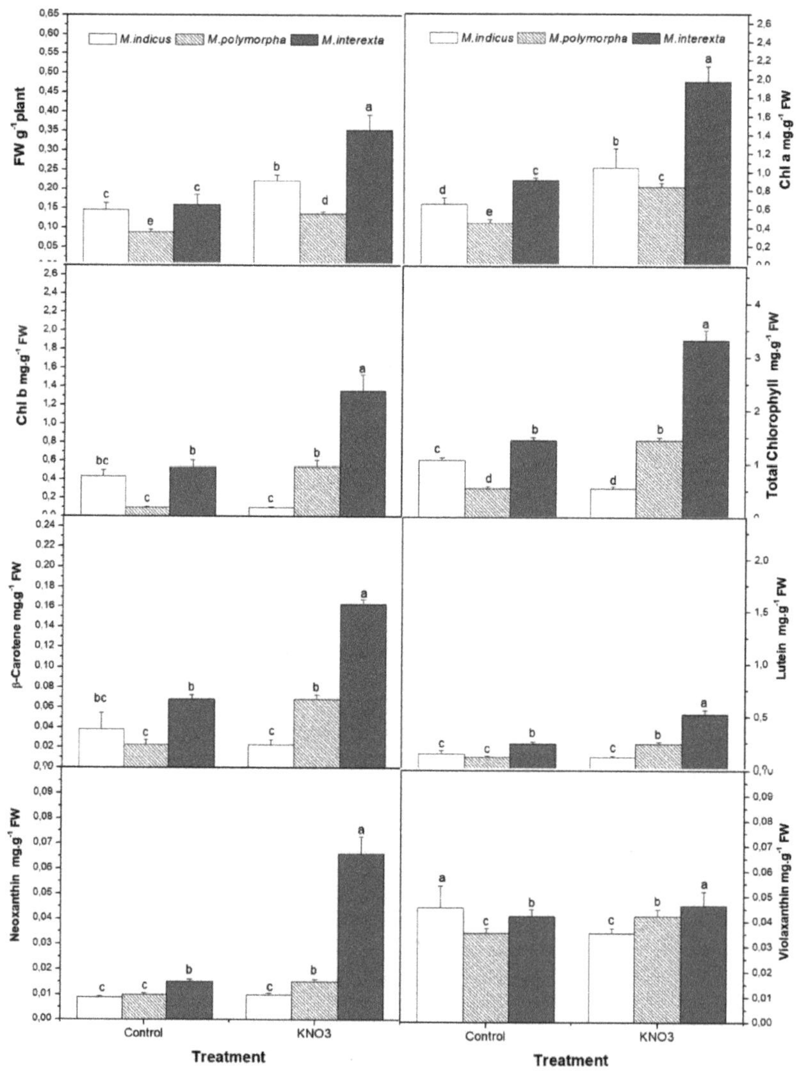

Figure 1. Effect of KNO_3 priming on total fresh weight and leaf pigments contents of three species of *Medicago* sprouts. Values are represented by mean ± standard deviation of at least three independent replicates. Within the same species, different letters on the bars indicate significant differences at $p < 0.05$.

3.2. Improvement of Nutritive Values: Mineral Content, Vitamins, and Antioxidant Activities

The Ca, K, Mg, Fe, Mn, and Zn concentrations, as well as Cu, were determined in dried *Medicago* sprouts. Regarding total nutrients analysis, the Ca, K, and Zn concentrations accounted for 24%, 22%, and 31%, respectively, on average in all species and treatment. However, Cu, Mg, Mn, and Fe were present in small amounts (Table 1). The mineral content of sprouts is very dependent on the sprouting conditions and species. The KNO_3 priming increased the Ca, Cu, Fe, K, and P in the three species. Indeed, the greatest accumulation of K and P, by 2- and 5-fold, respectively, was observed in *Medicago interexta* sprouts. Likewise, KNO_3 priming increased the content of Zn in *Medicago polymorpha* and *Medicago indicus* by 36% and 21%, respectively, as compared to control, while no significant change was observed in *Medicago interexta* sprouts. Furthermore, Mn and Mg content increased by 36% and 37% in *Medicago polymorpha*, and by 50% and 26% in *Medicago interexta*. In contrast, the Mn and Mg were not affected by KNO_3 priming in *Medicago indicus*.

Table 1. Effect of KNO_3 priming on nutriments content of three species of *Medicago* sprouts. Values are represented by mean $\pm$ standard deviation of at least three independent replicates. Means marked by different letters are significantly different than control at $p < 0.05$.

Parameters	Control			KNO_3 Priming		
	M. indicus	*M. polymorpha*	*M. interexta*	*M. indicus*	*M. polymorpha*	*M. interexta*
Ca mg·g^{-1} dw	17.57 $\pm$ 2.35 b	12.68 $\pm$ 1.87 b	15.79 $\pm$ 3.5 b	27.79 $\pm$ 6.69 a	19.63 $\pm$ 3.48 a	25.17 $\pm$ 0.46 a
Cu µg·g^{-1} dw	2.26 $\pm$ 0.71 a	2.42 $\pm$ 0.79 a	2.57 $\pm$ 1.06 b	2.87 $\pm$ 0.28 a	3.963 $\pm$ 0.86 a	4.38 $\pm$ 1.13 a
Fe µg·g^{-1} dw	3.99 $\pm$ 0.238 b	4.80 $\pm$ 1.06 b	3.15 $\pm$ 0.78 b	5.48 $\pm$ 1.02 a	6.798 $\pm$ 1.69 a	5.76 $\pm$ 0.44 a
Zn µg·g^{-1} dw	22.62 $\pm$ 2.08 b	13.49 $\pm$ 1.10 b	36.61 $\pm$ 3.25 a	35.8 $\pm$ 3.21 a	17.24 $\pm$ 1.50 a	35.71 $\pm$ 3.16 a
Mn µg·g^{-1} dw	0.25 $\pm$ 0.12 a	0.19 $\pm$ 0.13 b	0.13 $\pm$ 0.13 b	0.27 $\pm$ 0.12 a	0.30 $\pm$ 0.12 a	0.27 $\pm$ 0.12 a
Mg mg·g^{-1} dw	2.90 $\pm$ 0.160 a	2.38 $\pm$ 0.10 b	2.69 $\pm$ 0.14 b	2.67 $\pm$ 0.12 a	3.81 $\pm$ 0.22 a	3.65 $\pm$ 0.22 a
K mg·g^{-1} dw	15.7 $\pm$ 1.311 b	20.8 $\pm$ 1.76 b	12.0 $\pm$ 0.98 b	40.6 $\pm$ 3.57 a	60.5 $\pm$ 5.50 a	67.3 $\pm$ 6.02 a
P mg·g^{-1} dw	5.8 $\pm$ 0.57 b	7.6 $\pm$ 0.595 b	6.5 $\pm$ 0.46 b	10.4 $\pm$ 0.84 a	11.2 $\pm$ 0.93 a	13.6 $\pm$ 1.13 a

An analysis of vitamins was carried out for three species of *Medicago* sprouts (Table 2). The results revealed that Vit C presented the greatest accumulation by 41%, 32%, and 39% in *Medicago indicus*, *Medicago polymorpha*, and *Medicago interexta*, respectively (Table 1). KNO_3 priming enhanced the Vit C, Vit E, and riboflavin content in all three species. Obviously, Vit C content was higher in *Medicago interexta* in comparison to other species. Likewise, both Vit E and riboflavin showed the highest accumulation by 46% and 39%, respectively, in *Medicago polymorpha* compared to control and other species. In contrast, a slight increase in thiamine was recorded in *Medicago indicus* in a response to KNO_3 priming, and no significant change in *Medicago polymorpha*. While KNO_3 priming enhanced the thiamine content in *Medicago interexta* by 2-fold compared to control conditions.

Table 2. Effect of KNO_3 priming on vitamin content of three species of *Medicago* sprouts. Values are represented by mean $\pm$ standard deviation of at least three independent replicates. Means marked by different letters are significantly different than control at $p < 0.05$.

Parameters	Control			KNO_3 Priming		
	M. indicus	*M. polymorpha*	*M. interexta*	*M. indicus*	*M. polymorpha*	*M. interexta*
Vit C (mg·g^{-1} FW)	7.81 $\pm$ 1.35 b	7.67 $\pm$ 1.33 b	7.31 $\pm$ 1.26 b	8.15 $\pm$ 2.39 a	9.48 $\pm$ 1.37 a	13.92 $\pm$ 0.76 a
Vit E (mg·g^{-1} FW)	47.46 $\pm$ 1.17 a	38.05 $\pm$ 1.09 b	44.57 $\pm$ 1.58 b	48.47 $\pm$ 4.40 a	59.72 $\pm$ 2.43 a	61.92 $\pm$ 3.86 a

Table 2. *Cont.*

Parameters	Control			KNO$_3$ Priming		
	M. indicus	*M. polymorpha*	*M. interexta*	*M. indicus*	*M. polymorpha*	*M. interexta*
Thiamin (mg·g^{-1} FW)	0.10 ± 0.00 b	0.13 ± 0.02 a	0.06 ± 0.02 b	0.13 ± 0.02 a	0.13 ± 0.07 a	0.14 ± 0.06 a
Riboflavin (mg·g^{-1} FW)	0.30 ± 0.05 b	0.52 ± 0.09 b	0.75 ± 0.22 b	0.46 ± 0.22 a	0.87 ± 0.07 a	0.96 ± 0.46 a

3.3. Amino Acid Metabolism

HPLC analysis of amino acids in *Medicago* sprouts species revealed the presence of eighteen amino acids, with significantly different concentrations (Table 3). The highest value was recorded for glutamine in *Medicago interexta* (2.17 mg/g dry wt), followed by asparagine, glycine, phenylalanine, serine, proline, threonine, isoleucine, valine, leucine, lysine, tryptophane, cysteine, histidine, alanine, methionine, arginine, and tyrosine. KNO$_3$ priming improved the accumulation of glutamine, serine, arginine, alanine, proline, histidine, valine, methionine, cystine, isoleucine, leucine, phenylalanine, tyrosine, lysine, threonine, and tryptophan in the three species. Indeed, the greatest accumulation was recorded in glutamine by 49% in *Medicago interexta* sprout. Likewise, Arginine showed the highest increase by 60% and 58% in *Medicago indicus* and *Medicago polymorpha*, respectively, compared to control. In contrast, the rest of amino acids were not significantly affected by KNO$_3$ priming, depending on *Medicago* species. Compared to control sprouts, the N content was significantly affected by KNO$_3$ in the three species. Unlike, the activities of nitrate reductase (NR), GDH, GOGAT, and GS were changed in response to KNO$_3$ priming, depending on species (Table 3). Indeed, the highest increase of NR, GDH, GOGAT, and GS, by 26%, 32%, 32%, and 19%, respectively, was recorded in *Medicago interexta* sprouts. Furthermore, the protein content decreased in *Medicago indicus* and *Medicago polymorpha*, whereas it increased by 13% in *Medicago interexta*, compared to control.

Table 3. Effect of KNO$_3$ priming on total amino acids content, total N, proteins and NR, GDH and GOGAT activities of three species of *Medicago* sprouts, Values are represented by mean ± standard deviation of at least three independent replicates, Means marked by different letters are significantly different than control at $p < 0.05$.

Parameters	Control			KNO$_3$ Priming		
	M. indicus	*M. polymorpha*	*M. interexta*	*M. indicus*	*M. polymorpha*	*M. interexta*
Amino acids (mg·g^{-1} fw)						
Asparagine	1.64 ± 0.29 a	1.71 ± 0.38 b	1.9 ± 0.35 b	1.79 ± 1.52 a	1.99 ± 1.73 a	2.2 ± 1.93 a
Glutamine	1.86 ± 0.41 b	1.96 ± 0.45 b	2.17 ± 0.53 b	3.56 ± 3.32 a	4.07 ± 3.95 a	4.27 ± 4.23 a
Serine	1.37 ± 0.04 b	1.43 ± 0.15 a	1.59 ± 0.08 b	2.03 ± 2.16 a	1.25 ± 1.35 b	2.02 ± 2.48 a
Glycine	1.56 ± 0.28 a	1.68 ± 0.41 a	1.83 ± 0.28 b	1.18 ± 1.16 b	1.88 ± 1.62 a	2.02 ± 1.86 a
Arginine	0.31 ± 0.1 b	0.35 ± 0.15 b	0.37 ± 0.17 b	0.88 ± 0.78 a	0.84 ± 0.81 a	0.41 ± 0.61 a
Alanine	0.58 ± 0.1 a	0.71 ± 0.03 a	0.68 ± 0.12 a	0.51 ± 0.5 a	0.57 ± 0.5 a	0.65 ± 0.61 a
Proline	1.19 ± 0.44 b	1.34 ± 0.46 b	1.59 ± 0.59 b	2.67 ± 3.11 a	2.37 ± 3.26 a	3.09 ± 3.42 a
Histidine	0.74 ± 0.11 a	0.81 ± 0.11 a	0.89 ± 0.11 a	0.58 ± 0.58 b	0.41 ± 0.59 b	0.52 ± 0.74 b
Valine	0.89 ± 0.38 a	0.97 ± 0.43 a	1.05 ± 0.48 a	0.79 ± 0.54 a	0.65 ± 0.55 b	0.96 ± 0.71 b
Methionine	0.47 ± 0 b	0.55 ± 0 b	0.63 ± 0 b	0.85 ± 0.89 a	0.86 ± 0.89 a	0.86 ± 0.89 a
Cystine	0.87 ± 0.06 b	0.89 ± 0.03 b	0.92 ± 0.06 b	1.44 ± 1.17 a	1.45 ± 1.26 a	1.41 ± 1.37 a
Isoleucine	1.11 ± 0.72 b	1.12 ± 0.76 b	1.14 ± 0.8 b	1.39 ± 1.13 a	1.46 ± 1.17 a	1.28 ± 0.95 a

Table 3. *Cont.*

Parameters	Control			KNO₃ Priming		
	M. indicus	*M. polymorpha*	*M. interexta*	*M. indicus*	*M. polymorpha*	*M. interexta*
Leucine	1.04 ± 0.15 a	1.01 ± 0.29 a	0.98 ± 0.43 b	1.02 ± 0.86 a	1.15 ± 0.87 a	1.34 ± 1 a
Phenylalanine	1.98 ± 1.08 a	1.82 ± 1.17 a	1.65 ± 1.45 b	1.88 ± 2.14 a	1.13 ± 2.11 b	2.27 ± 2.44 a
Tyrosine	0.31 ± 0.08 b	0.32 ± 0.05 b	0.33 ± 0.05 b	0.4 ± 0.35 a	0.44 ± 0.34 a	0.47 ± 0.41 a
Lysine	0.77 ± 0.14 b	0.84 ± 0.1 b	0.92 ± 0.07 a	1.07 ± 0.92 a	1 ± 0.9 a	0.85 ± 0.82 b
Threonine	1.33 ± 0.28 a	1.39 ± 0.26 a	1.45 ± 0.24 a	1.47 ± 1.25 a	1.23 ± 1.26 a	1.59 ± 1.46 a
Tryptophan	0.78 ± 0.21 b	0.84 ± 0.23 b	0.9 ± 0.24 b	1.11 ± 0.92 a	1.32 ± 1.04 a	1.36 ± 1.1 a
Nitrogen content and metabolism						
Nitrogen (g/100 g)	23.39 ± 0.89 b	19.69 ± 0.82 b	15.72 ± 0.53 b	28.11 ± 1.24 a	24.95 ± 0.72 a	19.20 ± 0.8 a
Total protein (mg/g FW)	169.5 ± 1.9 a	153.61 ± 8.86 a	118.0 ± 3.16 b	99.6 ± 2.27 b	129.0 ± 8.47 b	136.8 ± 2.81 a
Nitrate reductase μmol nitrite/mg protein.min	45.24 ± 0.03 a	49.53 ± 2.47 b	86.19 ± 5.45 b	33.11 ± 2.23 b	56.36 ± 0.91 a	118 ± 11.27 a
GDH μmol NADH/mg protein.min	4.14 ± 0.21 a	4.10 ± 0.09 a	6.99 ± 0.48 b	4.14 ± 0.21 a	4.10 ± 0.09 a	10.33 ± 0.48 a
GOGAT μmol NADH/mg protein.min	7.83 ± 0.28 a	11.16 ± 0.45 b	14.35 ± 0.45 b	6.35 ± 0.29 b	14.70 ± 0.52 a	21.38 ± 1.83 a
GS μmol γ-glutamyl hydroxamate/mg protein.min	16.12 ± 0.91 b	23.09 ± 1.24 b	26.16 ± 0.45 b	23.00 ± 1.24 a	29.2 ± 0.77 a	32.55 ± 0.8 a

3.4. Antioxidant and Antidiabetic Avtivities

3.4.1. Antioxidant Metabolites and Free Radical Scavenging Activity of *Medicago* Sprouts

The present results revealed that the priming with KNO₃ during germination has induced changes in phenolic compounds concentrations in *Medicago* sprouts (Figure 2). KNO₃ priming significantly increased total phenolic and flavonoid contents in the three species.

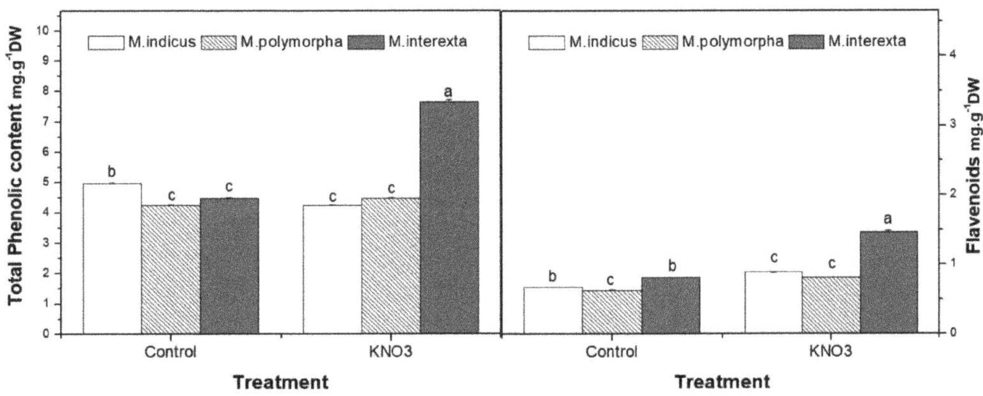

Figure 2. Effect of KNO₃ priming on total phenolic and flavonoid contents of three species of *Medicago* sprouts. Values are represented by mean ± standard deviation of at least three independent replicates. Within the same species, different letters on the bars indicate significant differences at $p < 0.05$.

Besides, KNO$_3$ priming enhanced the FRAP activities by 41%, 59%, and 35% for *Medicago indicus*, *Medicago polymorpha*, and *Medicago interexta*, respectively, in comparison to control (Figure 3). Furthermore, the reduced ascorbate and glutamine increased significantly ($p < 0.05$) in the three species as response to KNO$_3$ priming, and the highest increase was recorded in *Medicago interexta* (Figure 3).

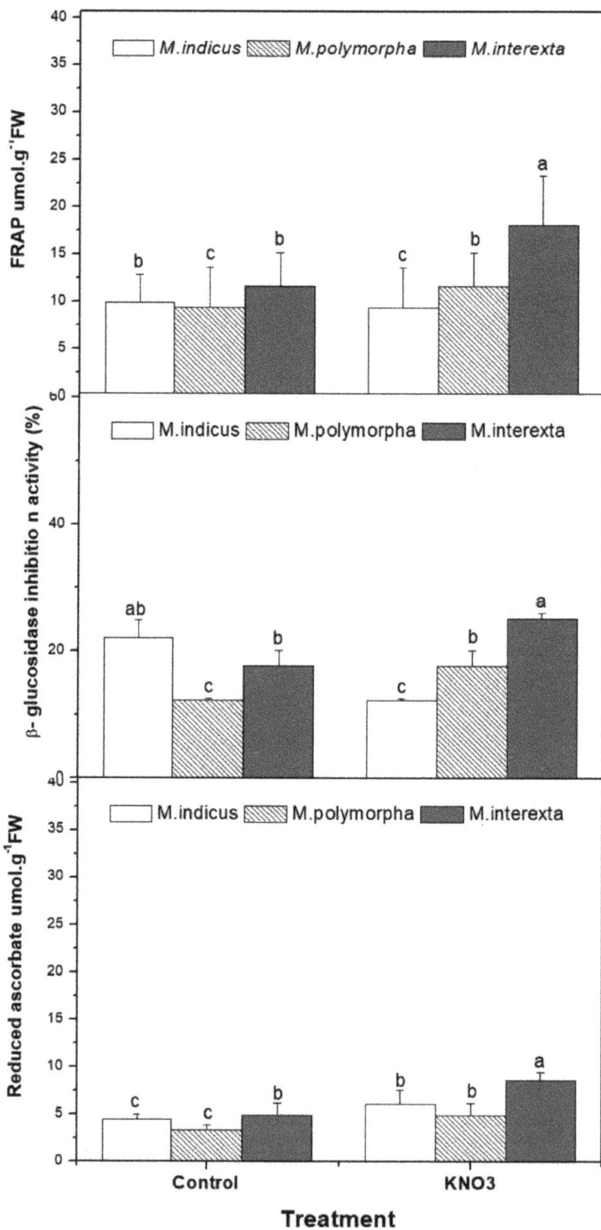

Figure 3. Effect of KNO$_3$ priming on antioxidant activity of three species of *Medicago* sprouts. Values are represented by mean $\pm$ standard deviation of at least three independent replicates. Within the same species, different letters on the bars indicate significant differences at $p < 0.05$.

3.4.2. Antidiabetic Activity

As shown in Figure 4, each species of *Medicago* sprouts exhibited antidiabetic activity. Under control conditions, *Medicago interexta* had the best inhibitory effect on α-amylase, while *Medicago indicus* had the best inhibitory effects on α-glucosidase. KNO_3 priming seemed to enhance the inhibition activity of α-amylase by 28% in *Medicago indicus*. Similarly, KNO_3 priming enhanced the inhibition activity of α-glucosidase by 30%, 40%, and 29% for *Medicago indicus*, *Medicago polymorpha*, and *Medicago interexta*, respectively, compared to control. The current results also demonstrated that KNO_3 priming caused a marked decrease in GI in *Medicago polymorpha* and *Medicago interexta* (lower than 70).

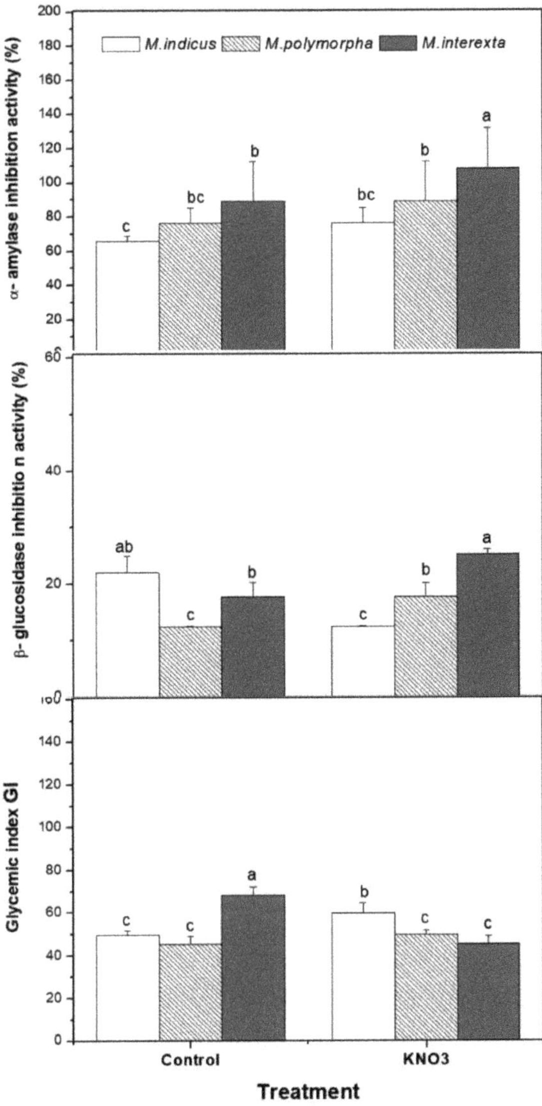

Figure 4. Effect of KNO_3 priming on antidiabetic activity of three species of *Medicago* sprouts. Values are represented by mean ± standard deviation of at least three independent replicates. Within the same species, different letters on the bars indicate significant differences at $p < 0.05$.

3.5. Principal Component Analysis (PCA)

PCA is a multivariate statistical analysis that can be used to examine and simplify complex and huge datasets. The pattern of variation in *Medicago* species was also analyzed using principal component analysis (PCA) to evaluate the variety of the species and their link with the observed traits based on the correlation between the traits and extracted clusters. The chemical profiles of plants that were not primed with KNO_3 were grouped and clearly separated from sprouts primed with KNO_3. In our dataset, two groups of traits were identified in the PCA biplot considering both PC1 and PC2 simultaneously (Figures 5 and 6). The FW, leaf pigments, mineral content, N metabolism, polyphenols, antioxidants enzymes, and glucosidase and amylase inhibitor enzymes linked with *Medicago interexta* and *polymorpha* primed with KNO_3.

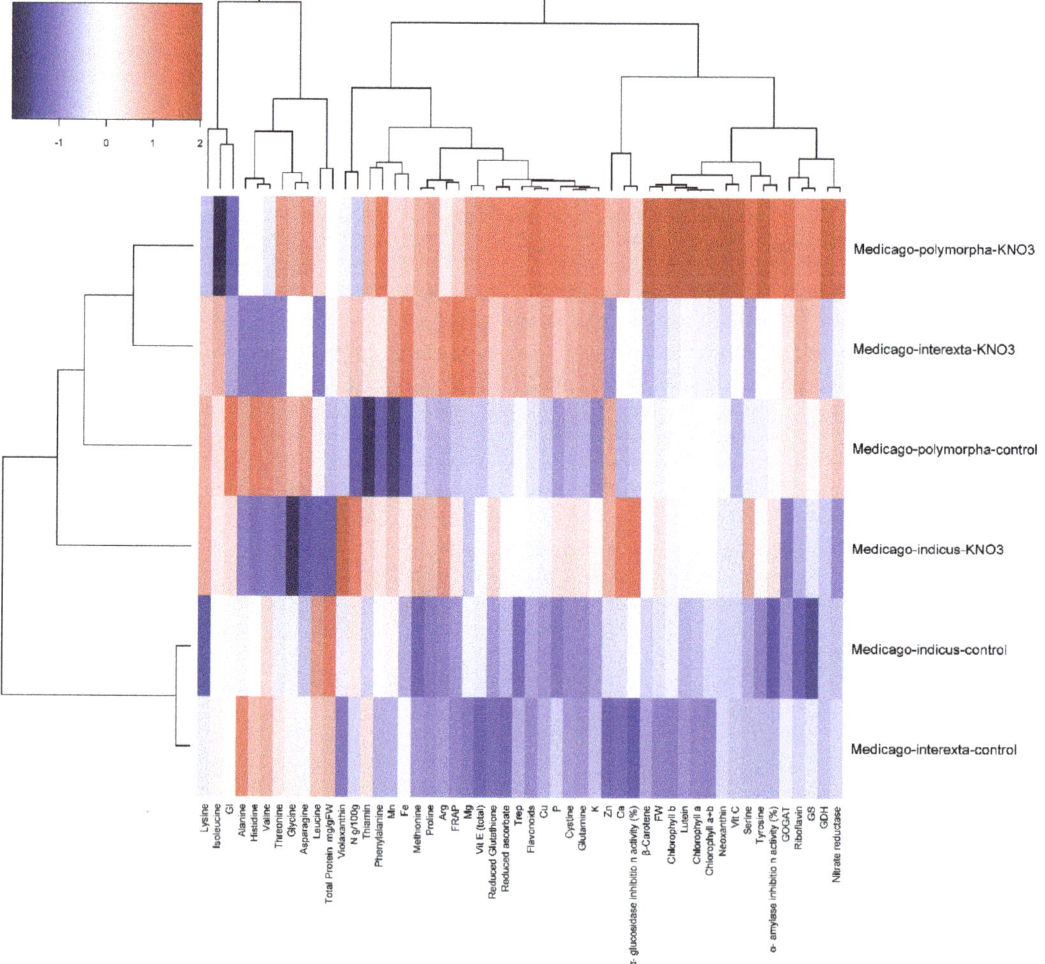

Figure 5. The cluster heatmap of primary and secondary metabolites of three *Medicago* species KNO_3 priming treatment. The graph's horizontal axis shows different treatments for each species, and the vertical axis shows different phytocompounds, amino acids, and nitrogen content. Color gradients represent the different values of contents under KNO_3 priming compared with that of control.

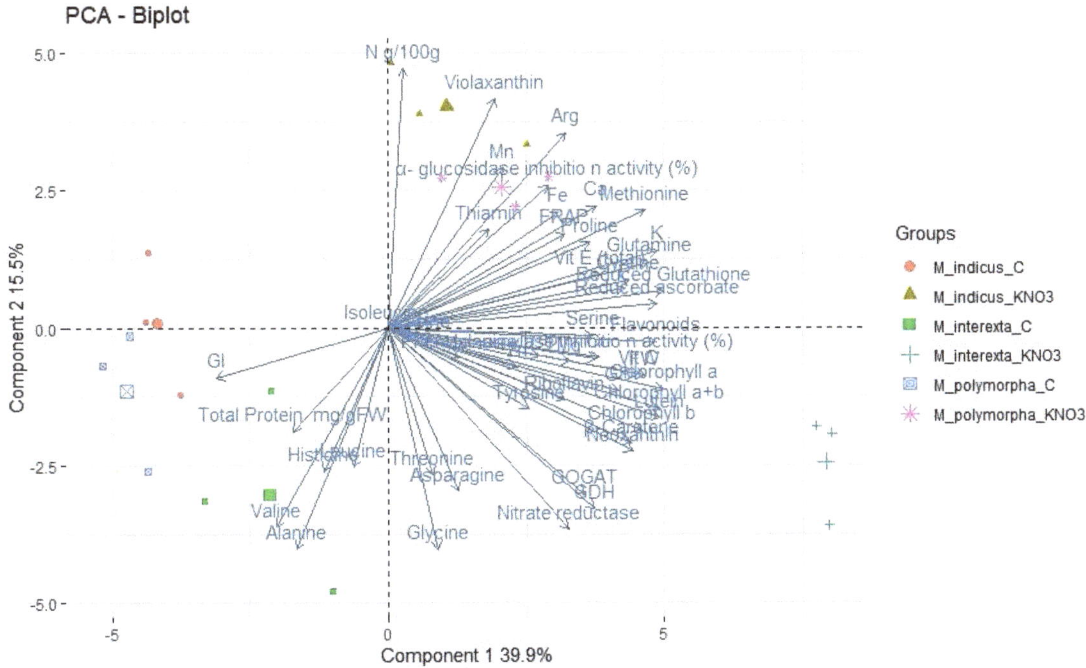

Figure 6. PCA-Biplot of *Medicago* species.

4. Discussion

4.1. KNO$_3$ Priming Increased Biomass Accumulation in Medicago Sprouts

Seed priming is used to ensure rapid and uniform seed germination and seedling emergence in order to improve agricultural production performance. Priming with KNO$_3$ has also been found to boost seedling germination, growth, establishment, and productivity in numerous studies. Seed priming has been extensively studied in terms of plant ecology, physiology, cellular biology, and molecular biology [28,29]. Increased seed quality has recently become a top focus in the agriculture business. Seed priming with KNO$_3$ increased the growth of three *Medicago* species in the current study. The results were similar to those of earlier studies, which reported that KNO$_3$ priming enhanced cucumber [30], white clover [31], and soybean [32] seedling fresh weights, when compared with unprimed seedlings. Similarly, seeds of *Medicago sativa* var. Anand-z primed with 0.1 percent MgCl$_2$ also had a high rate of seed germination and seedling growth [33]. Enhancement of *Medicago* growth might be due to increased cell division and elongation and activation of ROS scavenging enzymes in KNO$_3$ primed seeds [30]. KNO$_3$ may promote the growth of *Medicago* sprouts by acting as nutrients and initiators of crucial emergence and growth processes in sprouts. The improvement of growth by KNO$_3$ priming seemed to be varied among the species of *Medicago*. Indeed, *Medicago interexta* presented the highest biomass accumulation compared to other species. Differences in emergence and growth among species in response to germination conditions were reported in several research works [34]. The results could be confirmed by the photosynthetic pigments contents of *Medicago* sprouts, where the contents of chla, chlb, and total chlorophyll were generally significantly increased in response to KNO$_3$ priming sprouts. Furthermore, the enhancement of chlorophyll biosynthesis by KNO$_3$ was inconsistent with the increases of Mg^{2+} contents in *Medicago polymorpha* and *Medicago interexta*. The chlorophyll molecule containing Mg covalently linked with four nitrogen (N) atoms, and this might be the reason that KNO$_3$ priming enhanced nutrients uptake and resulted to enhanced chlorophyll contents in *Medicago* sprouts leaves. The results are in agreement with previous studies, which reported that biopriming significantly enhanced chlorophyll contents in wheat leaves [35]. Furthermore, many works reported that lucerne possesses detoxifying and anticancerogenic properties due to its high chlorophyll content [35]; thus, it seemed that KNO_3 priming could enhanced these proprieties in the three species. Moreover, the results revealed that the KNO_3 priming enhanced the two groups of carotenoids, including *β-carotene* and hydroxylated carotenoids, designated as xanthophyll pigments, such as lutein, neoxanthin, and violaxanthin. Many reports revealed that *Medicago* contains 400–500 mg total carotenoids/kg, with the majority being xanthophylls, such as lutein and zeaxanthin [36]. Furthermore, a review of xanthophylls' possible functions in disease prevention has revealed that they may have a preventive impact against some malignancies, coronary heart disease, and stroke. Besides, lutein, neoxanthin, and violaxanthin pigments possess anti-inflammatory and strong antioxidant properties, and they are very active against liver neoplasms [37].

4.2. KNO_3 Priming Increases Nutritive Values of Medicago Sprouts

Mineral and vitamin deficiencies have been linked to a variety of detrimental health impacts in humans. Sprouts have long been thought to be a good source of bioavailable minerals, such as Fe, Zn, Mn, Mg, Cu, and Ca [38]. As a result, using KNO_3 priming to increase the mineral element content of sprouts could improve the nutritional and health-promoting effects of *Medicago* sprouts. In the present study, KNO_3 priming resulted in a significant increment in minerals nutrition concentration (Ca, Cu, Fe, Mn, K, and P) in the three species. The mineral content of sprouts is highly dependent on the sprouting conditions in general; however, the mineral contents observed in the sprouts analyzed in this study are consistent with those described in the literature [1]. Supporting our results, KNO_3 priming enhanced leaf nutrient accumulation and significantly enhanced seedling growth in mung beans [29] and cucumber seedlings [30]. Furthermore, the increase in mineral content varied between species. *Medicago interexta* presented the highest increment in Ca content. The high content of Ca makes these products suitable for consumers with lactose intolerance [39]. Other minerals, such as Fe, were also quantified in *Medicago interexta*, being higher than the ones reported for other species. The high content of Fe could be helpful to prevent anemia caused by iron deficiency [39].

Concerning vitamins, this outcome revealed that the *Medicago* sprouts were rich in Vit E (tocopherol), and its content varied between species, since *Medicago polymorpha* and *Medicago interexta* showed the greatest levels. Besides, KNO_3 priming enhanced the accumulation of Vit E and riboflavin in the three species, but it decreased the level of thiamine in *Medicago polymorpha*. The positive effect of KNO_3 priming on vitamins content makes *Medicago* species a remarkable source of vitamins for supporting immunity.

4.3. KNO_3 Priming Promotes N assimilation in Medicago Sprouts

It is widely recognized that the essential amino acid content in plants is well acknowledged to have a substantial impact on their nutritional and health-promoting characteristics. Essential amino acids are critical for human health since they cannot be generated from scratch and act as building blocks for a variety of proteins that play critical roles in human health [40]. For instance, lucerne was approved for use in human nutrition [41]. The outcomes of the present study revealed that *Medicago* sprouts were rich in essential and semi-essential amino acids, such as glutamine, phenylalanine, threonine, asparagine, and glycine, which are considered as the most important bioactive components. Further, KNO_3 priming enhanced the accumulation of proline, histidine, valine methionine cystine, and isoleucine. Moreover, the effect of KNO_3 priming varied between the three species, since some amino acids greatly increased *Medicago polymorpha* and *Medicago interexta* more than *Medicago indicus*. In line with our findings, a previous study discovered that the response of plants to environmental conditions in primary N absorption differed by species [13]. The amino acid in *Medicago* sprouts is present in substantially larger concentrations than in eggs or wheat and was approved for use in human nutrition [41]. This increment in several amino acids in response to KNO_3 priming has resulted in increases in total N in the three species and total protein in *Medicago interexta*. The increase in the activity of important enzymes in N metabolism, including NR, GS, GDH, and GOGAT, could explain the changes in amino acid levels. We suggested that the nitrate supplemented by the priming with KNO_3 had a positive effect on NR. In fact, the increases in the activity of nitrate reductase (NR) increased the potential for nitrate reduction, resulting in increased capacity for amino acid synthesis, protein synthesis, and total N assimilation [12]. These results were in accordance with the results obtained in leaves of wheat seedlings [12] and Safflower [42] treated with KNO_3. Concerning the total protein, the increase in this compound in response to KNO_3 priming, except in *Medicago interexta*, may be due to the direct involvement of K in several steps of the translocation process, including the binding of RANt to ribosomes [13]. Interestingly, here, the KNO_3 priming effect on N level and GS, GDH, and GOGAT enzymes activities were varied with different species.

4.4. Antioxidant Metabolites Accumulation Increased Antioxidant Biological Activity of Medicago Sprouts Extracts

The bioactive compounds of *Medicago* have gotten a lot of interest because of their antibacterial, anti-inflammatory, anticancer, and antioxidant properties [7]. These biological actions, which include anti-inflammatory and antioxidant properties, may be due to the presence of phenolic and flavonoids chemicals, which function as free radical scavengers and/or metal chelators [43]. Caunii et al. (2012) [44] found that the lucerne extract included a number of free hydroxyl groups (hydrogen donors), giving the product a significant antioxidizing effect in the human body. As such, the outcomes of the current study revealed that KNO_3 enhanced the antioxidants properties of *Medicago* sprouts, and the increases depend on species. Indeed, *Medicago polymorpha* exhibit the highest antioxidant activities since flavonoids reduced glutathion and ascorbate content, and FRAP showed an increment by 2-fold, compared to control. This suggests that the obtained values of these compounds are impacted by species and germination processes.

As antidiabetic agents, the leaves of *Medicago* have been used traditionally to reduce plasma glucose levels in diabetic subjects [45,46]. In this investigation, the accumulation of nitrogen compounds enhanced by KNO_3 priming increased the antidiabetic activity of *Medicago* sprouts extracts, since the GI decrease sharply mainly in *Medicago interexta* compared to others sprouts in response to KNO_3 priming. Furthermore, the current study illustrated that KNO_3 priming enhanced the inhibitor effect against α-amylase and α-glucosidase in the three species of *Medicago* sprouts. Following the previous studies and examination, the anti-diabetic activity of *Medicago* was tested against α-amylase [46]. The accumulation of N leads to an increase in amino acid synthesis. Hence, amino acids may regulate insulin secretion in several ways, including the production of metabolic coupling factors, plasma membrane depolarization, and mitochondrial function augmentation [40]. Furthermore, the enhancement in several enzymes of N metabolism caused a high production of glutamine. This former amino acid has been postulated to play a role in nutrient-induced stimulus-secretion coupling as an additive factor in the glucose-stimulated insulin secretion amplification pathway [40]. Overall, as antidiabetic agents, accumulation of nitrogenous compounds can explain the increased antidiabetic activity of *Medicago* sprouts extracts.

4.5. Species-Specific Responses to KNO_3 Priming

According to hierarchical clustering, the effect of KNO_3 seemed to be related to *Medicago* species (Figure 3). The genetic diversity of *Medicago* has been identified by numerous studies [47]. Besides, *Medicago* was considered as the genetically complex species [48]. These findings indicate that there is a genotypic difference in seed priming efficacy, which is in accordance with other studies [47,48]. The differences between the three species may be due to ontogeny and species diversity. *Medicago indicus* sprouts primed by KNO_3 showed the highest antioxidant and antidiabetics activities. Besides, it seemed that KNO_3 improved the nutritive value of these farmer sprouts by enhancing the accumulation of vitamins (Vit E and Vit C), proteins, mineral nutrients, and nitrogen, followed by *Medicago interexta* and *Medicago polymorpha*, which responded to KNO_3 priming by enhancing the accumulation of the major amino acids.

5. Conclusions

The use of KNO_3 priming to improve the biological and nutritional qualities of *Medicago* sprouts has been proven to be effective. Hence, *Medicago* sprouts are increasingly being used as an alternative source of natural antioxidant and mineral components in ready-to-eat fresh products or the manufacture of new safe functional foods. At the species level, *Medicago interexta* and *Medicago polymorpha* were more responsive to the KNO_3 positive effect, being better than other species (*Medicago indicus*), since it gave the highest antioxidant and antidiabetic activities.

Author Contributions: Conceptualization, H.A.; methodology, F.H.; formal analysis, H.A.; investigation, A.S. and M.K.O.; writing—original draft preparation, A.Z.; writing—review and editing, A.Z. and H.A.; visualization, W.H.A.-Q., Y.A.A. and M.Y.H.; supervision, A.H.A.H.; project administration, H.A.; funding acquisition, W.H.A.-Q. and Y.A.A.; Validation, A.A.-H. All authors have read and agreed to the published version of the manuscript.

Funding: This research received no external funding.

Institutional Review Board Statement: Not applicable.

Informed Consent Statement: Not applicable.

Acknowledgments: The authors extend their appreciation to the Researchers Supporting Project number (RSP-2021/374) King Saud University, Riyadh, Saudi Arabia.

Conflicts of Interest: The authors declare no conflict of interest.

References

1. Plaza, L.; De Ancos, B.; Cano, M.P. Nutritional and health-related compounds in sprouts and seeds of soybean (*Glycine max*), wheat (*Triticum aestivum* L.) and alfalfa (*Medicago sativa*) treated by a new drying method. *Eur. Food Res. Technol.* **2003**, *216*, 138–144. [CrossRef]
2. Marton, M.; Mandoki, Z.; Caspo, J.; Caspo-Kiss, Z.; Marton, M.; Mándoki, Z.; Marton, M.; Mandoki, Z.; Caspo, J.; Caspo-Kiss, Z. The role of sprouts in human nutrition. A review. *Aliment. Hungarian Univ. Transylvania* **2010**, *3*, 81–117.
3. Gabrovská, D.; Paulíčková, I.; Mašková, E.; Fiedlerová, V.; Kocurová, K.; Průchová, J.; Strohalm, J.; Houška, M. Changes in selected vitamins, microorganism counts, and sensory quality during storage of pressurised sprouted seed of alfalfa (*Medicago sativa* L.). *Czech. J. Food Sci.* **2005**, *23*, 246–250. [CrossRef]
4. Mattioli, S.; Dal Bosco, A.; Martino, M.; Ruggeri, S.; Marconi, O.; Sileoni, V.; Falcinelli, B.; Castellini, C.; Benincasa, P. Alfalfa and flax sprouts supplementation enriches the content of bioactive compounds and lowers the cholesterol in hen egg. *J. Funct. Foods* **2016**, *22*, 454–462. [CrossRef]
5. Wrona, O.; Rafińska, K.; Walczak-Skierska, J.; Mozeński, C.; Buszewski, B. Extraction and Determination of Polar Bioactive Compounds from Alfalfa (*Medicago sativa* L.) Using Supercritical Techniques. *Molecules* **2019**, *24*, 4608. [CrossRef] [PubMed]
6. Apostol, L.; Iorga, S.; Mosoiu, C.; Racovita, R.C.; Niculae, O.M.; Vlasceanu, G. Alfalfa Concentrate—A Rich Source of Nutrients for Use in Food Products. *Int. Sci. Publ.* **2017**, *5*, 66–73.
7. Pandey, S.; Shenmare, K. Review on nutritional profile of *Medicago sativa* seeds. *Asian J. Adv. Res.* **2021**, *7*, 28–31.
8. Hasanuzzaman, M.; Bhuyan, M.H.M.B.; Nahar, K.; Hossain, M.S.; Al Mahmud, J.; Hossen, M.S.; Masud, A.A.C.; Moumita; Fujita, M. Potassium: A vital regulator of plant responses and tolerance to abiotic stresses. *Agronomy* **2018**, *8*, 31. [CrossRef]
9. Zrig, A.; Tounekti, T.; BenMohamed, H.; Abdelgawad, H.; Vadel, A.M.; Valero, D.; Khemira, H. Differential response of two almond rootstocks to chloride salt mixtures in the growing medium. *Russ. J. Plant Physiol.* **2016**, *63*, 143–151. [CrossRef]
10. Nacry, P.; Bouguyon, E.; Gojon, A. Nitrogen acquisition by roots: Physiological and developmental mechanisms ensuring plant adaptation to a fluctuating resource. *Plant Soil* **2013**, *370*, 1–29. [CrossRef]
11. Reddy, M.M.; Ulaganathan, K. Nitrogen Nutrition, Its Regulation and Biotechnological Approaches to Improve Crop Productivity. *Am. J. Plant Sci.* **2015**, *6*, 2745–2798. [CrossRef]
12. Balotf, S.; Kavoosi, G. Differential nitrate accumulation, nitrate reduction, nitrate reductase activity, protein production and carbohydrate biosynthesis in response to potassium and sodium nitrate. *Afr. J. Biotechnol.* **2011**, *10*, 17973–17980. [CrossRef]
13. Yang, S.; Zu, Y.; Li, B.; Bi, Y.; Jia, L.; He, Y.; Li, Y. Response and intraspecific differences in nitrogen metabolism of alfalfa (*Medicago sativa* L.) under cadmium stress. *Chemosphere* **2019**, *220*, 69–76. [CrossRef] [PubMed]

14. Khan, S.; Yu, H.; Li, Q.; Gao, Y.; Sallam, B.N.; Wang, H.; Liu, P.; Jiang, W. Exogenous application of amino acids improves the growth and yield of lettuce by enhancing photosynthetic assimilation and nutrient availability. *Agronomy* **2019**, *9*, 266. [CrossRef]
15. Zhou, H.X. Protein folding and binding in confined spaces and in crowded solutions. *J. Mol. Recognit.* **2004**, *17*, 368–375. [CrossRef] [PubMed]
16. Tanha, A.; Golzardi, F.; Mostafavi, K. Seed Priming to Overcome Autotoxicity of Alfalfa (*Medicago sativa*). *World J. Environ. Biosci.* **2017**, *6*, 1–5.
17. Lara, T.S.; Lira, J.M.S.; Rodrigues, A.C.; Rakocevic, M.; Alvarenga, A.A. Potassium Nitrate Priming Affects the Activity of Nitrate Reductase and Antioxidant Enzymes in Tomato Germination. *J. Agric. Sci.* **2014**, *6*, 72. [CrossRef]
18. Ali, M.M.; Javed, T.; Mauro, R.P.; Shabbir, R.; Afzal, I.; Yousef, A.F. Effect of seed priming with potassium nitrate on the performance of tomato. *Agriculture* **2020**, *10*, 498. [CrossRef]
19. Abdelgawad, H.; De Vos, D.; Zinta, G.; Domagalska, M.A.; Beemster, G.T.S.; Asard, H. Grassland species differentially regulate proline concentrations under future climate conditions: An integrated biochemical and modelling approach. *New Phytol.* **2015**, *208*, 354–369. [CrossRef]
20. Hu, X.; Tanaka, A.; Tanaka, R. Simple extraction methods that prevent the artifactual conversion of chlorophyll to chlorophyllide during pigment isolation from leaf samples. *Plant Methods* **2013**, *1*, 9–19. [CrossRef]
21. Velioglu, Y.S.; Mazza, G.; Gao, L.; Oomah, B.D. Antioxidant Activity and Total Phenolics in Selected Fruits, Vegetables, and Grain Products. *J. Agric. Food Chem.* **1998**, *46*, 4113–4117. [CrossRef]
22. Kamal, J. Quantification of alkaloids, phenols and flavonoids in sunflower (*Helianthus annuus* L.). *Afr. J. Biotechnol.* **2011**, *10*, 3149–3151. [CrossRef]
23. Sykes, M.; Croucher, J.; Smith, R.A. Proficiency testing has improved the quality of data of total vitamin B2 analysis in liquid dietary supplement. *Anal. Bioanal. Chem.* **2011**, *400*, 305–310. [CrossRef] [PubMed]
24. Bligh, E.G.; Dyer, W.J. Canadian Journal of Biochemistry and Physiology. *Can. J. Biochem. Physiol.* **1959**, *37*, 911–917. [CrossRef]
25. Cataldo, D.A.; Haroon, M.H.; Schrader, L.E.; Youngs, V.L. Rapid colorimetric determination of nitrate in plant tissue by nitration of salicylic acid. *Commun. Soil Sci. Plant Anal.* **1975**, *6*, 71–80. [CrossRef]
26. Hamad, I.; Abdelgawad, H.; Al Jaouni, S.; Zinta, G.; Asard, H.; Hassan, S.; Hegab, M.; Hagagy, N.; Selim, S. Metabolic analysis of various date palm fruit (*Phoenix dactylifera* L.) cultivars from Saudi Arabia to assess their nutritional quality. *Molecules* **2015**, *20*, 13620–13641. [CrossRef]
27. Zinta, G.; Abdelgawad, H.; Peshev, D.; Weedon, J.T.; Van Den Ende, W.; Nijs, I.; Janssens, I.A.; Beemster, G.T.S.; Asard, H. Dynamics of metabolic responses to periods of combined heat and drought in Arabidopsis thaliana under ambient and elevated atmospheric CO_2. *J. Exp. Bot.* **2018**, *69*, 2159–2170. [CrossRef]
28. Farooq, M.; Irfan, M.; Aziz, T.; Ahmad, I.; Cheema, S.A. Seed Priming with Ascorbic Acid Improves Drought Resistance of Wheat. *J. Agron. Crop Sci.* **2013**, *199*, 12–22. [CrossRef]
29. Shah, S.A.; Zeb, A.; Masood, T.; Noreen, N.; Abbas, S.J.; Samiullah, M.; Alim, M.A.; Muhammad, A. Effects of sprouting time on biochemical and nutritional qualities of Mungbean varieties. *Afr. J. Agric. Res.* **2011**, *6*, 5091–5098. [CrossRef]
30. Anwar, A.; Yu, X.; Li, Y. Seed priming as a promising technique to improve growth, chlorophyll, photosynthesis and nutrient contents in cucumber seedlings. *Not. Bot. Horti Agrobot. Cluj-Napoca* **2020**, *48*, 116–127. [CrossRef]
31. Ahmadvand, G.; Soleimani, F.; Saadatian, B.; Pouya, M. Effect of Seed Priming with Potassium Nitrate on Germination and Emergence Traits of Two Soybean Cultivars under Salinity Stress Conditions. *Am. J. Agric. Environ. Sci.* **2012**, *12*, 769–774. [CrossRef]
32. Iskender, T.; Mustafa, K.A.; Mahmut, K. Rapid and enhanced germination at low temperature of alfalfa and white clover seeds following osmotic priming. *Trop. Grasslands* **2009**, *43*, 171–177.
33. Brooks, S.; Athinuwat, D.; Chiangmai, P.N. Enhancing germination and seedling vigor of upland rice seed under salinity and water stresses by osmopriming. *Sci. Technol. Asia* **2020**, *25*, 63–74. [CrossRef]
34. Mapongmetsem, P.M.; Duguma, B.; Nkongmeneck, B.A.; Selegny, E. The effect of various seed pretreatments to improve germination in eight indigenous tree species in the forests of Cameroon. *Ann. For. Sci.* **1999**, *56*, 679–684. [CrossRef]
35. Gaweł, E. Chemical composition of lucerne leaf extract (EFL) and its applications as a phytobiotic in human nutrition. *Acta Sci. Pol. Technol. Aliment.* **2012**, *11*, 303–309.
36. Yanar, M.; Erçen, Z.; Özlüer Hunt, A.; Büyükçapar, H.M. The use of alfalfa, *Medicago sativa* as a natural carotenoid source in diets of goldfish, Carassius auratus. *Aquaculture* **2008**, *284*, 196–200. [CrossRef]
37. Sathasivam, R.; Ki, J.S. A review of the biological activities of microalgal carotenoids and their potential use in healthcare and cosmetic industries. *Mar. Drugs* **2018**, *16*, 26. [CrossRef]
38. Kim, H.Y.; Shin, H.S.; Park, H.; Kim, Y.C.; Yun, Y.G.; Park, S.; Shin, H.J.; Kim, K. In vitro inhibition of coronavirus replications by the traditionally used medicinal herbal extracts, Cimicifuga rhizoma, Meliae cortex, Coptidis rhizoma, and Phellodendron cortex. *J. Clin. Virol.* **2008**, *41*, 122–128. [CrossRef] [PubMed]
39. Soto-Zarazúa, M.G.; Bah, M.; Costa, A.S.G.; Rodrigues, F.; Pimentel, F.B.; Rojas-Molina, I.; Rojas, A.; Oliveira, M.B.P.P. Nutraceutical Potential of New Alfalfa (*Medicago sativa*) Ingredients for Beverage Preparations. *J. Med. Food* **2017**, *20*, 1039–1046. [CrossRef] [PubMed]
40. Newsholme, P.; Brennan, L.; Bender, K. Amino acid metabolism, β-cell function, and diabetes. *Diabetes* **2006**, *55*, 39–47. [CrossRef]

41. Gaweł, E.; Grzelak, M.; Janyszek, M. Lucerne (*Medicago sativa* L.) in the human diet—Case reports and short reports. *J. Herb. Med.* **2017**, *10*, 8–16. [CrossRef]

42. Jabeen, N.; Ahmad, R. Foliar application of potassium nitrate affects the growth and nitrate reductase activity in sunflower and safflower leaves under salinity. *Not. Bot. Horti Agrobot. Cluj-Napoca* **2011**, *39*, 172–178. [CrossRef]

43. León-López, L.; Escobar-Zúñiga, Y.; Salazar-Salas, N.Y.; Mora Rochín, S.; Cuevas-Rodríguez, E.O.; Reyes-Moreno, C.; Milán-Carrillo, J. Improving Polyphenolic Compounds: Antioxidant Activity in Chickpea Sprouts through Elicitation with Hydrogen Peroxide. *Foods* **2020**, *9*, 1791. [CrossRef] [PubMed]

44. Caunii, A.; Pribac, G.; Grozea, I.; Gaitin, D.; Samfira, I. Design of optimal solvent for extraction of bio-active ingredients from six varieties of *Medicago sativa*. *Chem. Cent. J.* **2012**, *6*, 1–8. [CrossRef]

45. Salih, B.A.; Azeez, K.O. Antidiabetic Action of Alfalfa (*Medicago sativa*) Leaves Powder on Type II Diabetic Patients. *Polytech. J.* **2019**, *9*, 23–25. [CrossRef]

46. Helal, E.G.E.; Abd El-Wahab, S.M.; Atia, T.A. Hypoglycemic Effect of the Aqueous Extracts of Lupinus Albus, *Medicago sativa* (Seeds) and their Mixture on Diabetic Rats. *Egypt. J. Hosp. Med.* **2013**, 685–698. [CrossRef]

47. Julier, B.; Huyghe, C.; Ecalle, C. Within- and Among-Cultivar Genetic Variation in Alfalfa: Forage Quality, Morphology, and Yield. *Crop Sci.* **2000**, *369*, 365–369. [CrossRef]

48. Basbag, M.; Demirel, R.; Avci, M. Determination of some agronomical and quality properties of wild alfalfa (*Medicago sativa* L.) clones in Turkey Determination of some agronomical and quality properties of wild alfalfa (*Medicago sativa* L.) clones in Turkey. *J. Food Agric. Environ.* **2009**, *7*, 357–359.

Article

Innovating the Synergistic Assets of β-Amino Butyric Acid (BABA) and Selenium Nanoparticles (SeNPs) in Improving the Growth, Nitrogen Metabolism, Biological Activities, and Nutritive Value of *Medicago interexta* Sprouts

Samy Selim [1,*,†], Nosheen Akhtar [2,†], Eman El Azab [3], Mona Warrad [3], Hassan H. Alhassan [1], Mohamed Abdel-Mawgoud [4,†], Soad K. Al Jaouni [5] and Hamada Abdelgawad [6,*]

1 Department of Clinical Laboratory Sciences, College of Applied Medical Sciences, Jouf University, Sakaka 72341, Saudi Arabia; h.alhasan@ju.edu.sa
2 Department of Biological Sciences, National University of Medical Sciences, Rawalpindi 46000, Pakistan; nosheenakhtar@numspak.edu.pk
3 Department of Clinical Laboratory Sciences, College of Applied Medical Sciences at Al-Quriat, Jouf University, Al-Quriat 77454, Saudi Arabia; efelazab@ju.edu.sa (E.E.A.); mfwarad@ju.edu.sa (M.W.)
4 Department of Medicinal and Aromatic Plants, Desert Research Centre, Cairo 11753, Egypt; Mohamed_drc@yahoo.com
5 Hematology/Pediatric Oncology, Yousef Abdulatif Jameel Scientific Chair of Prophetic Medicine Application, Faculty of Medicine, King Abdulaziz University, Jeddah 21589, Saudi Arabia; saljaouni@kau.edu.sa
6 Botany and Microbiology Department, Faculty of Science, Beni-Suef University, Beni-Suef 62511, Egypt
* Correspondence: sabdulsalam@ju.edu.sa (S.S.); hamada.abdelgawad@uantwerpen.be (H.A.)
† These authors contributed equally to this work.

Citation: Selim, S.; Akhtar, N.; El Azab, E.; Warrad, M.; Alhassan, H.H.; Abdel-Mawgoud, M.; Al Jaouni, S.K.; Abdelgawad, H. Innovating the Synergistic Assets of β-Amino Butyric Acid (BABA) and Selenium Nanoparticles (SeNPs) in Improving the Growth, Nitrogen Metabolism, Biological Activities, and Nutritive Value of *Medicago interexta* Sprouts. *Plants* **2022**, *11*, 306. https://doi.org/10.3390/plants11030306

Academic Editors: Angelica Galieni and Beatrice Falcinelli

Received: 18 November 2021
Accepted: 17 January 2022
Published: 24 January 2022

Publisher's Note: MDPI stays neutral with regard to jurisdictional claims in published maps and institutional affiliations.

Abstract: In view of the wide traditional uses of legume sprouts, several strategies have been approved to improve their growth, bioactivity, and nutritive values. In this regard, the present study aimed at investigating how priming with selenium nanoparticles (SeNPs, 25 mg L^{-1}) enhanced the effects of β-amino butyric acid (BABA, 30 mM) on the growth, physiology, nitrogen metabolism, and bioactive metabolites of *Medicago interexta* sprouts. The results have shown that the growth and photosynthesis of *M. interexta* sprouts were enhanced by the treatment with BABA or SeNPs, being higher under combined treatment. Increased photosynthesis provided the precursors for the biosynthesis of primary and secondary metabolites. In this regard, the combined treatment had a more pronounced effect on the bioactive primary metabolites (essential amino acids), secondary metabolites (phenolics, GSH, and ASC), and mineral profiles of the investigated sprouts than that of sole treatments. Increased amino acids were accompanied by increased nitrogen metabolism, i.e., nitrate reductase, glutamate dehydrogenase (GDH), glutamate synthase (GOGAT), glutamine synthase (GS), cysteine synthesis serine acetyltransferase, arginase, threonine synthase, and methionine synthase. Further, the antioxidant capacity (FRAP), the anti-diabetic activities (i.e., α-amylase and α-glucosidase inhibition activities), and the glycemic index of the tested sprouts were more significantly improved by the combined treatment with BABA and SeNPs than by individual treatment. Overall, the combined effect of BABA and SeNPs could be preferable to their individual effects on plant growth and bioactive metabolites.

Keywords: *Medicago interexta*; sprouts; SeNPs; BABA; nutritious metabolites; anti-diabetic

1. Introduction

Phytochemicals are important metabolic compounds that confer the capability of plants to combat environmental stress, boost their defense systems, and protect them from pathogens and insects. Secondary metabolites also play key roles as health-promoting enzymes; accordingly, they are an essential part of human health. These metabolites, especially the phenolic compounds [1] and glucosinolates [2], have been reported for their protective effects against the oxidative process and provide protection against different diseases, such as cardiovascular diseases, neurodegenerative diseases, and cancer [3]. Currently, the development of new strategies to improve plant growth and boost the production of secondary metabolites is one of the fascinating fields of research. Application of elicitors and nanoparticles can enhance the production of bioactive metabolites in plants, including their qualitative value in producing fresh produce, enriched foods, or raw ingredients for feed/food and pharmaceutical products [4,5].

Elicitors mimic the action of plant signaling and increase the production of reactive oxygen species (ROS) and reactive nitrogen species (RNS), upregulate the defense-related genes, change the potential of plasma membrane cells, and enhance ion fluxes (Cl^- and K^+ efflux and Ca^{2+} influxes) [6,7]. They also induce changes in protein phosphorylation and lipid oxidation, and activate the de novo biosynthesis of transcription factors, which directly regulate the expression of genes involved in secondary metabolites production [6,7]. β-aminobutyric acid (BABA) is a nonprotein amino acid that is considered as one of the plant activators that induce resistance in many different plant species against a wide range of abiotic and biotic stresses. BABA, which was found to present naturally at low concentrations in plant tissues, can be increased 5-fold to 10-fold under stress conditions [8]. The broad spectrum protective effect of BABA against numerous plant diseases has been well documented [9,10] and is attributable to enhanced phenolics content or related compounds. For example, research has shown that BABA induces changes in the response of leaf antioxidants to UV-B [11,12]. Moreover, BABA interacts with several hormones, such as salicylic acid (SA), abscisic acid (ABA), and ethylene [8] and thereby takes part in the growth of plants, including development, photosynthesis, transpiration, and ion uptake and transport.

Furthermore, in the context to plant growth, nanoparticles have unique physicochemical properties and the potential to boost plant metabolism, and thus the production of secondary metabolites [13]. Application of nanoparticles (NPs) is currently an interesting area for minimizing the use of chemical fertilizers and improving the growth and yield of plants [14,15]. The unique physicochemical properties of NPs have potentially opened up new paradigms, and the introduction of NPs to plants might have a significant impact; therefore, they can be used in agricultural applications for better growth and yield.

Among different nanoparticles, selenium nanoparticles (SeNPs) have precedence over other nanoparticles because of the significant role of selenium in activating plants' defense systems. Several studies have demonstrated that Se may exert diverse beneficial effects at low concentrations as an antioxidant and as a growth-promoting agent in higher plants. Moreover, some plants are able to accumulate large amounts of Se as an essential element [16].

Se uptake by plants depends on some environmental factors, such as soil pH, salinity, and concentration of competing ions. Usually, the stems and leaves accumulate higher Se levels than do the roots [17]. It has also been demonstrated that Se might affect plant growth and many metabolic processes. For instance, Se might contribute to maintaining the water potential of plants under drought conditions [18]. Se could enhance light harvesting, thereby increasing the available energy for plants [17]. On the other hand, the phytoxicity of Se might be related to an interaction with sulfur; consequently, sulfur-containing amino acids might be replaced by Se-containing amino acids [19].

The toxicity of Se depends on its chemical form as well as on plant age. Se toxicity could be observed at a concentration of ≥ 2 mg/kg dry weight. The maximum Se content (safest concentration) in the medium without growth inhibition was found to be 1, 10, 0.25, and 0.25 mg/L for radish, sunflower, alfalfa, beetroot, respectively [20]. On the other hand, SeNPs have a more enhancing effect on plants, with low toxicity, when compared with the bulk form [21]. In addition, the use of biogenic SeNPs is known to be an environmentally friendly and ecologically biocompatible approach in enhancing crop production by alleviating biotic and abiotic stresses [22]. Moreover, SeNPs enhance photosynthetic pigment activity, nutrient status, antioxidant activity, and total phenolic content under drought stress. Surprisingly, at a minimal dose, Se is highly effective against salinity stress by maintaining turgor pressure, controlling the accumulation of total sugars, amino acids, and potential antioxidant enzymes, and improving the transpiration rate [22]. Se also decreases chloride ion contents, ROS species, and membrane damage. In addition, Se decreases sodium-ion accumulation and increases potassium-ion accumulation, thereby reducing the detrimental effects of salt stress on plants [23].

Legumes are valued worldwide as a sustainable and inexpensive meat alternative and are considered the second most important food source after cereals. Legumes are a rich source of many nutrient components, including starch, protein, certain fatty acids, and micronutrients such as vitamins, minerals, and bioactive compounds [24–26]. Medicago is the genius of leguminous plants and *Medicago interexta* (*M. interexta*) is an important member, reported to be the source of proteins and tannins [26]. Regarding the significance of BABA and SeNPs in triggering the production of phytochemicals, we hypothesized that the application of both can have additive effects and can enhance the nutritional and pharmacological value of *M. interexta* by improving the production of primary and secondary metabolites. Thus, the present study aimed to evaluate the impact of BABA, SeNPs, and their combined effects on *M. interexta* sprouts. We evaluated the impacts on growth, mineral content, the vitamin and amino acid profile, nitrogen, and phenolic metabolism, as well as on the concentrations of several phytochemical compounds. We further examined the role of SeNPs and/or BABA in the enhancement of the antioxidant and antidiabetic potential of *M. interexta*. Overall, our study contributes to an understanding of the biochemical basis of BABA, SeNPs, and their combination in *M. interexta*.

2. Results

2.1. Enhanced Growth of M. interexta Sprouts under Sole and Combined Treatments with BABA and/or SeNPs

The present investigation revealed that the treatment of *M. interexta* with β-amino butyric acid (BABA) led to a significant increase in biomass accumulation (expressed as fresh weight FW, dry weight DW), photosynthesis, and respiration by approximately 40%, in comparison to control sprouts (Figure 1). The addition of SeNPs to the target sprouts also induced a higher increase in growth and photosynthesis of *M. interexta* sprouts (by about 50–90%), in comparison to the non-treated plants. Interestingly, the combined effect of BABA and SeNPs resulted in a much higher increment in growth parameters, by approximately 200% when compared with the control sprouts. Thus, the growth of *M. interexta* sprouts was enhanced by the sole and combined treatment with BABA and/or SeNPs, with higher enhancement under the combined treatment.

Regarding pigment content, the sole treatment of *M. interexta* sprouts with BABA significantly increased almost all carotenoids (i.e., chl a, b, β-carotene, lutein, neoxanthin, and violaxanthin) (Table 1). In addition, when *M. interexta* sprouts were grown under individual treatment with SeNPs, there were significant increments in all the detected carotenoids, except for neoxanthin. Moreover, the combined treatment of *M. interexta* sprouts with BABA and SeNPs increased all the detected carotenoids, when compared with the control sprouts.

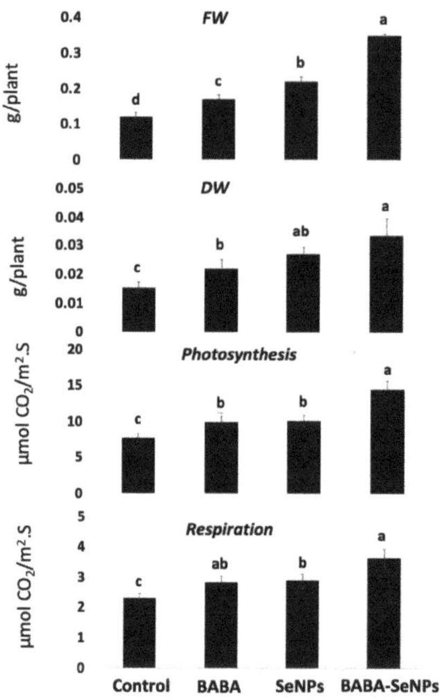

Figure 1. Biomass; fresh weight (FW) (mg g^{-1} FW) and dry weight (DW) (mg g^{-1} FW); photosynthesis (μmol CO$_2$ m^{-2} s^{-1}); and respiration of control in BABA- and/or SeNPs-treated *M. interexta* sprouts. Data are represented by the means of four replicates $\pm$ standard deviations. Different small letter superscripts (a–d) within a row indicate significant differences between control and BABA and/or SeNPs samples. One-way analysis of variance (ANOVA) was performed. Tukey's test was used as the post hoc test for the separation of means ($p < 0.05$).

Table 1. Pigment content (chlorophyll *a* + *b*) (mg g^{-1} FW) of control and BABA- and/or Se NPs-treated *M. interexta* sprouts. Data are represented by the means of four replicates $\pm$ standard deviations.

	Control	BABA	SeNPs	BABA-SeNPs
Chl a	0.65 $\pm$ 0.06 c	0.92 $\pm$ 0.02 b	1.05 $\pm$ 0.2 b	1.97 $\pm$ 0.17 a
Chl b	0.43 $\pm$ 0.069 c	0.53 $\pm$ 0.08 bc	0.59 $\pm$ 0.116 b	1.18 $\pm$ 0.19 a
β-Carotene	0.04 $\pm$ 0.01 c	0.07 $\pm$ 0.004 b	0.07 $\pm$ 0.017 b	0.11 $\pm$ 0.01 a
Lutein	0.14 $\pm$ 0.03 c	0.24 $\pm$ 0.02 b	0.23 $\pm$ 0.02 b	0.53 $\pm$ 0.03 a
Neoxanthin	0.02 $\pm$ 0.01 c	0.02 $\pm$ 0.003 b	0.01 $\pm$ 0.001 c	0.05 $\pm$ 0.007 a
Violaxanthin	0.05 $\pm$ 0.01 c	0.04 $\pm$ 0.003 b	0.07 $\pm$ 0.009 a	0.05 $\pm$ 0.001 b

Different small letters (a–c) within a row indicate significant differences between control and BABA- and/or SeNPs-samples. One-way analysis of variance (ANOVA) was performed. Tukey's test was used as the post hoc test for the separation of means ($p < 0.05$).

2.2. Combined Treatment of M. interexta Sprouts with BABA and SeNPs Induced a More Pronounced Effect on Mineral and Vitamin Profiles than That of a Sole Treatment

In the current study, the mineral and vitamin profiles were investigated in *M. interexta* under the different effects of BABA and/or SeNPs (Table 2). Under control conditions, eight mineral elements (Ca, Cu, Fe, Zn, Mn, Mg, K, and P) were detected in *M. interexta* sprouts, whereas Zn had the highest content, followed by Ca and K. When *M. interexta* sprouts were treated individually with BABA, there was a significant increase only in Zn (by about 70%), in addition to a significant decrease in Mn, while no changes were observed for Cu, Fe, Ca, or K. In the case of sole treatment of *M. interexta* sprouts with SeNPs, there were remarkable increases in Ca, Fe, Zn (increased by 60–80%), K, and P (increased by approximately 100–150%), while no significant changes were reported for Cu and Mn. On the other hand, the combined treatment of *M. interexta* sprouts with BABA and SeNPs induced enhancing effects on the contents of Ca (elevated by 50%), Fe, Zn, Cu (increased by 80–100%), K (450%), and P (increased by about 170%). It was observed that Mn was not affected by any of the treatments used.

Table 2. Mineral elements (mg g^{-1} FW) and vitamins (mg g^{-1} FW) of control and BABA- and/or Se NPs-treated *M. interexta* sprouts. Data are represented by the means of four replicates $\pm$ standard deviations.

Parameters	Control	BABA	SeNPs	BABA-SeNPs
Elements				
Ca	17.57 $\pm$ 2.3 b	15.79 $\pm$ 3.5 b	27.79 $\pm$ 6.7 a	25.17 $\pm$ 0.47 a
Cu	2.26 $\pm$ 0.71 b	2.57 $\pm$ 1.07 b	2.87 $\pm$ 0.28 b	4.38 $\pm$ 1.1 a
Fe	3.99 $\pm$ 0.23 b	3.15 $\pm$ 0.78 b	5.48 $\pm$ 1.02 a	5.76 $\pm$ 0.44 a
Zn	22.62 $\pm$ 2.0 b	36.62 $\pm$ 3.3 a	35.88 $\pm$ 3.2 a	35.72 $\pm$ 3.2 a
Mn	0.25 $\pm$ 0.03 a	0.13 $\pm$ 0.1 b	0.28 $\pm$ 0.13 a	0.27 $\pm$ 0.1 a
K	15.60 $\pm$ 1.3 c	11.95 $\pm$ 3 c	40.60 $\pm$ 3.6 b	67.29 $\pm$ 6 a
P	5.81 $\pm$ 0.6 c	6.48 $\pm$ 0.5 c	10.44 $\pm$ 0.8 b	13.56 $\pm$ 1.1 a
Vitamins				
Vit C	7.81 $\pm$ 1.3 b	7.31 $\pm$ 1.2 b	8.15 $\pm$ 2.4 b	13.92 $\pm$ 0.7 a
Vit E	47.47 $\pm$ 1.2 b	44.57 $\pm$ 1.6 cb	48.47 $\pm$ 4.4 b	61.92 $\pm$ 3.9 a
Thiamin	0.10 $\pm$ 0 b	0.07 $\pm$ 0 b	0.13 $\pm$ 0.02 a	0.14 $\pm$ 0.06 a
Riboflavin	0.35 $\pm$ 0.3 b	0.51 $\pm$ 0.75 a	0.24 $\pm$ 0.47 b	0.49 $\pm$ 0.96 a

Different small letters (a–c) within a row indicate significant differences between control and BABA and/or Se NPs samples. One-way analysis of variance (ANOVA) was performed. Tukey's test was used as the post hoc test for the separation of means ($p < 0.05$).

Regarding vitamin content, four vitamins (Vit C, Vit E, thiamin, and riboflavin) were detected in *M. interexta* sprouts under control conditions, wherein Vit E was the predominant vitamin (Table 2). When treated individually with BABA, the target sprouts did not show significant changes in vitamin content, except for riboflavin (increased by 80%), in comparison to control plants. Similarly, there were no significant differences in vitamins, except for thiamin, in response to the sole treatment with SeNPs. Meanwhile, the interactive impact of both BABA and SeNPs has been reflected on increasing all vitamins detected in comparison to the control. Overall, the combined treatment with BABA and SeNPs had a more pronounced effect on the mineral and vitamin profiles of *M. interexta* sprouts than did a sole treatment.

2.3. M. interexta Sprouts Were More Responsive to the Combined Effect of BABA and SeNPs on Nitrogen Metabolism than to Individual Treatments

In the present investigation, amino acids have been analyzed in *M. interexta* sprouts grown under higher concentrations of BABA and/or SeNPs (Table 3). Under control conditions, 18 amino acids (i.e., asparagine, glutamine, serine, glycine, arginine, alanine, proline, histidine, valine, methionine, cystine, ornithine, leucine, phenylalanine, tyrosine, lysine, threonine, and tryptophane) were quantified in *M. interexta*, where glutamine had the highest percentage. From the current data, it is clear that *M. interexta* sprouts interacted differently to the effects of BABA and/or SeNPs. There were significant elevations in the contents of serine, glycine, alanine, proline, histidine, valine, ornithine, and phenylalanine, while no significant changes were observed for asparagine, glutamine, cystine, leucine, arginine, methionine, lysine, threonine, tryptophane, or tyrosine in *M. interexta* sprouts treated solely with BABA, when compared with the control sprouts.

Table 3. Amino acids (μg g^{-1} FW) of control and BABA- and/or SeNPs-treated *M. interexta* sprouts. Data are represented by the means of four replicates $\pm$ standard deviations.

Amino Acids	Control	BABA	SeNPs	BABA-SeNPs
Asparagine	1.53 ± 0.1 b	1.71 ± 0.06 b	1.76 ± 0.02 b	2.17 ± 0.01 a
Glutamine	1.89 ± 0.19 c	2.15 ± 0.25 c	3.39 ± 0.08 b	4.53 ± 0.12 a
Serine	1.18 ± 0.07 c	1.36 ± 0.13 ab	2.31 ± 0.13 b	2.66 ± 0.3 a
Glycine	1.40 ± 0.01 c	1.69 ± 0.07 b	1.16 ± 0.1 c	2.01 ± 0.01 a
Arginine	0.30 ± 0.05 c	0.38 ± 0.08 c	0.77 ± 0.05 a	0.57 ± 0.08 b
Alanine	0.54 ± 0.03 b	0.62 ± 0.03 a	0.51 ± 0 b	0.61 ± 0.02 a
Proline	0.93 ± 0.01 c	1.25 ± 0.03 b	2.54 ± 0.06 a	2.71 ± 0.18 a
Histidine	0.75 ± 0.05 b	0.90 ± 0.05 a	0.67 ± 0.04 b	0.72 ± 0.09 b
Valine	0.76 ± 0.15 b	0.91 ± 0.2 a	0.61 ± 0.09 b	0.73 ± 0.11 b
Methionine	0.66 ± 0.09 c	0.75 ± 0.05 c	0.97 ± 0.01 b	1.17 ± 0.1 a
Cystine	0.99 ± 0.14 b	0.79 ± 0.15 b	1.47 ± 0.08 a	1.56 ± 0.04 a
Ornithine	1.17 ± 0.18 c	2.10 ± 0.21 b	1.72 ± 0.04 b	3.06 ± 0.1 a
Leucine	0.98 ± 0.06 a	0.86 ± 0.18 a	0.86 ± 0.07 a	1.07 ± 0.12 a
Phenylalanine	1.42 ± 0.22 b	1.87 ± 0.23 a	1.65 ± 0.11 b	2.04 ± 0.11 a
Tyrosine	0.31 ± 0.04 a	0.30 ± 0 a	0.42 ± 0.01 ab	0.45 ± 0.01 ab
Lysine	0.70 ± 0.02 b	0.88 ± 0.02 b	1.02 ± 0.03 b	1.91 ± 0.03 a
Threonine	1.18 ± 0.05 b	1.32 ± 0.03 b	1.68 ± 0.08 a	1.78 ± 0.09 a
Treptophane	0.72 ± 0.08 b	0.83 ± 0.1 b	1.06 ± 0.02 ab	1.28 ± 0.04 a

Different small letters (a–c) within a row indicate significant differences between control and BABA and/or SeNPs samples. One-way analysis of variance (ANOVA) was performed. Tukey's test was used as the post hoc test for the separation of means ($p < 0.05$).

The individual treatment of *M. interexta* sprouts with SeNPs markedly induced the contents of asparagine, glutamine, serine, arginine, proline, methionine, cystine, ornithine, tyrosine, threonine, and tryptophane, but there were no significant changes in the levels of glycine, alanine, histidine, valine, leucine, lysine, or phenylalanine, when compared with the control sprouts. Moreover, the interaction between BABA and SeNPs led to significant increments in most of the detected amino acids in *M. interexta* sprouts, except for histidine, valine, and leucine, when compared with control plants.

Regarding nitrogen metabolism, *M. interexta* sprouts interacted differently to the effects of BABA and/or SeNPs on N, total protein, nitrate reductase, GDH, GOGAT, GS, cysteine synthesis serine acetyltransferase, arginase, threonine synthase, and methionine synthase (Table 4). When *M. interexta* sprouts were grown under the individual impact of BABA, there were remarkable increases in N content, nitrate reductase, GDH, GOGAT, GS, cysteine synthesis serine acetyltransferase, arginase, threonine synthase and methionine synthase, as well as significant reductions in total protein, in comparison to the control sprouts. In the case of treatment individually with SeNPs, the tested sprouts tended to display notable increases in N, GOGAT, GS, cysteine synthesis serine acetyltransferase, arginase, and methionine synthase, in addition to significant decreases in total protein, while no changes were reported for nitrate reductase, GDH, or threonine synthase, when compared to control sprouts. On the other hand, the combined treatment of *M. interexta* sprouts with BABA and SeNPs positively influenced the levels of all the measured related N-parameters, except for total proteins, which were significantly decreased when compared with control sprouts. It could be noted that the interaction between BABA and SeNPs exerted a more pronounced effect on the nitrogen metabolism of *M. interexta* than their individual treatments.

Table 4. Nitrogen (g 100 g^{-1} FW), protein content (g 100 g^{-1} FW), and nitrogen-related enzymes (umol mg^{-1} protein. min) of control and BABA- and/or SeNPs-treated *M. interexta* sprouts. Data are represented by the means of four replicates $\pm$ standard deviations.

	Control	BABA	SeNPs	BABA-SeNPs
Nitrogen	23.3 $\pm$ 0.8 b	35.7 $\pm$ 0.5 a	28.1 $\pm$ 1.2 b	41.2 $\pm$ 0.8 a
Total Protein	169.5 $\pm$ 1.9 a	118.0 $\pm$ 3.1 d	99.6 $\pm$ 2.2 c	136 $\pm$ 2.8 b
Nitrate reductase	45.2 $\pm$ 0.03 c	86.1 $\pm$ 5.4 b	43.1 $\pm$ 2.2 c	118 $\pm$ 11 a
GDH	4.14 $\pm$ 0.2 c	6.99 $\pm$ 0.48 b	4.9 $\pm$ 0.21 c	10 $\pm$ 0.48 a
GOGAT	7.8 $\pm$ 0.28 d	14.35 $\pm$ 0.4 b	10.3 $\pm$ 0.2 c	21 $\pm$ 1.8 a
GS	16.12 $\pm$ 0.9 d	26.10 $\pm$ 0.4 c	23.0 $\pm$ 1 b	32 $\pm$ 0.8 a
Cyst syn ser acetyltransferase	6.7 $\pm$ 0.28 d	11.05 $\pm$ 0.0 b	9.0 $\pm$ 0.4 c	14.2 $\pm$ 0.38 a
Arginase	4.01 $\pm$ 0.02 d	7.7 $\pm$ 0.46 b	5.9 $\pm$ 0.2 cd	10.7 $\pm$ 0.9 a
Threonine synthase	1.0 $\pm$ 0.02 c	1.70 $\pm$ 0.1 b	0.9 $\pm$ 0.04 c	2.6 $\pm$ 0.17 a
Methionine synthase	2.0 $\pm$ 0.01 c	4.30 $\pm$ 0.05 a	3.40 $\pm$ 0.1 b	4.4 $\pm$ 0.2 a

Different small letters (a–d) within a row indicate significant differences between control and BABA and/or Se NPs-samples. One-way analysis of variance (ANOVA) was performed. Tukey's test was used as the post hoc test for the separation of means ($p < 0.05$).

2.4. Antioxidants of M. interexta Sprouts Were Improved by the Sole and Combined Treatments with BABA and/or SeNPs

The levels of antioxidants (i.e., phenolics, FRAP, CAT, POX, GSH, and ASC) were measured in the target sprouts under the impact of BABA and/or SeNPs (Table 5). The individual treatment of *M. interexta* sprouts with BABA resulted in significant increases in flavonoids, phenols, FRAP (by about 90%), and GSH (by about 20%), as well as in ASC content (by about 80%), in comparison to control plants. Meanwhile, the sole treatment of the target sprouts with SeNPs also increased the levels of flavonoids, phenolics, antioxidant activity (by about 90%), GSH (by about 20%), and ASC content (by 20%), when compared with the control sprouts. Interestingly, highly significant increases in flavonoids, phenols, FRAP (by about 130%), GSH, and ASC (by about 100%) were obtained in *M. interexta* sprouts when treated with the combination of BABA and SeNPs. Thus, the levels of antioxidants of *M. interexta* were enhanced by the sole and combined treatments with BABA and/or Se NPs, with higher enhancement under the combined treatment.

Table 5. Flavonoids (mg g^{-1} FW), phenolic acids (mg g^{-1} FW), antioxidant capacity (FRAP) (µmol trolox g $^{-1}$ FW), GSH (mg g^{-1} FW), and ASC (mg g^{-1} FW) of control and BABA- and/or SeNPs-treated *M. interexta* sprouts. Data are represented by the means of four replicates $\pm$ standard deviations.

	Control	BABA	SeNPs	BABA-SeNPs
FRAP	11.9 $\pm$ 1.19 c	18.3 $\pm$ 2.5 b	18.8 $\pm$ 3.6 b	24.0 $\pm$ 5.6 a
Phenolics	3.54 $\pm$ 0.01 c	5.7 $\pm$ 0.02 b	6.7 $\pm$ 0.04 b	8.9 $\pm$ 0.04 a
Flavonoids	0.58 $\pm$ 0.01 c	0.81 $\pm$ 0.01 b	0.89 $\pm$ 0 b	1.47 $\pm$ 0.02 a
Reduced GSH	0.85 $\pm$ 0.11 b	1.03 $\pm$ 0.3 b	1.1 $\pm$ 0.24 a	1.56 $\pm$ 0.19 a
Reduced ASC	4.22 $\pm$ 0.47 b	7.19 $\pm$ 0.69 a	5.4 $\pm$ 0.56 b	8.56 $\pm$ 0.38 a

Different small letters (a–c) within a row indicate significant differences between control and BABA and/or SeNPs samples. One-way analysis of variance (ANOVA) was performed. Tukey's test was used as the post hoc test for the separation of means ($p < 0.05$).

2.5. Anti-Diabetic Activity of M. interexta Sprouts Was More Improved by the Combined Treatment with BABA and Se NPs than by Individual Treatments

In the present study, anti-diabetic activity (i.e., α-amylase and α-glucosidase inhibition activities, and the glycemic index GI) was investigated in *M. interexta* sprouts in response to the different effects of BABA and/or SeNPs (Figure 2). When treated individually with BABA, *M. interexta* sprouts showed more increases in α-amylase and α-glucosidase inhibition activity (by about 40% and 20%, respectively), as well as a significant decrease in GI (by about 50%) in comparison to the control. Meanwhile, the sole treatment of *M. interexta* sprouts with SeNPs induced significant increments in both α-amylase and α-glucosidase inhibition activities (increased by 20% and 10% respectively), but it decreased the GI (by about 30%). Interestingly, the interactive impact imposed by BABA and SeNPs has induced the levels of α-amylase and α-glucosidase inhibition activities, (by about 50%, and 90%, respectively), but decreased the GI (by about 30%), when compared with control plants. Thus, the anti-diabetic activity of *M. interexta* sprouts was more improved by the combined treatment with BABA and SeNPs than by an individual treatment.

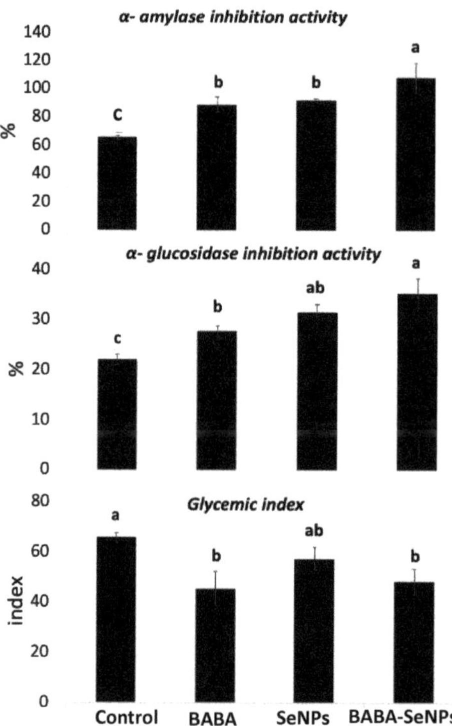

Figure 2. α-amylase and α-glucosidase inhibition activities, and the glycemic index (GI) of control and BABA- and/or SeNPs-treated *M. interexta* sprouts. Data are represented by the means of four replicates ± standard deviations. Different small letters (a–c) within a row indicate significant differences between control and BABA and/or SeNPs samples. One-way analysis of variance (ANOVA) was performed. Tukey's test was used as the post hoc test for the separation of means ($p < 0.05$).

3. Discussion

The present study was conducted to explore the collective effects of BABA and SeNPs on *M. interexta* sprouts in enhancing resistance against infections and increasing nutritional and pharmacological values. The effects of BABA and SeNPs on biosynthetic pathways and on the biological activities of *M. interexta* sprouts were evaluated, both alone and in combination. SeNPs and BABA have emerged as part of an effective class of elicitors that induce a defense mechanism that enhances the production of valuable bioactive metabolites. Our results indicated that the intervention comprised of a combined BABA and SeNPs treatment had a more significant impact on the endogenous biosynthetic pathways of *M. interexta* sprouts, as compared to individual treatments.

3.1. Improved Growth of M. Interexta Sprouts

In the current study, significant increases in the biomass production and photosynthetic activity of *M. interexta* sprouts were observed under treatment with BABA and SeNPs alone; however, the increases were remarkable when both agents were used in combination. The increases might be attributed to the additive effects of BABA and SeNPs that elicited a vigorous increase of metabolism and mineral content, as measured in our study. Many previous studies described the growth-modulating effects of BABA on different plants. Jisha et al. reported that BABA seed-priming increased seedling growth, under both unstressed and stressed conditions in rice [27]. BABA has been thought to enhance nitrogen metabolism, which consequently provides precursors needed for the biosynthesis of amino acids and protein and increases photosynthesis, growth rates, and biomass accumulation. In addition, the improved photosynthetic pigments under treatment with BABA, as reported herein, are directly related to the photosynthesis process and to the efficiency of photosynthesis. We observed that the increase of biomass using SeNPs was higher, as compared to using BABA. However, a remarkable increase in plant growth was observed when a combined treatment used both BABA and SeNPs. The positive effects of SeNPs on the growth of different plants support our data on the increased growth of *M. interexta* when SeNPs were used, either alone or in combination.

Previous studies showed that the use of SeNPs indicated growth-promoting effects in cowpea yield [28], efficiently upregulated selenoenzymes, and exhibited less toxicity [29]. Previous studies have also shown that SeNPs could enhance the photosynthetic efficiency of some plants, such as tomato. Such positive effects could also be reflected in increasing pigment contents, as reported in our study. This might be due to the small size of NPs, enabling them to easily move through plant parts [30]. In tomato, SeNPs improved the parameters of plant growth at low concentration (1 μM) [31]. Similarly, SeNPs at 400 mg improved the growth of the cluster bean [32].

3.2. Improved Pigment Content of M. interexta Sprouts

Interesting patterns were observed in the pigment contents of sprouts, using individual and combined treatment groups of *M. interexta*. The differential patterns indicated that the combined treatment targeted multiple pathways that were not affected when a single agent was used. For example, the combined treatment increased Chl a, Chl b, and neoxanthin, while BABA alone also increased these pigments. At a concentration of 25 mg L^{-1}, SeNPs' suspension-priming significantly reduced neoxanthin when used alone, while in sprouts subjected to combined treatment, neoxanthin was observed to be increased. Similarly, the combined intervention and the sole treatment with SeNPs or BABA resulted in significant increases in violaxanthin.

Previous studies reported the effects of SeNPS and BABA on photosynthetic pigments. SeNPs at a low concentration of 6.25 μM were found to be effective in increasing total photosynthetic pigments in the leaves of cowpea [28]. Similarly, in tomato leaves, application of SeNPs at 1 μM improved the chlorophyll content by 27.5% [31]. Contrary to our results, the priming of seeds with BABA is reported to have positive effects on pigment content. For example, rice seed-priming with BABA increased the photosynthetic pigment content of leaves, modified the Chl a fluorescence, and enhanced the photosystem activities of seedlings [27].

Our study results are also contrary to the reported finding that BABA exhibited an undesirable side effect, i.e., that it reduces plant growth [33]; however, we observed that BABA alone also enhanced photosynthesis and plant growth. This is attributable to the fact that different plant species employ different defense mechanisms and, accordingly, differential effects of the same elicitor can be observed among species. Our results showed that the combined treatment of *M. interexta* could increase the content of Chl a, Chl b, and carotene significantly, indicating that it could strengthen MI by enhancing the photosynthetic system.

3.3. Improved Mineral Content and Vitamin Profile of M. interexta Sprouts

Plant-derived foods have the potential to serve as dietary sources for all human-essential minerals. The essential minerals include N, S, P, K, Ca, Cl, Fe, Zn, Mn, Cu, B, Mo, and Ni. Among these, Ca, Zn, Ca, Cu, Fe, K, Mn, K, and P were detected in *M. interext* sprouts, from which Zn, Ca, and K were present in higher amounts. We evaluated the effects of treatments on mineral content and the results revealed that Zn concentration was increased by BABA while SeNPs increased K and P. The combined treatment resulted in a robust increase in the concentration of K, P, Fe, Zn Cu, and Ca. These increases in minerals might lead to remarkable increases in the growth of sprouts of *M. interexta*, as the minerals, especially K, modulate various biochemical and physiological processes that are responsible for plant growth and development. Also, the BABA-induced increases in minerals could be due to increased root growth that, in turn, triggers nutrient uptake by plants [34,35]. BABA upregulated mineral transporters [36,37]. Moreover, improved nitrogen nutrition by BABA treatment could enhance root uptake, root-to-shoot translocation, and remobilization of Zn [38]. In this context, the positive effects of nitrogen and Zn uptake and translocation can be explained by upregulating the transporter proteins and nitrogenous chelators involved in these processes. Consequentially, an increased level of Zn is needed for biosynthesis and for the structural and functional integrity of proteins and amino acid metabolism [39].

Regarding the effect of SeNPs on mineral uptake, the study in [40] indicated that exposure to Se significantly upregulated the expressions of the phosphate transporter (PHT), the potassium channel protein (KCP), and the potassium transporter protein (KTP). In agreement with our results, Se application was found to enhance the mineral content (e.g., Zn, Mn, Cu, Ca, Mg, Na, and K) of alfalfa and radish [20]. It was found that the mineral content (P, K, Ca, and Mg) of garlic was significantly reduced under Se treatment [41]. Furthermore, our results showed that *M. interexta* sprouts are a rich source of vitamins, especially vitamin E, which were further increased by the combined treatment with BABA and SeNPs. High N availability under BABA treatment can also promote plants' Se absorption, and Se can then be further metabolized into seleno-proteins. In this regard, N fertilizer promotes growth, thereby promoting the absorption of P, K, S, and other mineral elements, including Se, by the root system [42].

Interestingly, neither BABA nor SeNPs had an effect on the concentrations of vitamins when used alone. Our study indicated that mutual intervention was more effective in triggering the multiple defense pathways that consequently enhanced the concentration of vitamins.

3.4. Improved Nitrogen Metabolism of M. interexta Sprouts

It is known that the nitrogen source, either nitrate or ammonium, affects the levels of amino acids and proteins, and consequently the rate of growth and biomass accumulation. Nitrogen metabolism is thought to be involved in the conversion of amino acids via nitrate reduction [43]. BABA is thought to enhance nitrogen metabolism, which consequently provides precursors that are needed for the biosynthesis of amino acids and protein. Previous reports have also shown that priming could increase nitrogen metabolism by enhancing the contents of amino acids and total protein, as well as nitrate reductase activity [44].

In our study, the individual and combined treatments with BABA and/or SeNPs have positively affected almost all the measured N-related parameters. In line with our results, priming has been shown to increase the production of GDH and GOGAT [44]. In this regard, the GS/GOGAT pathway is thought to assimilate ammonia at normal intracellular concentrations, while GDH plays a role in the assimilation of ammonia into amino acids. Similarly, γ-aminobutyric acid (GABA) has been previously found to promote total nitrate reductase activity [45].

Arginase is known to be involved in the conversion of arginine into ornithine, so it might contribute to increasing the ornithine content in the sprouts treated by BABA and/or SeNPs, as reported in our study. Consequently, ornithine could act as a precursor for the synthesis of polyamines and some amino acids, such as glutamate and proline, which are incorporated into many physiological processes, particularly under stress conditions [46]. In addition, arginase plays a role in increasing some other amino acids by providing the carbon and nitrogen skeleton required for their biosynthesis [47]. Further, the hydrolysis of arginine by arginase results in formation of urea, which in turn is hydrolyzed into ammonia. Finally, ammonia is involved in the glutamine synthetase/glutamine oxoglutarate aminotransferase (GS/GOGAT) cycle [46].

3.5. Improved Antioxidants of M. interexta Sprouts

Previous studies showed that BABA enhances a variety of plant metabolites and their associated mechanisms, and thus strengthens the defense systems of plants. BABA promotes the synthesis of phenolics and anthocyanins, and elevates the production of the enzymes associated with ROS [33]. Zhong et al. reported that BABA enhanced the activation of defense enzymes in soybean [48]. BABA also has been reported to potentiate different defence-signaling pathways under biotic and abiotic stresses [49]. Similarly, biogenic SeNPs improve the antioxidant defensive system of plants under abiotic stress [50]. SeNPs were also reported to be significantly involved in quenching ROS due to enhanced production of antioxidant enzymes, including guaiacol peroxidase (GPX), superoxide dismutase (SOD), proline oxidase (POX), and catalase (CAT) [51,52].

In the present study, we observed that both BABA and SeNPs increased the concentrations of phenolics and flavonoids, which might be attributed to enhanced antioxidant activity, as indicated by the results of FRAP assay. However, the increase was more significant under the combined treatment. Previous studies have shown the ability of BABA and/or Se to increase the levels of phenolics in plants grown under stress conditions [53,54]. Such induced increments might be due to activation of phenylalanine ammonia-lyas (PAL), which is a key enzyme in the phenylpropanoid pathway, as it is responsible for the biosynthesis of phenolic compounds [53,54]. Similarly, phenolic compounds were previously enhanced in potato when treated with BABA [55]. In addition, the PAL content of garlic has been found to be significantly increased under Se treatment, thereby enhancing its phenolic content [54]. In addition, the induced photosynthetic activity under treatment such as BABA and/or SeNPs could significantly increase the carbon skeleton necessary for the biosynthesis of different classes of secondary metabolites, such as phenolic compounds [56–58]. Moreover, the remarkable rise in GSH, the key non-enzymatic antioxidant, was measured in the combined treatment. The ameliorated ratio of GSH/GSSG is required for the generation of ascorbate (ASC) and the stimulation of numerous CO_2-fixing enzymes in the chloroplasts [59], ensuring the availability of $NADP^+$ to accept electrons from the photosynthetic electron transport chain.

3.6. Improved Antidiabetic Activity of M. interexta Sprouts

As the pharmacological properties of plants are correlated to their phytochemical content, we explored the enhanced phytochemical content that is attributed to the enhancement of the antidiabetic potential of *M. interexta* sprouts. We evaluated α-amylase and α-glucosidase inhibition activities, and the glycemic index of *M. interexta* sprouts. Results indicated that BABA increased the α-amylase inhibition activity of MI, while the GI of *M. interexta* sprouts was significantly decreased. SeNPs had positive effects in increasing the inhibition activity against α-amylase and α-glucosidase. Notably, the combined treatment increased the inhibitory effects against both enzymes but decreased the GI. A large variety of α-amylase and α-glucosidase inhibitors have been reported from various plants [60]. The reported inhibitory enzymatic activity in our study may be due to the presence of potentially bioactive compounds, such as polyphenols, alkaloids, flavonoids, tannins, and glycosides, which can enhance the combined treatment of BABA and SeNPs, leading to an increased antidiabetic potential of *M. interexta* sprouts.

3.7. Species-Specific Response to BABA and/or SeNPs

To better understand the BABA- and/or SeNPs-induced effects on *M. interexta* sprouts, we performed a principal component analysis (PCA) of the chemical composition and biological activities of the tested sprouts (Figure 3). There was a clear separation between the treatment parameters along the PC1, which explains 67% of the total variation. Obviously, the combined treatment of BABA and SeNPs induced the accumulation of amino acids, vitamins, and many components, as well as antidiabetic activity. There was also a clear separation between the parameters of the individually treated BABA or SeNPs sprouts along PC2 (representing 29% of the total variation). The sole treatment of *M. interexta* sprouts with SeNPs enhanced higher amounts of amino acids, vitamins, and other components, compared with sole treatment with BABA. Overall, the present data showed that *M. interexta* sprouts were differentially grouped, indicating the specificity of the accumulation of nutritive metabolites in response to the individual and/or the combined treatments with BABA and/or SeNPs.

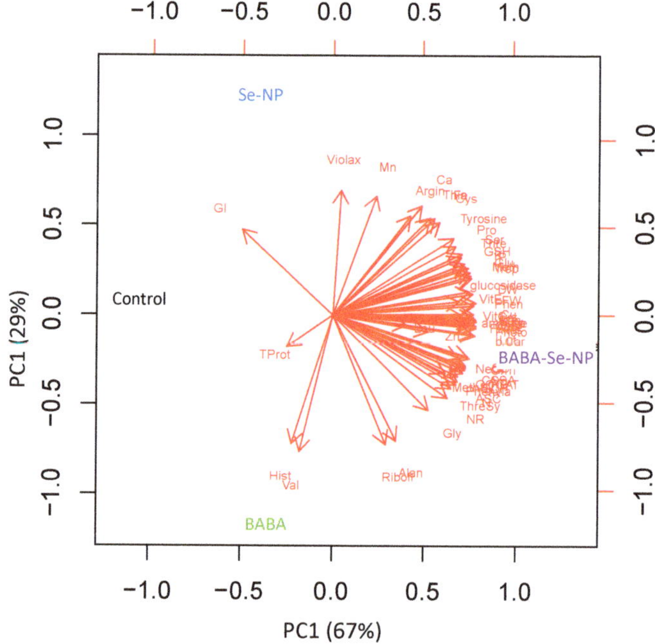

Figure 3. Principal component analysis (PCA) of chemical compositions and biological activities of control and BABA- and/or SeNPs-treated *M. interexta* sprouts.

4. Materials and Methods

4.1. Experimental Setup

Seeds of *M. interexta* were collected from the Agricultural Research Centers, where they were collected during filed trips to different locations in Egypt (Giza and Ismailia) and Saudi Arabia (Riyadh, Saudi Arabia). Seeds of *M. interexta* were collected from Dr. Mohammad K. Okla, Botany and Microbiology Department, College of Science, King Saud University, Riyadh, Saudi Arabia. The seeds were soaked for 1 h in 5 g L^{-1} of sodium hypochlorite for disinfection, and then they were washed with distilled water. The plant seeds were divided into two groups: the first group was primed with suspension containing 25 mg L^{-1} of selenium nanoparticles (SeNPs) for 10 h with continuous shaking (shaker (IKA KS 501 shaker, Staufen, Germany) at room temperature (24 °C). Then, the seeds were washed thrice with distilled water for 2 min. For sprouting processes, the seeds of both groups

(200 seeds per group) were distributed on trays (3 trays/treatment) filled with vermiculite and irrigated with 200 mL of 30 mM β-amino butyric acid (BABA) solution. The control trays were irrigated with Milli-Q water. Then, the seeds were evenly transferred to trays and covered. The applied concentrations of BABA and SeNPs were selected according to pilot experiments, where six concentrations of BABA (0 (distilled water) and 5, 15, 30, 60, and 90 mM) and 5 concentrations of SeNPs (0 (distilled water), 10, 25, 50, 75 mg L^{-1}) were tested. The growth conditions were adjusted to 150 μmol (photosynthetically active radiation) PAR m^{-2} s^{-1}, 23/18.5 °C air temperature, 63% humidity, and 16/8 h day/night photoperiod. Each experiment was replicated two times, and for all assays, four biological replicates (two biological replicates from each experiment) were used; accordingly, 16 samples in total were analyzed per each measurement. Moreover, each replicate corresponded to a group of 25 sprouts harvested from a certain tray. The sprout tissues (leaves and stems) from each treatment were harvested after 9 days. After fresh weight (FW) and dry weight (DW) measurements, the sprouts were frozen in liquid nitrogen and kept at −80 °C for biochemical analysis.

4.2. Selenium Nanoparticles Characterization

Selenium nanoparticles (SeNPs) were purchased from American Elements (Los Angeles, CA, USA) (https://www.americanelements.com/selenium-nanoparticles-7782-49-2, accessed on 25 February 2017). They are gray to black solids of a size of 20 and a specific surface area of 40 m^2/g, purity of 99.99%, and a density of 4.79 g/cm^3, according to the manufacturer's data. The morphological features were validated by using a scanning electron microscope (SEM manufacturered by JEOL JSM-6510, LA, Japan). To avoid coarse aggregation of SeNPs in aqueous solution, NPs were sonicated.

4.3. Determination of Photosynthetic Rate

Photosynthesis (μmol CO_2 m^{-2} s^{-1}) and dark respiration (μmol CO_2 m^{-2} s^{-1}) of the treated sprouts were detected by using an EGM-4 infrared gas analyzer (PP Systems, Hitchin, UK). Photosynthesis dark respiration was determined from 180 s measurements of net CO_2 exchange (NE).

4.4. Pigment Analysis

For homogenization of sprout samples, a MagNALyser (Roche, Vilvoorde, Belgium) was used for 1 min at 7000 rpm, then centrifugation was done for 20 min at 4 °C and 14,000× *g*. The supernatant was filtered through an Acrodisc GHP filter (0.45 μm 13 mm) (Gelman, Ann Arbor, MI, USA) and was further analyzed by HPLC (Shimadzu SIL10-ADvp, Kyoto, Japan, reversed-phase, at 4 °C). Pigments were separated on a C18 silica column (Waters Spherisorb, 5 μm ODS1, 4.6 × 250 mm, at 40 °C), using a mobile phase, as follows: (A) 81:9:10 acetonitrile/methanol/water and solvent; (B) 68:32 methanol/ethyl acetate, at a flow rate of 1.0 mL/min at room temperature [61]. A diode-array detector (Shimadzu SPD-M10Avp, Kyoto, Japan) was used for detection of chlorophyll *a* and *b*, and β-carotene at 420, 440, and 462 nm. Shimadzu Lab Solutions Lite software was used for the calculation of concentrations.

4.5. Analysis of Mineral Contents

Detection of mineral elements was carried out according to [62,63], whereas 200 mg from treated and control plants grown were digested by using an HNO_3/H_2O solution (5:1). Thereafter, macro- and micro-elements were evaluated by using inductively coupled plasma mass spectrometry (ICP-MS, Finnigan Element XR, and Scientific, Bremen, Germany). Nitric acid (1%) was used as a standard.

4.6. Determination of Phenolic, Flavonoid Contents, and Vitamins Levels

To extract phenolics and flavonoids, 150 mg of sprout material were extracted in 2 mL 80% methanol. Then, it was homogenized by a MagNALyser (Roche, Vilvoorde, Belgium;

7000 rpm/1 min). The extraction was performed three times. After each extraction, samples were centrifuged at 4 °C 20 min at 10,000× *g*, then the supernatants were transferred to clean tubes. The resulting supernatants were combined and centrifuged again at 4 °C for 30 min at 10,000× *g* to remove suspended particles. Prior to analysis, the samples were diluted 1:2 in 80% methanol, and 10 µL was used. The phenolic content was determined by using a Folin–Ciocalteu assay, where gallic acid was used as a standard [58]. The flavonoid content was evaluated following the modified aluminum chloride colorimetric method, where quercetin was applied as a standard [58]. The levels of phenolic and flavonoid compounds were identified by HPLC methods using the standards and their relative retention times, whereas the peak area of each standard could be used as an indication of the amount of each compound. For detection of the target compounds, approximately 50 mg samples were mixed with acetone/water (4:1). The HPLC system (SCL-10 AVP, Kyoto, Japan) was provided with a Lichrosorb Si-60, 7 µm, 3 mm × 150 mm column and a diode array detector. The mobile phase was a mixture of (90:10) water/formic acid, as well as (85:10:5) acetonitrile/water/formic acid, at a flow rate of 0.8 mL/min. The binary solvent system utilized in the mobile phase consisted of the following: (A) 1 percent acetic acid/water, and (B) methanol, with the gradient being 0 min 40% B, 5 min 65 percent B, 10 min 90% B, and 15 min 40% B until 17 min, as modified from the reference. The eluate was tested for UV absorbance at 260, 280, and 330 nm. Compounds were found by comparing retention times, absorbance spectrum profiles, and running samples, after pure standards had been added to known concentrations of each discovered compound to internal standards. Meanwhile, the internal standard was 3,5-dichloro-4- hydroxybenzoic.

Detection of vitamins in treated and control sprouts was carried out via HPLC, according to [58,64]. The contents of thiamine and riboflavin were determined in sprouts, by using UV and/or fluorescence detectors [58]. Separation was performed on a reverse-phase (C18) column (HPLC, methanol/water). Ascorbate (Vit C) was extracted in 1 mL of 6% (*w/v*) meta-phosphoric acid at 4 °C and was separated by reverse-phase HPLC coupled with a UV detector (100 mm × 4.6 mm Polaris C18-A, 3 lm particle size; 40 °C, isocratic flow rate: 1 mL min^{-1}, elution buffer: 2 mM KCl, pH 2.5 with O-phosphoric acid). Tocopherol (vit E) was separated on Particil Pac 5 µm column material (length 250 mm, i.d. 4.6 mm) and quantified by HPLC (Shimadzu's Hertogenbosch, s-Hertogenbosch, The Netherlands, normal phase conditions), coupled with a fluorometric detector (excitation at 290 nm and emission at 330 nm). Riboflavin and thiamine were extracted by homogenizing samples in ethanol solvent through a MagNALyser (Roche, Vilvoorde, Belgium, 1 min, 7000 rpm), then centrifuged for 20 min at 14,000× *g*, 4 °C. The supernatant was taken and filtered (Acrodisc GHP filter, 0.45 µm 13 mm). Then, the solution was analyzed by using HPLC (Shimadzu SIL10-ADvp, reverse-phased, at 4 °C), where the target compounds were separated on a reverse-phase (C18) column (HPLC, methanol/water as a mobile phase and fluorescence as a detector) [62].

4.6.1. Total Antioxidant Capacity (FRAP)

Total antioxidant capacity was determined by using the ferric-reducing antioxidant power (FRAP) method. The extraction of samples was performed by using 80% ethanol; then, the extracts were centrifuged for 20 min at 4 °C and 14,000× *g*. The FRAP reagent was prepared by adding FeCl$_3$ (20 mM) to the acetate buffer (0.25 M). Thereafter, the FRAP reagent (approximately 0.25 mL) was mixed with 0.1 mL of extracts, and the reading was taken at 593 nm, as previously outlined in [65]. The values were expressed as µmol trolox/g FW.

4.6.2. Amino Acid Analysis

For amino acid analysis, the method described in [66] was used, in which 100 mg of each plant was homogenized in 5 mL of 80% ethanol at 5000 rpm for 1 min. After centrifugation (14,000× *g* for 25 min), the supernatant was resuspended in 5 mL of chloroform. Thereafter, 1 mL of H$_2$O was used for the residue extraction. The supernatant and pellet were resuspended in chloroform and centrifuged at 8000× *g* for 10 min. A total of 15 amino acids (0.05 µmoles mL^{-1} for each one) were used as reference standards for determination of the retention time of each amino acid. An internal standard α-aminobutyric was also used for amino acid detection. Then, the extracts were centrifuged for 10 min at 20,000× *g* and the aqueous phase was filtered by Millipore micro-filters (0.2-lm pore size). The amino acids were quantified (using a Waters Acquity UPLC TQD device coupled to a BEH amide column, 2.1 mm × 50 mm). The elution (A, 84% ammonium formate, 6% formic acid, and 10% acetonitrile, *v*/*v*, and B, acetonitrile and 2% formic acid, *v*/*v*) resulted in amino acid peak integration. Star Chromatography (version 5.51) software was applied.

4.7. Determination of Nitrogen Content and Metabolism

Total nitrogen (N) content was determined by digestion of the sprout samples (0.2 g) in H_2SO_4 at 260 °C; the amount of N was detected by using a CN element analyzer (NC-2100, Carlo Erba Instruments, Milan, Italy). For enzyme assays, the samples (100 mg) were extracted with 400 µL of extraction buffer (50 mM HEPES-KOH pH 7.5, 10% (*v*/*v*) glycerol, 0.1% Triton X-100, 10 mM $MgCl_2$, 1 mM EDTA, 1 mM benzamidine, 1 mM ε-aminocapronic acid, 1 mM DTT, and 20 µM flavin adenine dinucleotide). The samples were centrifuged at 4 °C 13,000× *g* for 5 min, and the supernatant was used in the reactions. The determination of glutamine synthetase (GS) and glutamine 2-oxoglutarate aminotransferase (GOGAT) was conducted as indicated by the reduction of NADH at A_{340}. Glutamate dehydrogenase (GDH) was determined by 2-oxoglutarate-dependent NADH oxidation. Determination of GS activity was performed by monitoring γ-glutamyl hydroxamate at A_{340}. Estimation of GOGAT activity was achieved according to glutamine-dependent NADH oxidation at A340. Nitrate reductase (NR) activity was determined by measuring nitrite-dependent NADH oxidation (A_{340}) [44,45]. Arginase was determined according to [67], based on the formation of urea from arginine, where the reaction mixture consisted of 1 mM $MnCl_2$, 10 mM Tris (pH 9.5), and 125 mM L-arginine (pH 9.5), in addition to the enzyme solution, to make a total volume of 10 mL. Then, incubation was carried out for 30 min at 37 °C. The reaction was started by adding the enzyme and terminated by the addition of 0.1 mL 50% TCA. Protein removal was performed by centrifugation, and the urea content in the supernatant was colorimetrically measured, where one unit was defined as the amount of enzyme producing 1 umol urea per min. The arginase activity was detected as a linear function of both incubation time and concentration under these conditions. Boiled enzyme preparations were used as the control [67]. Total proteins were detected by using Lowery methods [68].

4.8. Determination of Antidiabetic Activity

For sample homogenization, a MagNALyser and a phosphate buffer (1 mL, 50 mM, pH 5.2) were used. Then, centrifugation was carried out for 5 min at 4 °C and 14,000× *g*. The α-amylase inhibitory activity of the extracts and fractions was carried out according to a standard method, with minor modification [69]. In a 96-well plate, the reaction mixture containing a 50 µL phosphate buffer (100 mM, pH = 6.8), 10 µL α–amylase (2 U/mL), and 20 µL of varying concentrations of the extracts (0.1, 0.2, 0.3, 0.4, and 0.5 mg/mL) was preincubated at 37 °C for 20 min. Then, 20 µL of 1% soluble starch (100 mM phosphate buffer pH 6.8) was added as a substrate and incubated further at 37 °C for 30 min; 100 µL of the DNS color reagent was then added and boiled for 10 min. The absorbance of the resulting mixture was measured at 540 nm using a multiplate reader. Acarbose at various concentrations (0.1–0.5 mg/mL) was used as a standard. A without-test substance was set up in parallel as a control, and each experiment was performed in triplicate.

4.9. Statistical Analyses

Statistical analyses were completed, using an SPSS statistical package (SPSS Inc., Chicago, IL, USA). Replication of each experiment was performed twice. Four replicates were used for all assays and each replicate corresponded to a group of 25 sprouts harvested from a certain tray. One-way analysis of variance (ANOVA) was carried out, where Tukey's test was used as the post hoc test for the separation of means ($p < 0.05$). Principal component analysis (PCA) was generated by a multi-experimental viewer (TM4 software package, http://mev.tm4.org, accessed on 18 November 2021).

5. Conclusions

Based on the above results, it could be concluded that the application of BABA and/or SeNPs could be a useful technique to enhance the growth and photosynthetic activity of sprouts. As a result, the combined treatment had a more pronounced effect on the bioactive primary metabolites (essential amino acids), secondary metabolises (phenolics, GSH, ASC), mineral profiles, and nitrogen metabolism of the investigated sprouts than that of sole treatments. Concomitantly, the antioxidant (FRAP), the anti-diabetic activities (i.e., α-amylase and α-glucosidase inhibition activities) and the glycemic index) of the tested sprouts were more significantly improved by the combined treatment with BABA and SeNPs than by individual treatment. Thus, this study represents the first report that supports the use of the combined treatment of BABA and SeNPs to increase plant growth and bioactive metabolites.

Author Contributions: S.S., N.A., E.E.A. and H.H.A. planned and designed the research; S.S., S.K.A.J., M.W., H.H.A., E.E.A., N.A., H.A. and M.A.-M. performed the experiments; S.S., S.K.A.J., M.W., H.H.A., E.E.A., N.A., H.A. and M.A.-M. analyzed the data; S.S., N.A., H.H.A. and S.S. contributed to the reagents/chemicals; S.S., S.K.A.J., M.W., H.H.A., E.E.A., N.A. and M.A.-M. provided a draft version of the manuscript, and S.S., S.K.A.J., M.W., H.H.A., E.E.A. and H.A. revised and finalized the manuscript. All authors have read and agreed to the published version of the manuscript.

Funding: The authors extend their appreciation to the Deanship of Scientific Research at Jouf University for funding this work through research grant No. DSR-2021-01-0122.

Institutional Review Board Statement: Not applicable.

Informed Consent Statement: Not applicable.

Data Availability Statement: Data is contained within the article.

Acknowledgments: The authors extend their appreciation to the Deanship of Scientific Research at Jouf University for funding this work through research grant no. (DSR-2021-01-0122).

Conflicts of Interest: The authors declare no conflict of interest.

References

1. Haminiuk, C.W.I.; Maciel, G.M.; Plata-Oviedo, M.S.V.; Peralta, R.M. Phenolic compounds in fruits–an overview. *Int. J. Food Sci. Technol.* **2012**, *47*, 2023–2044. [CrossRef]
2. Dinkova-Kostova, A.T.; Kostov, R.V. Glucosinolates and isothiocyanates in health and disease. *Trends Mol. Med.* **2012**, *18*, 337–347. [CrossRef] [PubMed]
3. Björkman, M.; Klingen, I.; Birch, A.N.E.; Bones, A.M.; Bruce, T.J.A.; Johansen, T.J.; Meadow, R.; Mølmann, J.; Seljåsen, R.; Smart, L.E. Phytochemicals of Brassicaceae in plant protection and human health–Influences of climate, environment and agronomic practice. *Phytochemistry* **2011**, *72*, 538–556. [CrossRef] [PubMed]
4. Poulev, A.; O'Neal, J.M.; Logendra, S.; Pouleva, R.B.; Timeva, V.; Garvey, A.S.; Gleba, D.; Jenkins, I.S.; Halpern, B.T.; Kneer, R. Elicitation, a new window into plant chemodiversity and phytochemical drug discovery. *J. Med. Chem.* **2003**, *46*, 2542–2547. [CrossRef]
5. Smetanska, I. Production of secondary metabolites using plant cell cultures. *Food Biotechnol.* **2008**, *111*, 187–228.
6. Zhao, J.; Davis, L.C.; Verpoorte, R. Elicitor signal transduction leading to production of plant secondary metabolites. *Biotechnol. Adv.* **2005**, *23*, 283–333. [CrossRef]
7. Ferrari, S. Biological elicitors of plant secondary metabolites: Mode of action and use in the production of nutraceutics. In *Bio-Farms for Nutraceuticals*; Springer: Berlin/Heidelberg, Germany, 2010; pp. 152–166.
8. Thevenet, D.; Pastor, V.; Baccelli, I.; Balmer, A.; Vallat, A.; Neier, R.; Glauser, G.; Mauch-Mani, B. The priming molecule β-aminobutyric acid is naturally present in plants and is induced by stress. *New Phytol.* **2017**, *213*, 552–559. [CrossRef]
9. Jakab, G.; Cottier, V.; Toquin, V.; Rigoli, G.; Zimmerli, L.; Métraux, J.-P.; Mauch-Mani, B. β-Aminobutyric acid-induced resistance in plants. *Eur. J. Plant Pathol.* **2001**, *107*, 29–37. [CrossRef]
10. Cohen, Y.R. β-aminobutyric acid-induced resistance against plant pathogens. *Plant Dis.* **2002**, *86*, 448–457. [CrossRef]

11. Shaw, A.K.; Bhardwaj, P.K.; Ghosh, S.; Roy, S.; Saha, S.; Sherpa, A.R.; Saha, S.K.; Hossain, Z. β-aminobutyric acid mediated drought stress alleviation in maize (*Zea mays* L.). *Environ. Sci. Pollut. Res.* **2016**, *23*, 2437–2453. [CrossRef]
12. Mátai, A.; Jakab, G.; Hideg, É. Single-dose β-aminobutyric acid treatment modifies tobacco (*Nicotiana tabacum* L.) leaf acclimation to consecutive UV-B treatment. *Photochem. Photobiol. Sci.* **2019**, *18*, 359–366. [CrossRef] [PubMed]
13. Giraldo, J.P.; Landry, M.P.; Faltermeier, S.M.; McNicholas, T.P.; Iverson, N.M.; Boghossian, A.A.; Reuel, N.F.; Hilmer, A.J.; Sen, F.; Brew, J.A.; et al. Plant nanobionics approach to augment photosynthesis and biochemical sensing. *Nat. Mater.* **2014**, *13*, 400–408. [CrossRef] [PubMed]
14. Majumder, D.D.; Ulrichs, C.; Majumder, D.; Mewis, I.; Thakur, A.R.; Brahmachary, R.L.; Banerjee, R.; Rahman, A.; Debnath, N.; Seth, D. Current status and future trends of nanoscale technology and its impact on modern computing, biology, medicine and agricultural biotechnology. In Proceedings of the 2007 International Conference on Computing: Theory and Applications (ICCTA'07), Kolkata, India, 5–7 March 2007; pp. 563–573.
15. Siddiqui, M.H.; Al-Whaibi, M.H. Role of nano-SiO$_2$ in germination of tomato (*Lycopersicum esculentum* seeds Mill.). *Saudi J. Biol. Sci.* **2014**, *21*, 13–17. [CrossRef] [PubMed]
16. Terry, N.; Zayed, A.M.; De Souza, M.P.; Tarun, A.S. Selenium in higher plants. *Annu. Rev. Plant Biol.* **2000**, *51*, 401–432. [CrossRef] [PubMed]
17. Germ, M.; Stibilj, V. Selenium and plants. *Acta Agric. Slov.* **2007**, *89*, 65–71. [CrossRef]
18. Kuznetsov, V.V.; Kholodova, V.P.; Kuznetsov, V.V.; Yagodin, B.A. Selenium regulates the water status of plants exposed to drought. In *Doklady Biological Sciences*; Consultants Bureau: New York, NY, USA, 2003; pp. 266–268.
19. Ferrarese, M.; Sourestani, M.M.; Quattrini, E.; Schiavi, M.; Ferrante, A. Biofortification of spinach plants applying selenium in the nutrient solution of floating system. *Veg. Crop. Res. Bull.* **2012**, *2012*, 127–136. [CrossRef]
20. Guevara Moreno, O.D.; Acevedo Aguilar, F.J.; Yanez Barrientos, E. Selenium uptake and biotransformation and effect of selenium exposure on the essential and trace elements status: Comparative evaluation of four edible plants. *J. Mex. Chem. Soc.* **2018**, *62*, 247–258. [CrossRef]
21. Li, X.; Xu, H.; Chen, Z.-S.; Chen, G. Biosynthesis of nanoparticles by microorganisms and their applications. *J. Nanomater.* **2011**, *2011*, 1–16. [CrossRef]
22. Zohra, E.; Ikram, M.; Omar, A.A.; Hussain, M.; Satti, S.H.; Raja, N.I.; Ehsan, M. Potential applications of biogenic selenium nanoparticles in alleviating biotic and abiotic stresses in plants: A comprehensive insight on the mechanistic approach and future perspectives. *Green Process. Synth.* **2021**, *10*, 456–475. [CrossRef]
23. Rady, M.O.A.; Semida, W.M.; Abd El-Mageed, T.A.; Howladar, S.M.; Shaaban, A. Foliage applied selenium improves photosynthetic efficiency, antioxidant potential and wheat productivity under drought stress. *Int. J. Agric. Biol* **2020**, *24*, 1293–1300.
24. Chang, K.C.; Satterlee, L.D. Chemical, nutritional and microbiological quality of a protein concentrate from culled dry beans. *J. Food Sci.* **1979**, *44*, 1589–1593. [CrossRef]
25. Bakoğlu, A.; Bağcı, E.; Koçak, A.; Yüce, E. Fatty acid composition of some *Medicago* L. (Fabaceae) species from Turkey. *Asian J. Chem.* **2010**, *2*, 651–656.
26. Kokten, K.; Bakoglu, A.; Kocak, A.; Bagci, E.; Akcura, M.; Kaplan, M. Chemical composition of the seeds of some Medicago species. *Chem. Nat. Compd.* **2011**, *47*, 619–621. [CrossRef]
27. Jisha, K.C.; Puthur, J.T. Seed priming with beta-amino butyric acid improves abiotic stress tolerance in rice seedlings. *Rice Sci.* **2016**, *23*, 242–254. [CrossRef]
28. El Lateef Gharib, F.A.; Zeid, I.M.; Ghazi, S.M.; Ahmed, E.Z. The response of cowpea (*Vigna unguiculata* L) plants to foliar application of sodium selenate and selenium nanoparticles (SeNPs). *J. Nanomater. Mol. Nanotechnol.* **2019**, *8*, 1000272.
29. Wang, H.; Zhang, J.; Yu, H. Elemental selenium at nano size possesses lower toxicity without compromising the fundamental effect on selenoenzymes: Comparison with selenomethionine in mice. *Free Radic. Biol. Med.* **2007**, *42*, 1524–1533. [CrossRef]
30. Morales-Espinoza, M.C.; Cadenas-Pliego, G.; Pérez-Alvarez, M.; Hernández-Fuentes, A.D.; Cabrera de la Fuente, M.; Benavides-Mendoza, A.; Valdés-Reyna, J.; Juárez-Maldonado, A. Se nanoparticles induce changes in the growth, antioxidant responses, and fruit quality of tomato developed under NaCl Stress. *Molecules* **2019**, *24*, 3030. [CrossRef]
31. Haghighi, M.; Abolghasemi, R.; Teixeira da Silva, J.A. Low and high temperature stress affect the growth characteristics of tomato in hydroponic culture with Se and nano-Se amendment. *Sci. Hortic.* **2014**, *178*, 231–240. [CrossRef]
32. Ragavan, P.; Ananth, A.; Rajan, M.R. Impact of selenium nanoparticles on growth, biochemical characteristics and yield of cluster bean Cyamopsis tetragonoloba. *Int. J. Environ. Agric. Biotechnol.* **2017**, *2*, 238983. [CrossRef]
33. Wu, C.C.; Singh, P.; Chen, M.C.; Zimmerli, L. L-Glutamine inhibits beta-aminobutyric acid-induced stress resistance and priming in Arabidopsis. *J. Exp. Bot.* **2010**, *61*, 995–1002. [CrossRef]
34. Choudhary, A.; Kumar, A.; Kaur, H.; Balamurugan, A.; Padhy, A.K.; Mehta, S. Plant Performance and Defensive Role of β-Amino Butyric Acid Under Environmental Stress. In *Plant Performance under Environmental Stress*; Springer: Berlin/Heidelberg, Germany, 2021; pp. 249–275.
35. Shafique, H.; Jamil, Y.; ul Haq, Z.; Mujahid, T.; Khan, A.U.; Iqbal, M.; Abbas, M. Low power continuous wave-laser seed irradiation effect on Moringa oleifera germination, seedling growth and biochemical attributes. *J. Photochem. Photobiol. B Biol.* **2017**, *170*, 314–323.
36. Cao, S.; Jiang, L.; Yuan, H.; Jian, H.; Ren, G.; Bian, X.; Zou, J.; Chen, Z. β-Amino-butyric acid protects Arabidopsis against low potassium stress. *Acta Physiol. Plant.* **2008**, *30*, 309–314. [CrossRef]

37. Jiang, L.; Yang, R.Z.; Lu, Y.F.; Cao, S.Q.; Ci, L.K.; Zhang, J.J. β-aminobutyric acid-mediated tobacco tolerance to potassium deficiency. *Russ. J. Plant Physiol.* **2012**, *59*, 781–787. [CrossRef]

38. Erenoglu, E.B.; Kutman, U.B.; Ceylan, Y.; Yildiz, B.; Cakmak, I. Improved nitrogen nutrition enhances root uptake, root-to-shoot translocation and remobilization of zinc (65Zn) in wheat. *New Phytol.* **2011**, *189*, 438–448. [CrossRef] [PubMed]

39. Broadley, M.R.; White, P.J.; Hammond, J.P.; Zelko, I.; Lux, A. Zinc in plants. *New Phytol.* **2007**, *173*, 677–702. [CrossRef] [PubMed]

40. Pandey, C.; Gupta, M. Selenium amelioration of arsenic toxicity in rice shows genotypic variation: A transcriptomic and biochemical analysis. *J. Plant Physiol.* **2018**, *231*, 168–181. [CrossRef]

41. Poldma, P.; Tonutare, T.; Viitak, A.; Luik, A.; Moor, U. Effect of selenium treatment on mineral nutrition, bulb size, and antioxidant properties of garlic (*Allium sativum* L.). *J. Agric. Food Chem.* **2011**, *59*, 5498–5503. [CrossRef] [PubMed]

42. Chen, Y.; Liang, D.; Song, W.; Lei, L.; Yu, D.; Miao, S. Effects of nitrogen application on selenium accumulation, translocation and distribution of winter wheat at different growth periods. *J. Plant Nutr. Fert.* **2012**, *22*, 395–402.

43. Helali, S.M.; Nebli, H.; Kaddour, R.; Mahmoudi, H.; Lachaâl, M.; Ouerghi, Z. Influence of nitrate—Ammonium ratio on growth and nutrition of *Arabidopsis thaliana*. *Plant Soil* **2010**, *336*, 65–74. [CrossRef]

44. Okla, M.K.; Akhtar, N.; Alamri, S.A.; Al-Qahtani, S.M.; Ismail, A.; Abbas, Z.K.; Al-Ghamdi, A.A.; Qahtan, A.A.; Soufan, W.H.; Alaraidh, I.A. Potential importance of molybdenum priming to metabolism and nutritive value of *Canavalia* spp. sprouts. *Plants* **2021**, *10*, 2387. [CrossRef]

45. Janse van Rensburg, H.C.; Van den Ende, W. Priming with γ-aminobutyric acid against Botrytis cinerea reshuffles metabolism and reactive oxygen species: Dissecting signalling and metabolism. *Antioxidants* **2020**, *9*, 1174. [CrossRef] [PubMed]

46. Winter, G.; Todd, C.D.; Trovato, M.; Forlani, G.; Funck, D. Physiological implications of arginine metabolism in plants. *Front. Plant Sci.* **2015**, *6*, 534. [CrossRef] [PubMed]

47. Siddappa, S.; Marathe, G.K. What we know about plant arginases? *Plant Physiol. Biochem.* **2020**, *156*, 600–610. [CrossRef] [PubMed]

48. Zhong, Y.; Wang, B.; Yan, J.; Cheng, L.; Yao, L.; Xiao, L.; Wu, T. DL-β-aminobutyric acid-induced resistance in soybean against Aphis glycines Matsumura (Hemiptera: *Aphididae*). *PLoS ONE* **2014**, *9*, e85142. [CrossRef]

49. Baccelli, I.; Mauch-Mani, B. Beta-aminobutyric acid priming of plant defense: The role of ABA and other hormones. *Plant Mol. Biol.* **2016**, *91*, 703–711. [CrossRef]

50. Jóźwiak, W.; Politycka, B. Effect of selenium on alleviating oxidative stress caused by a water deficit in cucumber roots. *Plants* **2019**, *8*, 217. [CrossRef]

51. Menon, S.; Devi, K.S.S.; Agarwal, H.; Shanmugam, V.K. Efficacy of biogenic selenium nanoparticles from an extract of ginger towards evaluation on anti-microbial and anti-oxidant activities. *Colloid Interface Sci. Commun.* **2019**, *29*, 1–8. [CrossRef]

52. Zahedi, S.M.; Abdelrahman, M.; Hosseini, M.S.; Hoveizeh, N.F.; Tran, L.-S.P. Alleviation of the effect of salinity on growth and yield of strawberry by foliar spray of selenium-nanoparticles. *Environ. Pollut.* **2019**, *253*, 246–258. [CrossRef]

53. Baysal, Ö.; Gürsoy, Y.Z.; Örnek, H.; Duru, A. Induction of oxidants in tomato leaves treated with DL-β-Amino butyric acid (BABA) and infected with Clavibacter michiganensis ssp. michiganensis. *Eur. J. Plant Pathol.* **2005**, *112*, 361–369. [CrossRef]

54. Astaneh, R.K.; Bolandnazar, S.; Nahandi, F.Z.; Oustan, S. Effect of selenium application on phenylalanine ammonia-lyase (PAL) activity, phenol leakage and total phenolic content in garlic (*Allium sativum* L.) under NaCl stress. *Inf. Process. Agric.* **2018**, *5*, 339–344. [CrossRef]

55. Altamiranda, E.A.G.; Andreu, A.B.; Daleo, G.R.; Olivieri, F.P. Effect of β-aminobutyric acid (BABA) on protection against Phytophthora infestans throughout the potato crop cycle. *Australas. Plant Pathol.* **2008**, *37*, 421–427. [CrossRef]

56. Habeeb, T.H.; Abdel-Mawgoud, M.; Yehia, R.S.; Khalil, A.M.A.; Saleh, A.M.; AbdElgawad, H. Interactive Impact of Arbuscular Mycorrhizal Fungi and Elevated CO_2 on Growth and Functional Food Value of Thymus vulgare. *J. Fungi* **2020**, *6*, 168. [CrossRef]

57. Saleh, A.M.; Abdel-Mawgoud, M.; Hassan, A.R.; Habeeb, T.H.; Yehia, R.S.; AbdElgawad, H. Global metabolic changes induced by arbuscular mycorrhizal fungi in oregano plants grown under ambient and elevated levels of atmospheric CO_2. *Plant Physiol. Biochem.* **2020**, *151*, 255–263. [CrossRef] [PubMed]

58. Almuhayawi, M.S.; Hassan, A.H.A.; Abdel-Mawgoud, M.; Khamis, G.; Selim, S.; Al Jaouni, S.K.; AbdElgawad, H. Laser light as a promising approach to improve the nutritional value, antioxidant capacity and anti-inflammatory activity of flavonoid-rich buckwheat sprouts. *Food Chem.* **2020**, *345*, 128788. [CrossRef] [PubMed]

59. Crawford, N.M.; Brian, G.F. Molecular and developmental biology of inorganic nitrogen nutrition. In *The Arabidopsis Book/American Society of Plant Biologists 1*; SSPA Sweden AB: Gothenburg, Sweden, 2002; p. e0011. [CrossRef]

60. Ghosh, P.; Das, C.; Biswas, S.; Nag, S.K.; Dutta, A.; Biswas, M.; Sil, S.; Hazra, L.; Ghosh, C.; Das, S.; et al. Phytochemical composition analysis and evaluation of in vitro medicinal properties and cytotoxicity of five wild weeds: A comparative study. *F1000Research* **2020**, *9*, 493. [CrossRef] [PubMed]

61. Thayer, S.S.; Björkman, O. Leaf Xanthophyll content and composition in sun and shade determined by HPLC. *Photosynth. Res.* **1990**, *23*, 331–343. [CrossRef] [PubMed]

62. AbdElgawad, H.; Okla, M.K.; Al-Amri, S.S.; Al-Hashimi, A.; Al-Qahtani, W.H.; Al-Qahtani, S.M.; Abbas, Z.K.; Al-Harbi, N.A.; Abd Algafar, A.; Almuhayawi, M.S. Effect of elevated CO_2 on biomolecules' accumulation in caraway (*Carum carvi* L.) plants at different developmental stages. *Plants* **2021**, *10*, 2434. [CrossRef]

63. Almuhayawi, M.S.; Abdel-Mawgoud, M.; Al Jaouni, S.K.; Almuhayawi, S.M.; Alruhaili, M.H.; Selim, S.; AbdElgawad, H. Bacterial endophytes as a promising approach to enhance the growth and accumulation of bioactive metabolites of three species of Chenopodium Sprouts. *Plants* **2021**, *10*, 2745. [CrossRef]

64. Almuhayawi, M.S.; Mohamed, M.S.M.; Abdel-Mawgoud, M.; Selim, S.; Al Jaouni, S.K.; AbdElgawad, H. Bioactive potential of several actinobacteria isolated from microbiologically barely explored desert habitat, Saudi Arabia. *Biology* **2021**, *10*, 235. [CrossRef]

65. Almuhayawi, M.S.; Al Jaouni, S.K.; Almuhayawi, S.M.; Selim, S.; Abdel-Mawgoud, M. Elevated CO_2 improves the nutritive value, antibacterial, anti-inflammatory, antioxidant and hypocholestecolemic activities of lemongrass sprouts. *Food Chem.* **2021**, *357*, 129730. [CrossRef]

66. Sinha, A.K.; Giblen, T.; AbdElgawad, H.; De Rop, M.; Asard, H.; Blust, R.; De Boeck, G. Regulation of amino acid metabolism as a defensive strategy in the brain of three freshwater teleosts in response to high environmental ammonia exposure. *Aquat. Toxicol.* **2013**, *130–131*, 86–96. [CrossRef] [PubMed]

67. Kang, J.H.; Cho, Y.D. Purification and properties of arginase from soybean, Glycine max, axes. *Plant Physiol.* **1990**, *93*, 1230–1234. [CrossRef] [PubMed]

68. Lowry, O.H.; Rosebrough, N.J.; Farr, A.L.; Randall, R.J. Protein measurement with the Folin phenol reagent. *J. Biol. Chem.* **1951**, *193*, 265–275. [CrossRef]

69. Zinta, G.; AbdElgawad, H.; Peshev, D.; Weedon, J.T.; Van den Ende, W.; Nijs, I.; Janssens, I.A.; Beemster, G.T.S.; Asard, H. Dynamics of metabolic responses to periods of combined heat and drought in Arabidopsis thaliana under ambient and elevated atmospheric CO_2. *J. Exp. Bot.* **2018**, *69*, 2159–2170. [CrossRef] [PubMed]

Article

Ultraviolet-B Irradiation Increases Antioxidant Capacity of Pakchoi (*Brassica rapa* L.) by Inducing Flavonoid Biosynthesis

Juan Hao [1,†], Panpan Lou [1,†], Yidie Han [1], Lijun Zheng [1], Jiangjie Lu [1], Zhehao Chen [1], Jun Ni [1], Yanjun Yang [1] and Maojun Xu [1,2,*]

[1] Zhejiang Provincial Key Laboratory for Genetic Improvement and Quality Control of Medicinal Plants, Hangzhou Normal University, Hangzhou 311121, China; juanhao@hznu.edu.cn (J.H.); 2019111010029@stu.hznu.edu.cn (P.L.); 2020111010057@stu.hznu.edu.cn (Y.H.); 2021111010056@stu.hznu.edu.cn (L.Z.); lujj@hznu.edu.cn (J.L.); zhchen@hznu.edu.cn (Z.C.); nijun@hznu.edu.cn (J.N.); yjyang@hznu.edu.cn (Y.Y.)

[2] Key Laboratory of Hangzhou City for Quality and Safety of Agricultural Products, College of Life and Environmental Sciences, Hangzhou Normal University, Hangzhou 311121, China

* Correspondence: xumaojunhz@163.com; Tel.: +86-0571-2886-5335

† These authors contributed equally to this work.

Abstract: As an important abiotic stress factor, ultraviolet-B (UV-B) light can stimulate the accumulation of antioxidants in plants. In this study, the possibility of enhancing antioxidant capacity in pakchoi (*Brassica rapa* L.) by UV-B supplementation was assessed. Irradiation with 4 µmol·m^{-2}·s^{-1} UV-B for 4 h or 2 µmol·m^{-2}·s^{-1} UV-B for 24 h significantly increased the 1,1–diphenyl–2–picrylhydrazyl (DPPH) scavenging activity and total reductive capacity, as a result of inducing a greater accumulation of total polyphenols and flavonoids without affecting the plant biomass. A high performance liquid chromatography (HPLC) analysis showed that the concentrations of many flavonoids significantly increased in response to UV-B treatment. The activities of three enzymes involved in the early steps of flavonoid biosynthesis, namely phenylalanine ammonia-lyase (PAL), cinnamate-4-hydroxylase (C4H), and 4-coumarate: coenzyme A (CoA) ligase (4CL), were significantly increased after the corresponding UV-B treatment. Compared with the control, the expression levels of several flavonoid biosynthesis genes (namely *BrPAL*, *BrC4H*, *Br4CL*, *BrCHS*, *BrF3H*, *BrF3′H*, *BrFLS*, *BrDFR*, *BrANS*, and *BrLDOX*) were also significantly up–regulated in the UV-B treatment group. The results suggest that appropriate preharvest UV-B supplementation could improve the nutritional quality of greenhouse-grown pakchoi by promoting the accumulation of antioxidants.

Keywords: pakchoi; greenhouse; UV-B; antioxidant activity; flavonoids; biosynthetic pathway

Citation: Hao, J.; Lou, P.; Han, Y.; Zheng, L.; Lu, J.; Chen, Z.; Ni, J.; Yang, Y.; Xu, M. Ultraviolet-B Irradiation Increases Antioxidant Capacity of Pakchoi (*Brassica rapa* L.) by Inducing Flavonoid Biosynthesis. *Plants* **2022**, *11*, 766. https://doi.org/10.3390/plants11060766

Academic Editors: Beatrice Falcinelli and Angelica Galieni

Received: 30 January 2022
Accepted: 7 March 2022
Published: 13 March 2022

Publisher's Note: MDPI stays neutral with regard to jurisdictional claims in published maps and institutional affiliations.

1. Introduction

In recent years, consumers have become more aware of the importance of dietary nutrition. High-quality functional foods, combining health and safety, are desired by consumers. Secondary plant metabolites (SPM), which include flavonoids, can not only be used as sunscreens by plant leaves to protect inner cells from harmful radiation, but are also considered to be the major bioactive compounds in edible plants with respect to human health benefits due to their potent antioxidant capacity [1,2]. *Brassica* species are known for their high contents of SPM, many of which are appreciated for their health-promoting effects. Kale (*Brassica oleracea* L.) has high concentrations of the flavonol aglycones kaempferol and quercetin, which show different antioxidant activities dependent on their chemical structure [3,4]. Several antioxidant phenolic compounds including flavonoids have been investigated and identified in Chinese cabbage (*Brassica rapa* L.) leaves [5]. Cabbage (*Brassica oleracea* L.) heads have important antioxidant and anti-inflammatory properties due to their rich glucosinolates content [6]. Pakchoi (*Brassica rapa* L.) is rich in SPM and contains numerous antioxidants, including flavonoids, hydroxycinnamic acids, carotenoids, chlorophylls, and glucosinolates [7,8]. With increasing attention being paid to the quality and safety of food, *Brassica* vegetables rich in antioxidants are gradually finding their way into our diets.

The biosynthesis of antioxidants in plant-derived food is regulated by many factors, including the light environment [3]. Ultraviolet-B (UV-B; 280–315 nm) radiation is an intrinsic part of the solar radiation that reaches the earth's surface and plays an important role in regulating the growth, photosynthesis, and SPM of higher plants [9]. UV-B radiation resulted in changes in a number of antioxidants in different *Brassica* vegetables. Much evidence has indicated that the impact of UV-B radiation on plants depends upon the context, such as radiation dosage, exposure time, stress acclimation, nutritional status, and plant species [7,10,11]. Exposure to low doses of UV-B and UV during the late developmental stages of pakchoi resulted in higher concentrations of flavonoids, hydroxycinnamic acids, carotenoids, and chlorophylls [7]. Six leafy *Brassica* species were analyzed for their flavonoid glycoside accumulation after short-term UV-B treatment, which showed species-specific responses [12]. Blue light treatment after pre-exposure to UV-B stabilized the changes in flavonoid glycoside and led to a higher hydroxyl radical scavenging capacity in three different *Brassica* sprouts [4]. Moreover, the treatment of low, ecologically relevant UV-B levels did not result in adverse effects at the human cell level [13]. Cooking methods might affect the bioavailability and content of SPM. It was found that steaming retained more chlorophylls, glucosinolates, phenolic acids and flavonoid compounds than boiling in three different cultivars of pakchoi [8]. These findings suggested that the supplementation of white light with UV-B irradiation may be a sustainable tool for improving crop production quality and food safety. A crucial issue is the dosage of radiation necessary to optimize the biosynthesis of beneficial phytochemicals without affecting the times, quality, and quantity of the harvest.

Vegetables are the main source of antioxidants in the human diet and are essential in our daily lives. The consumption of diets high in vegetables has been associated with a lower risk of a number of chronic illnesses [14]. As a result of market demands and economic incentives, greenhouse vegetable production has been developed and rapidly expanded as an intensive form of agriculture, which provides consumers with sufficient vegetables in the on- and off-seasons in many developing countries [15]. However, most plastic films covering greenhouses or polytunnels almost completely absorb and hence block UV radiation (both UV-A and UV-B) reaching the plants, due mainly to the stabilizers used in the different materials to extend the longevity of the film [16]. Polycarbonate, polyethylene, and fiberglass are the most commonly used greenhouse covering materials, with the effect of excluding more than 90% of the incident UV-B radiation [17]. Therefore, most greenhouse-grown vegetables are basically protected from UV-B irradiation during the growth process, leading to a decrease in the content of antioxidants such as flavonoids. For example, the concentrations of flavonoid derivatives in the leaf blade of various pakchoi cultivars ranged from 15 to 39 mg·g^{-1} dry matter under field conditions, but only ranged from 4.7 to 16.7 mg·g^{-1} dry matter under greenhouse conditions. The concentrations of hydroxycinnamic acid derivatives were also significantly reduced [18,19]. So, it is of great significance to increase the accumulation of antioxidants in greenhouse vegetables by supplementation with UV-B.

Pakchoi is a leafy *Brassica* vegetable that is widely available in Asia and consumed in rising quantities in Europe with a high contents of antioxidants. Several studies have reported the effect of UV-B radiation on antioxidants in pakchoi as described earlier. However, to our knowledge, the direct correlation between antioxidants accumulation and antioxidant activity under different UV-B irradiation conditions has not yet been studied. The objective of this research is to identify the most appropriate UV-B treatment for improving the antioxidant capacity in pakchoi and to identify the antioxidants stimulated by UV-B radiation. The activities of the key enzymes and expression levels of the genes involved in flavonoid biosynthetic pathway were determined to explore the molecular mechanism of UV-B radiation in improving antioxidant capacity. Our results will provide a potential new tool by which to generate greenhouse vegetables enriched with antioxidants for either fresh consumption or as a source of functional foods.

2. Results and Discussion

2.1. Impacts of UV-B Radiation on Plant Growth and Biomass in Pakchoi

It is critical to identify the most appropriate UV-B radiation dosage and exposure period, which enhances antioxidant capacity without affecting the growth and morphology of pakchoi. We first assessed how pakchoi plant growth was impacted by two different doses (2 μmol·m^{-2}·s^{-1} and 4 μmol·m^{-2}·s^{-1}) of UV-B radiation over each off our different exposure periods (2 h, 4 h, 8 h and 24 h). Compared with the control, there was no significant difference in fresh weight and dry weight under any UV-B radiation fluence rates (Figure 1). Our results are consistent with previous studies reporting that low, ecologically relevant UV-B levels do not affect plant growth [7,12]. As such, we further explored the impact of supplementary UV-B radiation on nutritional components and secondary metabolites in pakchoi.

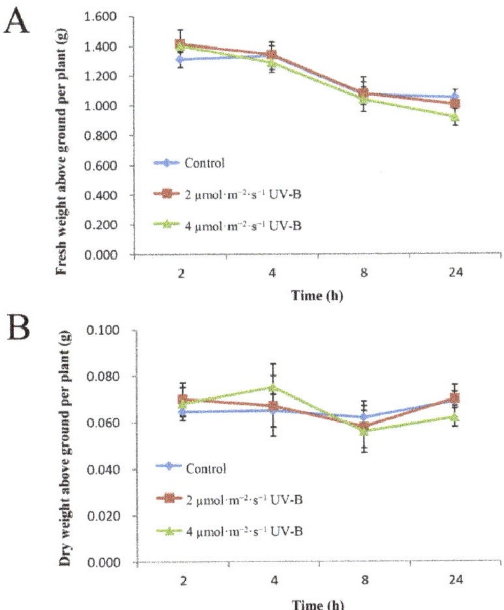

Figure 1. The effect of ultraviolet-B (UV-B) radiation on the above-ground biomass of pakchoi. (**A**) Fresh weight of 25-day-old seedlings treated with either 2 μmol·m^{-2}·s^{-1} or 4 μmol·m^{-2}·s^{-1} of UV-B irradiation for 2 h, 4 h, 8 h, or 24 h. (**B**) Dry weight of 25-day-old seedlings treated with either 2 μmol·m^{-2}·s^{-1} or 4 μmol·m^{-2}·s^{-1} of UV-B irradiation for 2 h, 4 h, 8 h, or 24 h. The plants without supplemental UV-B radiation served as controls. Data points are mean $\pm$ SE of three biological replicates. Significant differences between treated group and the control group at the same exposure period, identified by Student's *t*-test analysis.

2.2. UV-B Irradiation Effect on Total Antioxidant Capacity in Pakchoi

The total antioxidant capacity is often evaluated by 1,1–diphenyl–2–picrylhydrazyl (DPPH) scavenging activity, ferric reducing antioxidant power (FRAP), ABTS radical scavenging capacity, and oxygen radical absorption capacity assay in vegetables and fruits [20]. DPPH, as a stable free radical, has been widely employed to measure the radical scavenging effects of plant extracts [21]. The FRAP assay is a key method for assessing the total reduction capacity and offers a putative index of antioxidant capacity [22]. In the current study, the effect of UV-B radiation on total antioxidant capacity in pakchoi was assayed by measuring the DPPH scavenging activity and total reduction capacity.

There was no significant difference in DPPH scavenging activity between plants treated with either dose of UV-B radiation for 2 h and the control plants. Irradiation with $2\ \mu mol \cdot m^{-2} \cdot s^{-1}$ UV-B for 4 h or 8 h did not significantly increase the radical scavenging effects on DPPH. The DPPH scavenging activity of the plants irradiated with $2\ \mu mol \cdot m^{-2} \cdot s^{-1}$ UV-B for 24 h (81.12%) was significantly greater than that of the control (75.57%). The DPPH-scavenging activities of the plants irradiated with $4\ \mu mol \cdot m^{-2} \cdot s^{-1}$ UV-B for 4 h (87.84%; $p < 0.01$) or 8 h (84.95%; $p < 0.05$) were significantly greater than those of the controls (79.82% or 80.20%), although $4\ \mu mol \cdot m^{-2} \cdot s^{-1}$ UV-B irradiation for 24 h (74.64%) did not increase the scavenging effects on DPPH relative to the control (75.57%) (Figure 2A).

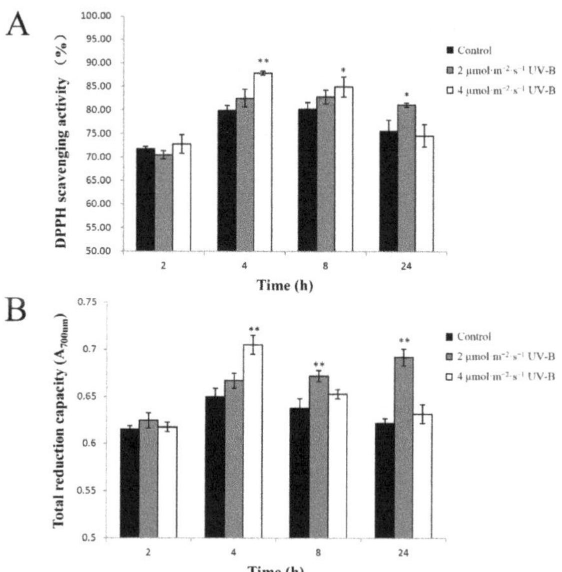

Figure 2. The effect of UV-B radiation on the total antioxidant capacity of pakchoi. (**A**) The 1,1–diphenyl-2-picrylhydrazyl (DPPH) scavenging activity of 25-day-old seedlings treated with either $2\ \mu mol \cdot m^{-2} \cdot s^{-1}$ or $4\ \mu mol \cdot m^{-2} \cdot s^{-1}$ of UV-B irradiation at 2 h, 4 h, 8 h, or 24 h. (**B**) The total reduction capacity of 25-day-old seedlings treated with either $2\ \mu mol \cdot m^{-2} \cdot s^{-1}$ or $4\ \mu mol \cdot m^{-2} \cdot s^{-1}$ of UV-B irradiation at 2 h, 4 h, 8 h, or 24 h. The plants not exposed to UV-B radiation served as controls. Three independent experiments were performed and data points represent the mean $\pm$ SE of three biological replicates. Asterisks indicate a significant difference (* $p < 0.05$; ** $p < 0.01$) to the corresponding control, using Student's *t*-test.

The effect of UV-B irradiation on total reduction capacity was similar to the effects on DPPH scavenging activity. The total reduction capacity of $2\ \mu mol \cdot m^{-2} \cdot s^{-1}$ UV-B radiation for either 8 h or 24 h and of $4\ \mu mol \cdot m^{-2} \cdot s^{-1}$ UV-B radiation for 4 h was significantly greater ($p < 0.01$) than that of the controls. No significant difference was found in the total reduction capacity between the plants exposed to other UV-B treatments and control plants (Figure 2B). These results indicated that the effect of UV-B radiation on total antioxidant capacity in pakchoi was dose-dependent, consistent with the previously reported results in the literature [7]. When the samples were collected immediately after the irradiation time-points, treatments with $2\ \mu mol \cdot m^{-2} \cdot s^{-1}$ UV-B radiation for 24 h or $4\ \mu mol \cdot m^{-2} \cdot s^{-1}$ UV-B radiation for 4 h had the greatest enhancement effect on antioxidant capacity in pakchoi. There was no significant stimulatory effect in response to $2\ \mu mol \cdot m^{-2} \cdot s^{-1}$ UV-B radiation for a shorter time or $4\ \mu mol \cdot m^{-2} \cdot s^{-1}$ UV-B radiation for a longer time. This effect may be related to different UV-B intensities activating particular signaling pathways, as described earlier [23]. However, the effect of UV-B radiation on antioxidant capacity varies over the collection time, which needs to be further studied.

In general, the effect of secondary metabolite accumulation induced by UV-B radiation lasts for some time. Su et al. reported that UV-B-induced anthocyanin accumulation in hypocotyls of radish sprouts could be sustained for a long time (more than 24 h) in the dark after irradiation [24]. To investigate whether a UV-B-induced increase of total antioxidant capacity continues after UV-B irradiation in pakchoi, the seedlings were first exposed to 4 $\mu mol \cdot m^{-2} \cdot s^{-1}$ UV-B for 4 h, and then transferred to darkness for 6, 12, 24, 36, and 48 h, respectively. The induction of DPPH scavenging activity could be maintained for 24 h in the dark following the radiation treatment (Figure 3A), although the enhancement effect was not apparent at 36 h or 48 h after treatment. As with DPPH scavenging activity, the induction of the total reduction capacity could also be maintained for 24 h in the dark after radiation (Figure 3B). Pakchoi is commonly consumed not only fresh (e.g., as salad), but also after cooking or fermentation. Furthermore, 21-day-old seedlings in three different cultivars of pakchoi were used to analyze the effect of domestic cooking methods (boiling and steaming) on secondary metabolites [8]. The production cycle of the pakchoi cultivar 'Can Bai' is 20–40 days, depending on growing temperature and consumer preference. The 25-day-old seedlings can be consumed for their high nutritional value, especially as baby salads. At the same time, there are some other ways to stabilize or further increase the enhancement effect of antioxidant capacity, such as blue light treatment after pre-exposure to UV-B as previously reported [4]. These might make the preharvest UV-B treatment an effective tool, allowing people to harvest more nutritious greenhouse-grown pakchoi in time. The decrease in DPPH scavenging activity from 6 h to 12 h might be due to the variable time of darkness in both the UV-B treatment and control groups. Dark treatments after UV-B irradiation can eliminate confounding factors such as incandescent and provide experimental evidence of energy efficiency and practical applications for enhancing the nutritional quality of pakchoi. The results showed that the energy could be saved for at least 6 h by the dark treatment.

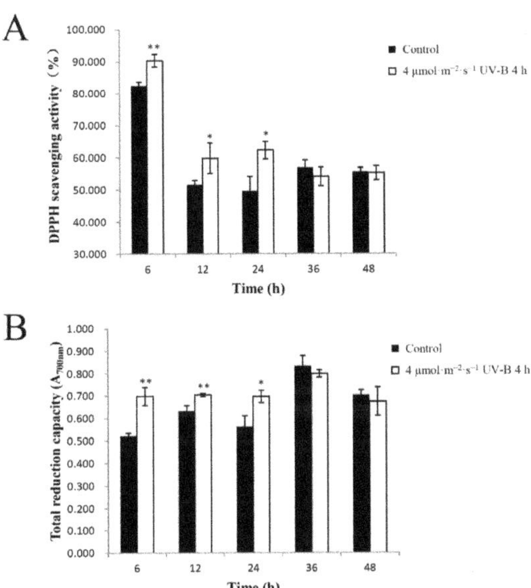

Figure 3. Duration of UV-B radiation effects on the antioxidant capacity of pakchoi. (**A**) The DPPH scavenging activity of 25-day-old seedlings at 6 h, 12 h, 24 h, 36 h, or 48 h after irradiation with 4 $\mu mol \cdot m^{-2} \cdot s^{-1}$ UV-B radiation for 4 h. (**B**) The total reduction capacity of 25-day-old seedlings at

6 h, 12 h, 24 h, 36 h, or 48 h after 4 μmol·m^{-2}·s^{-1} UV-B radiation for 4 h. Plants grown without UV-B radiation served as controls. Three independent biological replicate experiments were performed; data points represent the mean ± SE of the three biological replicates. Asterisks indicate a significant difference (* $p < 0.05$; ** $p < 0.01$) relative to the corresponding control, using Student's *t*-test.

2.3. Effects of UV-B Irradiation on Non-Enzymatic Antioxidants

Polyphenols, glutathione and ascorbate are considered to be potent non-enzymatic antioxidants in plants as they exhibit a high scavenging activity of harmful reactive oxygen species (ROS) [25–27]. Phenolic compounds are ubiquitous in the plant kingdom and constitute a large class of secondary metabolites, including phenolic acids, flavonoids, tannins, lignans, coumarins, and stilbenes [28]. Flavonoids are a biologically important group of phenolics, which have been recently suggested to contribute primary antioxidant functions in the responses of plants to a wide range of abiotic stresses, including UV-B radiation [1].

In the current study, the concentrations of total polyphenols, flavonoids, glutathione, and ascorbate in UV-B-treated and untreated pakchoi were determined and compared. As shown in Figure 4A, the concentration of total polyphenols increased very significantly after treatment with 2 μmol·m^{-2}·s^{-1} UV-B radiation for 24 h or 4 μmol·m^{-2}·s^{-1} UV-B radiation for 4 h compared with the control. The total polyphenol concentration increased from 13.94 mg·g^{-1} to 15.63 mg·g^{-1} ($p < $p0.01) after treatment with 2 μmol·m^{-2}·s^{-1} UV-B radiation for 24 h or from 13.65 mg·g^{-1} to 16.13 mg·g^{-1} ($p < 0.01$) after treatment with 4 μmol·m^{-2}·s^{-1} UV-B radiation for 4 h. No significant changes were observed in response to other UV-B irradiation treatments.

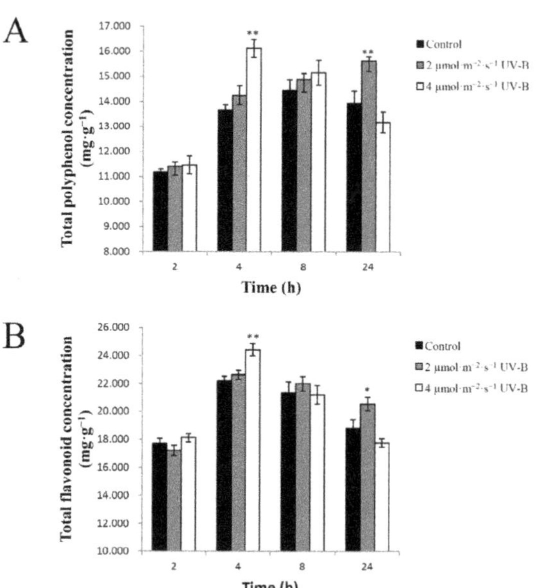

Figure 4. The effect of UV-B radiation on the antioxidant concentrations of pakchoi. (**A**) The total polyphenol concentration of 25-day-old seedlings treated with either 2 μmol·m^{-2}·s^{-1} or 4 μmol·m^{-2}·s^{-1} of UV-B radiation at 2 h, 4 h, 8 h, or 24 h. (**B**) The total flavonoid concentration of 25-day-old seedlings treated with either 2 μmol·m^{-2}·s^{-1} or 4 μmol·m^{-2}·s^{-1} of UV-B radiation at 2 h, 4 h, 8 h, or 24 h. The plants untreated with UV-B radiation served as controls. Three independent biological replicate experiments were performed; data points represent the mean ± SE of the three biological replicates. Asterisks indicate a significant difference (* $p < 0.05$; ** $p < 0.01$) relative to the corresponding control using Student's *t*-test.

As the main phenolic compounds, the response of total flavonoid concentration to UV-B radiation was similar to that of total polyphenol concentration. The total flavonoid concentration increased significantly from 18.81 mg·g^{-1} to 20.57 mg·g^{-1} ($p < 0.05$) after treatment with 2 μmol·m^{-2}·s^{-1} UV-B radiation for 24 h and from 22.20 mg·g^{-1} to 24.43 mg·g^{-1} ($p < 0.01$) after treatment with 4 μmol·m^{-2}·s^{-1} UV-B radiation for 4 h. There was no significant increase in the total flavonoid concentration in response to other UV-B irradiated conditions compared with the control. The results showed that the induction of total polyphenols and flavonoids in UV-B-treated pakchoi was dependent on the radiation dosage and time (Figure 4B).

There are many types of flavonoid, and changes in the concentrations of individual flavonoids in pakchoi between UV-B treatment and control groups were analyzed by high-performance liquid chromatography (HPLC). More than a dozen flavonoids were isolated from the pakchoi leaves based on their ultraviolet absorption spectrum and elution profile (Figure 5A). Among them, the peak areas of Peak 1, Peak 2, Peak 3, Peak 4, Peak 5, Peak 6 and Peak 9 increased very significantly ($p < 0.01$) in extracts of plants treated with 4 μmol·m^{-2}·s^{-1} UV-B radiation for 4 h. The peak areas of Peak 7 and Peak 8 increased significantly ($p < 0.05$) in the extracts of plants treated with 4 μmol·m^{-2}·s^{-1} UV-B radiation for 4 h. The peak area of Peak 10 did not change significantly (Figure 5B). In order to identify metabolic features, we carried out a liquid chromatography–mass spectrometry (LC–MS) analysis, and found that four of the peaks possibly representing flavonoids increased in response to UV-B (Figure S1). Due to the lack of suitable databases and standards, we could not confirm which specific type of flavonoids these peaks are. We speculate that they are most likely kaempferol glycosides according to the reported literature [8,11,12].

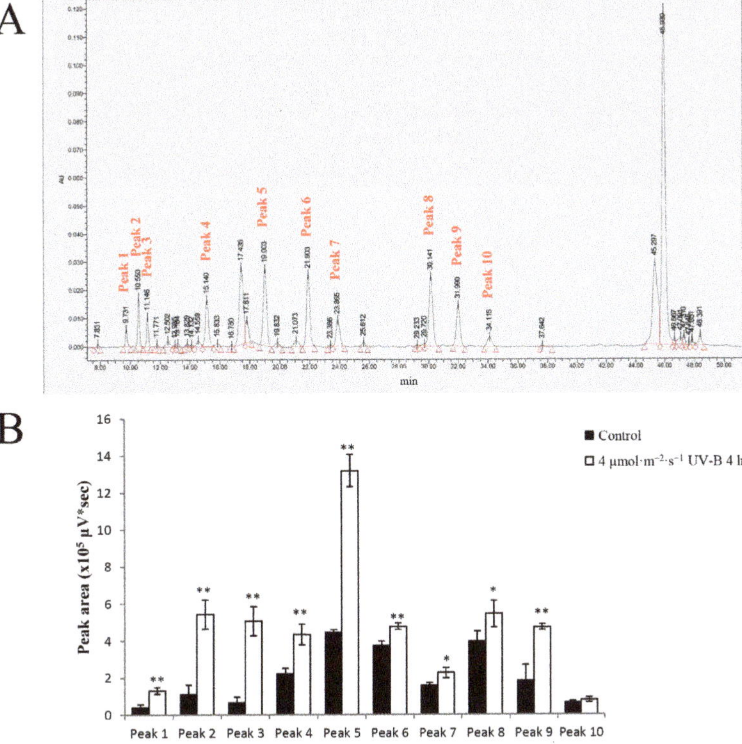

Figure 5. Determination of flavonoid concentrations in pakchoi in response to UV-B. (**A**) High

performance liquid chromatography (HPLC) chromatogram of the flavonoids in extracts of 25-day-old seedlings. (**B**) The peak areas fractioned by HPLC. Twenty-five-day-old seedlings treated with 4 $\mu mol \cdot m^{-2} \cdot s^{-1}$ UV-B for 4 h were harvested for extraction. Twenty-five-day-old seedlings not treated with UV-B radiation served as the control. Three independent biological replicate experiments were performed; data points represent the mean $\pm$ SE of the three biological replicates. Asterisks indicate a significant difference (* $p < 0.05$; ** $p < 0.01$) relative to the corresponding control, using Student's *t*-test.

On the other hand, there was no obvious enhancement effect of UV-B radiation on glutathione and ascorbate concentrations in pakchoi (Figure S2). Pakchoi synthesizes comparatively high amounts of glucosinolates, most of which were not affected by reduced UV-B conditions during the late developmental stages of pakchoi [7]. Flavonoids were still not evaluated and further investigations on glucosinolates are required. Nonetheless, these results revealed that the increase in non-enzymatic antioxidant activity was mainly due to the accumulation of phenolic compounds, especially flavonoids. This finding is in agreement with previous studies that showed that UV-B radiation can induce the biosynthesis of flavonoids in a range of plants [29–32].

2.4. Effects of UV-B Radiation on Flavonoid Biosynthesis Enzymes

The phenolic and flavonoid compound biosynthesis pathway is one of the most extensively studied areas of SPM. Flavonoids are synthesized via the shikimate-phenylpropanoid-flavonoid pathways in plants as documented in recent literature [33,34]. The phenylpropanoid pathway begins from the aromatic amino acids phenylalanine and tyrosine, which are synthesized by the shikimate pathway, to generate 4-coumaroyl-CoA, which is utilized in the flavonoid pathway. A number of important enzymes are involved in this process, such as PAL, C4H, 4CL, CHS, F3H, F3′H, FLS, DFR, and ANS. Among them, PAL, C4H and 4CL are three major enzymes in the phenylpropanoid pathway.

The effects of UV-B radiation on PAL, C4H and 4CL activities in pakchoi were examined in this study. PAL activity increased very significantly ($p < 0.01$) in response to 2 $\mu mol \cdot m^{-2} \cdot s^{-1}$ or 4 $\mu mol \cdot m^{-2} \cdot s^{-1}$ UV-B radiation for 4 h or 24 h compared with the control (Figure 6A). The activity of C4H increased very significantly ($p < 0.01$) after exposure to 2 $\mu mol \cdot m^{-2} \cdot s^{-1}$ or 4 $\mu mol \cdot m^{-2} \cdot s^{-1}$ UV-B radiation for 4 h compared with the control, whereas the increase after 24 h was only significant ($p < 0.05$) (Figure 6B). The activity of 4CL increased significantly ($p < 0.05$) in response to 2 $\mu mol \cdot m^{-2} \cdot s^{-1}$ or 4 $\mu mol \cdot m^{-2} \cdot s^{-1}$ UV-B radiation for 24 h compared with the control and very significantly ($p < 0.01$) after 2 $\mu mol \cdot m^{-2} \cdot s^{-1}$ or 4 $\mu mol \cdot m^{-2} \cdot s^{-1}$ UV-B radiation for 4 h compared with the control (Figure 6C). The results revealed that the observed stimulatory effect of UV-B radiation on the production of flavonoids could be explained by the induction of the activities of important enzymes in the flavonoid biosynthesis pathway, a finding which was consistent with previous reports from other plant species [24,26,35].

2.5. UV-B Effect on the Expression of Flavonoid Biosynthesis Genes

Anthocyanins and flavonols are the two major classes of flavonoid compounds, in terms of their role in protecting plants against abiotic and biotic stresses. A total of 73 anthocyanin biosynthetic genes in *Brassica rapa* have been identified using comparative genomic analyses between *Brassica rapa* and *Arabidopsis thaliana* [36]. The expression levels of some of these flavonoid biosynthesis genes in response to 2 or 4 $\mu mol \cdot m^{-2} \cdot s^{-1}$ UV-B irradiation for 4 h or 24 h were analyzed in pakchoi in the current study. The expression of each of the three major genes of the phenylpropanoid pathway, *BrPAL*, *BrC4H*, and *Br4CL*, was upregulated significantly ($p < 0.05$ or $p < 0.01$) in each of the UV-B treatment groups compared with the controls, a finding which was basically consistent with the results of the corresponding enzyme activity analysis. The expression of the early biosynthesis genes in the flavonoid pathway, *BrCHS*, *BrCHI*, *BrF3H*, *BrF3′H*, and *BrFLS*, and of the late biosynthesis genes in the flavonoid pathway, *BrDFR*, *BrANS*, *BrLDOX*, and *BrUFGT*, were upregulated significantly

($p < 0.05$) or very significantly ($p < 0.01$) after UV-B irradiation, compared with the control. Overall, the highest expression level occurred at 2 $\mu mol \cdot m^{-2} \cdot s^{-1}$ UV-B radiation for 24 h and 4 $\mu mol \cdot m^{-2} \cdot s^{-1}$ UV-B radiation for 4 h, findings which were consistent with the previous enzyme activity results (Figure 7). These results showed that the irradiation-induced increases in concentrations of flavonoids were associated with corresponding increases in the expression of flavonoid biosynthesis genes.

All of the above results indicated that the changes in gene expression, enzyme activity, and antioxidant concentration in response to supplementary UV-B radiation are basically consistent, as previously reported [37]. Moreover, the gene expression levels and biosynthetic enzyme activities are more sensitive to UV-B radiation than the flavonoid antioxidant concentration levels.

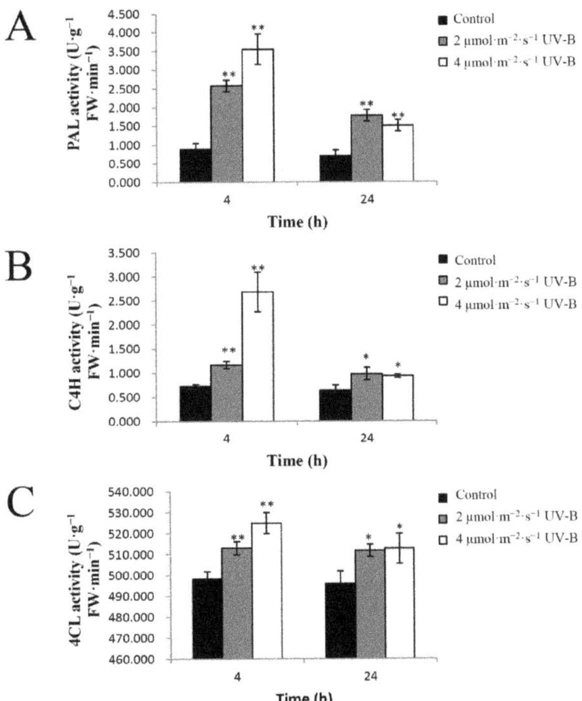

Figure 6. The effect of UV-B radiation on the activities of flavonoid biosynthesis enzymes in pakchoi. (**A**) The activity of PAL in 25-day-old seedlings treated with either 2 $\mu mol \cdot m^{-2} \cdot s^{-1}$ or 4 $\mu mol \cdot m^{-2} \cdot s^{-1}$ of UV-B radiation at 4 h or 24 h. (**B**) The activity of C4H in 25-day-old seedlings treated with either 2 $\mu mol \cdot m^{-2} \cdot s^{-1}$ or 4 $\mu mol \cdot m^{-2} \cdot s^{-1}$ of UV-B radiation at 4 h or 24 h. (**C**) The activity of 4CL in 25-day-old seedlings treated with either 2 $\mu mol \cdot m^{-2} \cdot s^{-1}$ or 4 $\mu mol \cdot m^{-2} \cdot s^{-1}$ of UV-B radiation at 4 h or 24 h. The plants without UV-B radiation served as controls. Three independent biological replicate experiments were performed; data points represent the mean $\pm$ SE of three biological replicates. Asterisks indicate a significant difference (* $p < 0.05$; ** $p < 0.01$) relative to the corresponding control, using Student's *t*-test.

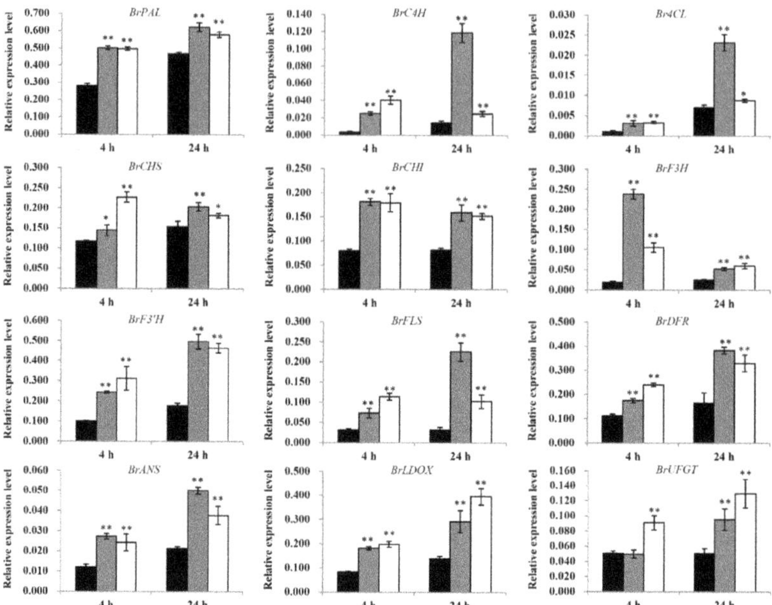

Figure 7. Relative expression levels of the genes related to flavonoid biosynthesis in response to UV-B radiation. Black columns represent control; gray columns represent 2 μmol·m^{-2}·s^{-1} UV-B; white columns represent 4 μmol·m^{-2}·s^{-1} UV-B. Gene expression values are relative to reference *BrActin2* expression; data points represent the mean ± SE of three biological replicates. Asterisks indicate a significant difference (* $p < 0.05$; ** $p < 0.01$) relative to the corresponding control using Student's *t*-test.

3. Materials and Methods

3.1. Plant Materials and Growth Conditions

A local commercial cultivar of pakchoi, 'Can Bai' (by Zhejiang Academy of Agricultural Sciences, Hangzhou, China), was used in the experiments. The plants were cultivated in an illuminated growth chamber (26 °C, 12 h/12 h light/dark cycle regime) in soil (peat, pH 5.5–6.5; Fafard, Saint-Bonaventure, QC, Canada). Water was supplied as required by the plants and fertilizer was administered weekly with Hoagland's nutrient solution. The light intensity of incandescent was 100 μmol·m^{-2}·s^{-1} (400–700 nm). For experimentation, 25-day-old seedlings were supplemented with (treatment group) or without (control group) UV-B irradiation and tissue was collected immediately after the irradiation time-points for analysis. Fifteen plants were pooled for one biological replicate, and all experiments were performed in triplicate.

3.2. Radiation Procedure

Pakchoi seedlings were placed on shelves and exposed to the supplementary UV-B radiation, at doses of 2 μmol·m^{-2}·s^{-1} (equals 0.7 W·m^{-2}) or 4 μmol·m^{-2}·s^{-1} (equals 1.4 W·m^{-2}) for 2 h, 4 h, 8 h, or 24 h. UV-B radiation was supplied by five fluorescent lamps (40 W 12RS, Beijing Lighting Research Institute, Beijing, China), whereby the UV-B emission peaked at 313 nm. The desired radiation dose was obtained by changing the number of UV-B lamps and the distance between the lamps and the plants. UV-B radiation was measured using an Optronics Model 720 spectroradiometer (Beijing Normal University Optronics Factory, Beijing, China), with a spectral range of 280 to 400 nm.

3.3. DPPH Scavenging Assay

The DPPH scavenging activity assay was performed according to the method reported by Alhaithloul et al. [38]. Aliquots (0.2 g) of the dried samples were extracted with 45 mL 70% methanol in a water bath at 70 °C for 60 min and centrifuged for 15 min at $4700\times g$. The supernatant was retained and used to determine DPPH scavenging activity. An aliquot (4 mL) of 2.0×10^{-4} mmol·L^{-1} DPPH solution in 70% ethanol was added to 1 mL of the supernatant. The mixture was allowed to incubate for 30 min at room temperature in the dark, after which the absorbance at 517 nm was measured.

3.4. Determination of Total Reduction Capacity

The FRAP assay was performed to determine the total reduction capacity according to the procedure reported previously [6]. Aliquots (0.2 g) of the dried samples were extracted with 45 mL 70% methanol in a water bath at 70°C for 60 min and centrifuged for 15 min at $4700\times g$. The supernatant was retained and used for assays. The FRAP reagent included a 300 mM acetate buffer (pH 3.6), 10 mM 2,4,6-tris(2-pyridyl)-1,3,5-triazine in 40 mM HCl, and 20 mM FeCl$_3$ in the ratio 10:1:1 (*v:v:v*). An aliquot (3 mL) of the FRAP reagent was mixed with 100 μL of the sample extract in a test tube, vortexed and incubated at 37 °C for 30 min in a water bath. The absorbance was measured at 700 nm.

The total phenolic concentration was measured using the Folin–Ciocalteu method as described previously with some modifications [39]. In brief, 0.5 mL of sample extract was mixed with 1.8 mL of 0.1 N Folin–Ciocalteu reagent (Sangon Biotech, Shanghai, China). After incubating for 5 min at room temperature, the reaction was stopped by the addition of 1.2 mL of an aqueous solution of 7.5% sodium carbonate. Then, the absorbance was measured at 765 nm. Gallic acid was used as the standard for a calibration curve, and the results were expressed as gallic acid equivalents.

The determination of total flavonoid concentration was performed as described previously with slight modifications [40]. Then, 2 mL of the sample extract was placed in a 10-mL volumetric flask and 0.5 mL of 5% NaNO$_2$ was added, following which, 0.5 mL of 10% AlCl$_3$ was added. After 6 min, 4 mL of 4% NaOH was added, and the total volume was 10 mL, with 70% ethanol. The solution was mixed well again, and the absorbance was measured at 510 nm. Rutin was used as the standard for a calibration curve, and the results were expressed as rutin equivalents.

3.5. Analysis of Flavonoids by HPLC and LC–MS

The HPLC and LC–MS analyses of flavonoids were carried out as described before with some modifications [41]. A tissue sample (0.1 g dry weight) was extracted in 1.5 mL of 80% methanol in the water bath for 60 min at 70 °C, centrifuged for 15 min at $4700\times g$ and then filtered through a 0.22-μm pore size filter (Millipore, Billerica, MA, USA) prior to analysis. An HPLC analysis was performed on a Waters 2695 Alliance HPLC system (Waters, Milford, MA, USA) equipped with a photodiode array detector. A C18 column (4.6 mm i.d. × 250 mm) (Waters, Milford, MA, USA) was used with a flow rate of 1 mL·min^{-1} at 25 °C. Gradient elution was employed using mobile phases of 0.1% trifluoroacetic acid (A) and acetonitrile (B) (Supplementary Table S1). Spectra were measured at a wavelength of 350 nm, and individual flavonoids were identified by comparing the retention time and UV spectra. LC–MS analyses were carried out using an LCQ ion trap mass spectrometer (Finnigan MAT, San Jose, CA, USA) equipped with an ESI source in the positive ion mode. Helium was used as the buffer gas and nitrogen was used as the dry gas (12 L·min^{-1}, 350 °C). The heated metal capillary temperature was 180 °C. The electrospray voltage was at 4.5 kV. The data were analyzed using a DataAnalysis Compass.

3.6. Assay of Flavonoid Biosynthesis Enzyme Activities

PAL activity was determined as described previously with some modifications [42]. Briefly, fresh samples (0.5 g) were homogenized in 10 mL of pre–cooled extractant solution (0.01 mol·L^{-1} boric acid buffer, pH 8.8; containing 5 mmol·L^{-1} β–mercaptoethanol) and centrifuged for 20 min at 4 °C at 9600× *g*. The supernatant was retained and used to determine the PAL activity. The reaction mixture consisted of 100 μL supernatant, 3.0 mL of 0.01 mol·L^{-1} sodium borate (pH 8.8; containing 0.005 mol·L^{-1} β–mercaptoethanol) and 700 μL of 0.01 mol·L^{-1} L–phenylalanine. The reaction mixture was incubated at 30 °C for 30 min and the reaction was stopped by adding 0.2 mL of 6 mol·L^{-1} HCl. PAL activity was spectrophotometrically measured by monitoring the absorbance at 290 nm. Then, C4H and 4CL activity were determined according to the procedures reported previously [43,44].

3.7. RNA Extraction and Quantitative Reverse Transcription PCR (qRT–PCR) Analysis

Total RNA was isolated with the OmniPlant RNA Kit (DNase I) CW2598 (CWBIO, Beijing, China) as previously described [45]. HiFiScript gDNA Removal cDNA Synthesis Kit CW2582 (CWBIO, Beijing, China) was used to achieve first-strand cDNA synthesis from approximately 1 μg of total RNA. qRT–PCR was performed using the iQ SYBR Green Supermix (Bio–Rad, Hercules, CA, USA) and run on the ABI Prism 7000 system (Applied Biosystems, Foster City, CA, USA). The sequences of the genes studied in this article (*BrActin*, Bra022356; *BrPAL*, Bra005221; *BrC4H*, Bra018311; *Br4CL*, Bra030429; *BrCHS*, Bra008792; *BrCHI*, Bra007142; *BrF3H*, Bra036828; *BrF3'H*, Bra009312; *BrFLS*, Bra009358; *BrDFR*, Bra027457; *BrANS*, Bra013652; *BrLDOX*, Bra019350; *BrUFGT*, Bra023954) were derived from a previously published paper [36] and *Brassica* database BRAD (http://brassicadb.cn, accessed on 20 January 2022) [46]. Furthermore, *BrActin2* was used as the reference housekeeping gene. The relative expression level of the target genes was normalized against the reference housekeeping gene [47]. The primers used in the qRT–PCR are listed in Supplementary Table S2.

4. Conclusions

Pakchoi is a very popular vegetable, rich in antioxidants with health benefits for consumers. However, greenhouse cultivation negatively affects the biosynthesis of antioxidants in pakchoi by interfering with incident UV-B. Preharvest UV-B supplementation has proved to be a very effective measure by which to improve the nutritional quality of pakchoi by promoting the accumulation of antioxidants in greenhouse-grown plants. Since the effects of UV-B radiation on plants depend on the radiation dose, exposure time, and plant species, we evaluated the effects of two different doses of UV-B radiation on pakchoi for four different irradiation periods. Our results showed that the appropriate UV-B irradiation treatments (4 μmol·m^{-2}·s^{-1} for 4 h or 2 μmol·m^{-2}·s^{-1} for 24 h) could significantly upregulate the expression of flavonoid biosynthesis genes (*BrPAL*, *BrC4H*, *Br4CL*, *BrCHS*, *BrF3H*, *BrF3'H*, *BrFLS*, *BrDFR*, *BrANS*, and *BrLDOX*), increase the activities of the most important enzymes (PAL, C4H and 4CL), promote the accumulation of flavonoids, and eventually lead to the improvement of antioxidant activity in pakchoi. This study provides a basis for future comprehensive studies on the metabolic mechanism of flavonoid biosynthesis, and new insight into an enhancement of the nutritional quality of greenhouse-grown vegetables.

Supplementary Materials: The following supporting information can be downloaded at: https://www.mdpi.com/article/10.3390/plants11060766/s1, Figure S1. Mass spectrometric analysis of the flavonoids which increased in response to 4 μmol·m^{-2}·s^{-1} UV-B radiation for 4 h. Figure S2: The effect of UV-B radiation on the antioxidant concentrations of pakchoi. (**A**) The glutathione concentration of 25-day-old seedlings treated with either dosage (2 μmol·m^{-2}·s^{-1} or 4 μmol·m^{-2}·s^{-1}) of UV-B radiation at 4 h or 24 h. (**B**) The ascorbate concentration of 25-day-old seedlings treated with either dosage (2 μmol·m^{-2}·s^{-1} or 4 μmol·m^{-2}·s^{-1}) of UV-B radiation at 4 h or 24 h. The plants without UV-B radiation served as controls. Three biologically independent replicate experiments were performed; data points represent the mean ± SE of three biological replicates. Asterisks indicate a significant difference (* $p < 0.05$; ** $p < 0.01$) relative to the corresponding control, using Student's *t*-test; Table S1: Gradient elution program for HPLC analysis; Table S2: Primers used in quantitative reverse transcription PCR (qRT–PCR).

Author Contributions: Conceptualization, J.H. and M.X.; validation, P.L.; formal analysis, Y.H.; investigation, L.Z.; resources, Z.C. and J.N.; data curation, Y.Y.; writing—original draft preparation, J.H.; writing—review and editing, J.L. and M.X. All authors have read and agreed to the published version of the manuscript.

Funding: This research was funded by the National Natural Science Foundation of China, grant number 31601343; Hangzhou Normal University Xinmiao Talent Program, grant number 2021R426077; Natural Science Foundation of Zhejiang Province, grant number LY19C150005 and LY19C020003; Major Increase or Decrease Program in The Central Finance Level, grant number 2060302; Zhejiang Provincial Key Research & Development Project, grant number 2018C02030.

Institutional Review Board Statement: Not applicable.

Informed Consent Statement: Not applicable.

Data Availability Statement: Not applicable.

Acknowledgments: We thank Xiaoqiong Ma (Zhejiang Provincial Hospital of Chinese Medicine) for LC–MS analysis.

Conflicts of Interest: The authors declare no conflict of interest.

References

1. Brunetti, C.; Di Ferdinando, M.; Fini, A.; Pollastri, S.; Tattini, M. Flavonoids as antioxidants and developmental regulators: Relative significance in plants and humans. *Int. J. Mol. Sci.* **2013**, *14*, 3540–3555. [CrossRef]
2. Del Rio, D.; Rodriguez-Mateos, A.; Spencer, J.P.; Tognolini, M.; Borges, G.; Crozier, A. Dietary (poly)phenolics in human health: Structures, bioavailability, and evidence of protective effects against chronic diseases. *Antioxid. Redox Signal.* **2013**, *18*, 1818–1892. [CrossRef]
3. Neugart, S.; Klaring, H.P.; Zietz, M.; Schreiner, M.; Rohn, S.; Kroh, L.W.; Krumbein, A. The effect of temperature and radiation on flavonol aglycones and flavonol glycosides of kale (*Brassica oleracea* var. *sabellica*). *Food Chem.* **2012**, *133*, 1456–1465. [CrossRef]
4. Neugart, S.; Majer, P.; Schreiner, M.; Hideg, É. Blue light treatment but not green light treatment after pre-exposure to UV-B stabilizes flavonoid glycoside changes and corresponding biological effects in three different Brassicaceae sprouts. *Front. Plant Sci.* **2021**, *11*, 611247. [CrossRef]
5. Seong, G.U.; Hwang, I.W.; Chung, S.K. Antioxidant capacities and polyphenolics of Chinese cabbage (*Brassica rapa* L. ssp. *Pekinensis*) leaves. *Food Chem.* **2016**, *199*, 612–618. [CrossRef]
6. Rokayya, S.; Li, C.J.; Zhao, Y.; Li, Y.; Sun, C.H. Cabbage (*Brassica oleracea* L. var. *capitata*) phytochemicals with antioxidant and anti-inflammatory potential. *Asian Pac. J. Cancer Prev.* **2014**, *14*, 6657–6662. [CrossRef]
7. Heinze, M.; Hanschen, F.S.; Wiesner-Reinhold, M.; Baldermann, S.; Gräfe, J.; Schreiner, M.; Neugart, S. Effects of developmental stages and reduced UVB and low UV conditions on plant secondary metabolite profiles in pak choi (*Brassica rapa* subsp. *chinensis*). *J. Agric. Food Chem.* **2018**, *66*, 1678–1692. [CrossRef] [PubMed]
8. Chen, X.M.; Hanschen, F.S.; Neugart, S.; Schreiner, M.; Vargas, S.A.; Gutschmann, B.; Baldermann, S. Boiling and steaming induced changes in secondary metabolites in three different cultivars of pak choi (*Brassica rapa* subsp. *chinensis*). *J. Food Compos. Anal.* **2019**, *82*, 103232. [CrossRef]
9. Jansen, M.A.; Bornman, J.F. UV-B radiation: From generic stressor to specific regulator. *Physiol. Plant* **2012**, *145*, 501–504. [CrossRef] [PubMed]
10. Reifenrath, K.; Muller, C. Species-specific and leaf-age dependent effects of ultraviolet radiation on two Brassicaceae. *Phytochemistry* **2007**, *68*, 875–885. [CrossRef] [PubMed]
11. Harbaum-Piayda, B.; Walter, B.; Bengtsson, G.B.; Hubbermann, E.M.; Bilger, W.; Schwarz, K. Influence of pre-harvest UV-B irradiation and normal or controlled atmosphere storage on flavonoid and hydroxycinnamic acid contents of pak choi (*Brassica campestris* L. ssp *chinensis* var. *communis*). *Postharvest Biol. Tech.* **2010**, *56*, 202–208. [CrossRef]
12. Neugart, S.; Bumke-Vogt, C. Flavonoid glycosides in *Brassica* species respond to UV-B depending on exposure time and adaptation time. *Molecules* **2021**, *26*, 494. [CrossRef] [PubMed]
13. Wiesner-Reinhold, M.; Gomas, J.V.D.; Herz, C.; Tran, H.T.T.; Baldermann, S.; Neugart, S.; Filler, T.; Glaab, J.; Einfeldt, S.; Schreiner, M.; et al. Subsequent treatment of leafy vegetables with low doses of UVB-radiation does not provoke cytotoxicity, genotoxicity, or oxidative stress in a human liver cell model. *Food Biosci.* **2021**, *43*, 101327. [CrossRef]
14. Hung, H.C.; Joshipura, K.J.; Jiang, R.; Hu, F.B.; Hunter, D.; Smith-Warner, S.A.; Colditz, G.A.; Rosner, B.; Spiegelman, D.; Willett, W.C. Fruit and vegetable intake and risk of major chronic disease. *J. Natl. Cancer Inst.* **2004**, *96*, 1577–1584. [CrossRef] [PubMed]
15. Yang, L.Q.; Huang, B.A.; Hu, W.Y.; Chen, Y.; Mao, M.C.; Yao, L.P. The impact of greenhouse vegetable farming duration and soil types on phytoavailability of heavy metals and their health risk in eastern China. *Chemosphere* **2014**, *103*, 121–130. [CrossRef]
16. Mormile, P.; Rippa, M.; Graziani, G.; Ritieni, A. Use of greenhouse-covering films with tailored UV-B transmission dose for growing 'medicines' through plants: Rocket salad case. *J. Sci. Food Agric.* **2019**, *99*, 6931–6936. [CrossRef]

17. Yoshida, H.; Nishikawa, T.; Hikosaka, S.; Goto, E. Effects of nocturnal UV-B irradiation on growth, flowering, and phytochemical concentration in leaves of greenhouse-grown red perilla. *Plants* **2021**, *10*, 1252. [CrossRef]
18. Harbaum, B.; Hubbermann, E.M.; Wolff, C.; Herges, R.; Zhu, Z.; Schwarz, K. Identification of flavonoids and hydroxycinnamic acids in pak choi varieties (*Brassica campestris* L. ssp. *chinensis* var. *communis*) by HPLC-ESI-MSn and NMR and their quantification by HPLC-DAD. *J. Agric. Food Chem.* **2007**, *55*, 8251–8260. [CrossRef]
19. Harbaum, B.; Hubbermann, E.M.; Zhu, Z.; Schwarz, K. Free and bound phenolic compounds in leaves of pak choi (*Brassica campestris* L. ssp. *chinensis* var. *communis*) and Chinese leaf mustard (*Brassica juncea* Coss). *Food chem.* **2008**, *110*, 838–846. [CrossRef]
20. Dudonne, S.; Vitrac, X.; Coutiere, P.; Woillez, M.; Merillon, J.M. Comparative study of antioxidant properties and total phenolic content of 30 plant extracts of industrial interest using DPPH, ABTS, FRAP, SOD, and ORAC assays. *J. Agric. Food Chem.* **2009**, *57*, 1768–1774. [CrossRef]
21. Nenadis, N.; Tsimidou, M. Observations on the estimation of scavenging activity of phenolic compounds using rapid 1,1-diphenyl-2-picrylhydrazyl (DPPH•) tests. *J. Am. Oil Chem. Soc.* **2002**, *79*, 1191. [CrossRef]
22. Benzie, I.F.F.; Strain, J.J. The ferric reducing ability of plasma (FRAP) as a measure of "antioxidant power": The FRAP assay. *Anal. Biochem.* **1996**, *239*, 70–76. [CrossRef] [PubMed]
23. Brown, B.A.; Jenkins, G.I. UV-B signaling pathways with different fluence-rate response profiles are distinguished in mature Arabidopsis leaf tissue by requirement for UVR8, HY5, and HYH. *Plant Physiol.* **2008**, *146*, 576–588. [CrossRef] [PubMed]
24. Su, N.N.; Lu, Y.W.; Wu, Q.; Liu, Y.Y.; Xia, Y.; Xia, K.; Cui, J. UV-B-induced anthocyanin accumulation in hypocotyls of radish sprouts continues in the dark after irradiation. *J. Sci. Food Agric.* **2016**, *96*, 886–892. [CrossRef]
25. Usano-Alemany, J.; Panjai, L. Effects of increasing doses of UV-B on main phenolic acids content, antioxidant activity and estimated biomass in lavandin (*Lavandula x intermedia*). *Nat. Prod. Commun.* **2015**, *10*, 1269–1272. [CrossRef]
26. Takshak, S.; Agrawal, S.B. Defence strategies adopted by the medicinal plant *Coleus forskohlii* against supplemental ultraviolet-B radiation: Augmentation of secondary metabolites and antioxidants. *Plant Physiol. Biochem.* **2015**, *97*, 124–138. [CrossRef]
27. Singh, V.P.; Srivastava, P.K.; Prasad, S.M. Differential effect of UV-B radiation on growth, oxidative stress and ascorbate-glutathione cycle in two cyanobacteria under copper toxicity. *Plant Physiol. Biochem.* **2012**, *61*, 61–70. [CrossRef]
28. Dai, J.; Mumper, R.J. Plant phenolics: Extraction, analysis and their antioxidant and anticancer properties. *Molecules* **2010**, *15*, 7313–7352. [CrossRef]
29. Nascimento, L.B.; Leal-Costa, M.V.; Menezes, E.A.; Lopes, V.R.; Muzitano, M.F.; Costa, S.S.; Tavares, E.S. Ultraviolet-B radiation effects on phenolic profile and flavonoid content of *Kalanchoe pinnata*. *J. Photochem. Photobiol. B.* **2015**, *148*, 73–81. [CrossRef]
30. Sun, M.Y.; Gu, X.D.; Fu, H.W.; Zhang, L.; Chen, R.Z.; Cui, L.; Zheng, L.H.; Zhang, D.W.; Tian, J.K. Change of secondary metabolites in leaves of *Ginkgo biloba* L. in response to UV-B induction. *Innov. Food Sci. Emerg. Technol.* **2010**, *11*, 672–676. [CrossRef]
31. Martinez-Luscher, J.; Torres, N.; Hilbert, G.; Richard, T.; Sanchez-Diaz, M.; Delrot, S.; Aguirreolea, J.; Pascual, I.; Gomes, E. Ultraviolet-B radiation modifies the quantitative and qualitative profile of flavonoids and amino acids in grape berries. *Phytochemistry* **2014**, *102*, 106–114. [CrossRef] [PubMed]
32. Avena-Bustillos, R.J.; Du, W.X.; Woods, R.; Olson, D.; Breksa, A.P., 3rd; McHugh, T.H. Ultraviolet-B light treatment increases antioxidant capacity of carrot products. *J. Sci. Food Agric.* **2012**, *92*, 2341–2348. [CrossRef] [PubMed]
33. Saito, K.; Yonekura-Sakakibara, K.; Nakabayashi, R.; Higashi, Y.; Yamazaki, M.; Tohge, T.; Fernie, A.R. The flavonoid biosynthetic pathway in Arabidopsis: Structural and genetic diversity. *Plant Physiol. Biochem.* **2013**, *72*, 21–34. [CrossRef] [PubMed]
34. Winkel-Shirley, B. Flavonoid biosynthesis. A colorful model for genetics, biochemistry, cell biology, and biotechnology. *Plant Physiol.* **2001**, *126*, 485–493. [CrossRef]
35. Jiao, J.; Gai, Q.Y.; Wang, W.; Luo, M.; Gu, C.B.; Fu, Y.J.; Ma, W. Ultraviolet radiation-elicited enhancement of isoflavonoid accumulation, biosynthetic gene expression, and antioxidant activity in *Astragalus membranaceus* hairy root cultures. *J. Agric. Food Chem.* **2015**, *63*, 8216–8224. [CrossRef]
36. Guo, N.; Cheng, F.; Wu, J.; Liu, B.; Zheng, S.N.; Liang, J.L.; Wang, X.W. Anthocyanin biosynthetic genes in *Brassica rapa*. *BMC genom.* **2014**, *15*, 426. [CrossRef]
37. Nguyen, C.T.T.; Lim, S.; Lee, J.G.; Lee, E.J. VcBBX, VcMYB21, and VcR2R3MYB transcription factors are involved in UV–B-induced anthocyanin biosynthesis in the peel of harvested blueberry fruit. *J. Agric. Food Chem.* **2017**, *65*, 2066–2073. [CrossRef]
38. Alhaithloul, H.A.; Soliman, M.H.; Ameta, K.L.; El-Esawi, M.A.; Elkelish, A. Changes in ecophysiology, osmolytes, and secondary metabolites of the medicinal plants of *Mentha piperita* and *Catharanthus roseus* subjected to drought and heat stress. *Biomolecules* **2020**, *10*, 43. [CrossRef]
39. Loayza, F.E.; Brecht, J.K.; Simonne, A.H.; Plotto, A.; Baldwin, E.A.; Bai, J.; Lon-Kan, E. A brief hot-water treatment alleviates chilling injury symptoms in fresh tomatoes. *J. Sci. Food Agric.* **2021**, *101*, 54–64. [CrossRef]
40. Jia, Z.S.; Tang, M.C.; Wu, J.M. The determination of flavonoid contents in mulberry and their scavenging effects on superoxide radicals. *Food Chem.* **1999**, *64*, 555–559. [CrossRef]
41. Ni, J.; Hao, J.; Jiang, Z.F.; Zhan, X.R.; Dong, L.X.; Yang, X.L.; Sun, Z.H.; Xu, W.Y.; Wang, Z.K.; Xu, M.J. NaCl induces flavonoid biosynthesis through a putative novel pathway in post-harvest Ginkgo leaves. *Front. Plant Sci.* **2017**, *8*, 920. [CrossRef] [PubMed]
42. Huang, J.; Gu, M.; Lai, Z.; Fan, B.; Shi, K.; Zhou, Y.H.; Yu, J.Q.; Chen, Z. Functional analysis of the Arabidopsis *PAL* gene family in plant growth, development, and response to environmental stress. *Plant Physiol.* **2010**, *153*, 1526–1538. [CrossRef] [PubMed]

43. Lamb, C.J.; Rubery, P.H. A spectrophotometric assay for trans-cinnamic acid 4-hydroxylase activity. *Anal. Biochem.* **1975**, *68*, 554–561. [CrossRef]
44. Knobloch, K.H.; Hahlbrock, K. Isoenzymes of p-coumarate: CoA ligase from cell suspension cultures of *Glycine max*. *Eur. J. Biochem.* **1975**, *52*, 311–320. [CrossRef]
45. Hao, J.; Lou, P.P.; Han, Y.D.; Chen, Z.H.; Chen, J.M.; Ni, J.; Yang, Y.J.; Jiang, Z.F.; Xu, M.J. GrTCP11, a cotton TCP transcription factor, inhibits root hair elongation by down-regulating jasmonic acid pathway in *Arabidopsis thaliana*. *Front. Plant Sci.* **2021**, *12*, 769675. [CrossRef]
46. Cheng, F.; Liu, S.Y.; Wu, J.; Fang, L.; Sun, S.L.; Liu, B.; Li, P.X.; Hua, W.; Wang, X.W. BRAD, the genetics and genomics database for Brassica plants. *BMC Plant Biol.* **2011**, *11*, 136. [CrossRef]
47. Long, L.; Yang, W.W.; Liao, P.; Guo, Y.W.; Kumar, A.; Gao, W. Transcriptome analysis reveals differentially expressed ERF transcription factors associated with salt response in cotton. *Plant Sci.* **2019**, *281*, 72–81. [CrossRef]

 *plants*

Article

Comparative Analysis of Policosanols Related to Growth Times from the Seedlings of Various Korean Oat (*Avena sativa* L.) Cultivars and Screening for Adenosine 5′-Monophosphate-Activated Protein Kinase (AMPK) Activation

Han-Gyeol Lee [1,2,†], So-Yeun Woo [3,†], Hyung-Jae Ahn [1,4], Ji-Yeong Yang [1], Mi-Ja Lee [1], Hyun-Young Kim [1], Seung-Yeob Song [1], Jin-Hwan Lee [5,*] and Woo-Duck Seo [1,*]

1   Crop Foundation Research Division, National Institute of Crop Science, Rural Development Administration, Jeollabuk-do, Wanju-gun 55365, Korea; gajae93@gmail.com (H.-G.L.); hengja112@gmail.com (H.-J.A.); yjy90@korea.kr (J.-Y.Y.); esilvia@korea.kr (M.-J.L.); hykim84@korea.kr (H.-Y.K.); s2y337@korea.kr (S.-Y.S.)

2   Division of Life Sciences, College of Natural Science, Jeonbuk National University, 567 Baekje-daero, Jeollabuk-do, Jeonju 54896, Korea

3   Natural Medicine Research Center, Korea Research Institute of Bioscience and Biotechnology, Chungcheongbuk-do, Cheongju-si 28116, Korea; woosy@kribb.re.kr

4   Department of Agbiotechnology and Natural Resources, College of Agriculture and Life Science, Gyeongsang National University, Jinju 52828, Korea

5   Department of Life Resources Industry, College of Natural Resources and Life Science, Dong-A University, 37, Nakdong-daero 550 beon-gil, Busan 49315, Korea

*   Correspondence: schem72@daum.net (J.-H.L.); swd2002@korea.kr (W.-D.S.); Tel.: +82-512-007-521 (J.-H.L.); +82-632-385-333 (W.-D.S.); Fax: +82-512-007-505 (J.-H.L.); +82-632-385-335 (W.-D.S.)

†   These authors contributed equally to this work as co-first authors.

Citation: Lee, H.-G.; Woo, S.-Y.; Ahn, H.-J.; Yang, J.-Y.; Lee, M.-J.; Kim, H.-Y.; Song, S.-Y.; Lee, J.-H.; Seo, W.-D. Comparative Analysis of Policosanols Related to Growth Times from the Seedlings of Various Korean Oat (*Avena sativa* L.) Cultivars and Screening for Adenosine 5′-Monophosphate-Activated Protein Kinase (AMPK) Activation. *Plants* **2022**, *11*, 1844. https://doi.org/10.3390/plants11141844

Academic Editors: Beatrice Falcinelli and Angelica Galieni

Received: 21 June 2022
Accepted: 12 July 2022
Published: 14 July 2022

Publisher's Note: MDPI stays neutral with regard to jurisdictional claims in published maps and institutional affiliations.

Abstract: The objectives of this research were to evaluate the policosanol profiles and adenosine-5′-monophosphate-activated protein kinase (AMPK) properties in the seedlings of Korean oat (*Avena sativa* L.) cultivars at different growth times. Nine policosanols in the silylated hexane extracts were detected using GC-MS and their contents showed considerable differences; specifically, hexacosanol (6) exhibited the highest composition, constituting 88–91% of the total average content. Moreover, the average hexacosanol (6) contents showed remarkable variations of 337.8 (5 days) → 416.8 (7 days) → 458.9 (9 days) → 490.0 (11 days) → 479.2 (13 days) → 427.0 mg/100 g (15 days). The seedlings collected at 11 days showed the highest average policosanol content (541.7 mg/100 g), with the lowest content being 383.4 mg/100 g after 5 days. Interestingly, policosanols from oat seedlings grown for 11 days induced the most prevalent phenotype of AMPK activation in HepG2 cells, indicating that policosanols are an excellent AMPK activator.

Keywords: oat seedling; policosanol; hexacosanol; AMPK; growth times; GC-MS

1. Introduction

Policosanols are long-chain aliphatic primary alcohols that were first extracted from sugar cane (*Saccharum officinarum* L.) at Dalmer Laboratories in Cuba and mainly composed of policosanol, octacosanol (C-28), triacontanol (C-30), and hexacosanol (C-26) [1,2]. These metabolites were also contained minor compositions including tetracosanol (C-24), heptacosanol (C-27), nonacosanol (C-29), dotriacontanol (C-32), and tetratriacontanol (C-34) [1]. Moreover, policosanol monotherapy reduces low-density lipoprotein (LDL)-cholesterol levels by increasing the expression of LDL receptors and increases high-density lipoprotein (HDL)-cholesterol levels [3,4]. Policosanol has recently shown several pharmacological activities such as reduction of platelet aggregation and antiulcer, antioxidant, and anti-inflammatory activities [1,5–8]. Specifically, these components and their natural sources have increased interest in the food and medical industries for use in emulsion, capsule, and tablet delivery systems, owing to their bioavailability and efficacy [1]. It is well established that the policosanols in crops shows significant differences according to various factors including environmental conditions and genetics, as reported in the literatures [9–12]. Moreover, many researchers have reported that the wheat grains contain policosanols and phytosterols, and their contents as well as compositions are positively correlated with the growth conditions and management [9,13,14]. Recently, several studies have demonstrated that natural plants and crops are adenosine-5′-monophosphate-activated protein kinase (AMPK) activators and the policosanol content plays an important role in determining the rate of AMPK activation [12,15]. Notably, AMPK regulates energy metabolism such as glucose and cholesterol as well as hepatic lipids, and AMPK activation blocks ATP-consuming anabolic pathways, including fatty acid, cholesterol, and protein syntheses [12]. Moreover, Policosanol is known to regulate AMPK-mediated cholesterol synthesis reduction and 3-hydroxy-glutaryl-CoA reductase (HMGCR) reductase activity in HepG2 cells [16,17]. In recent, the extracts of barley and wheat seedlings containing high levels of policosanols can reduce plasma cholesterol concentrations via the activity of the AMPK-dependent phosphorylation inhibition-limiting enzyme in cholesterol biosynthesis, HMGCR. Especially, hexacosanol (C26-OH), a major policosanol in barley seedlings, displayed considerable AMPK activation abilities [15].

Among diverse crops, oat (*Avena sativa* L.), belonging to the family *Gramineae*, is one of the most popular healthy foods worldwide and contains several biological metabolites, including carbohydrates, sterols, lipids, proteins, alkaloids, saponins, and flavonoids. This crop is of great important in the prevention and treatment of diseases in herbal remedies because it contains esters, phospholipids, triglycerides, and fatty acids [18,19]. For these above reasons, several researchers have recently documented the various biological properties such as antioxidant, antimicrobial, antidiabetic, anti-inflammatory, antiplatelet, and antiparasitic effects [20,21]. In addition, we have recently reported that the OSs exhibited anti-osteoporotic activity [22]. However, to the best of our knowledge, no studies have performed comparative analyses of policosanol derivatives and biological abilities in various cultivars of OSs. Therefore, we evaluated the metabolite compositions and AMPK activations in the seedlings of various oat cultivars at different growth times.

The purpose of this present work were to compare the policosanols and AMPK activation properties from OSs of Korean cultivars in growth times. Nine policosanols in the silylated hexane extracts of OSs were characterized by gas chromatography coupled with a single quadrupole mass spectrometry (GC-MS). We also evaluated potential cultivar and optimal conditions with high policosanol content to enhance the functional value of this species. In addition, our study was the first to document the degree of the viability and AMPK activation in HepG2 cells under abundant policosanols of OSs at different growth times.

2. Materials and Methods

2.1. Plant Material and Chemical Reagents

Fifteen Korean oat cultivars, namely, Gwanghan, Dahan, Donghan, Samhan, Shinhan, Okhan, Johan, Taehan, Punghan, Dakyung, Dajo, Jopung, Darkhorse, Hi-early, and High Speed were used in this research (Figure 1A). All cultivars of oat were planted in 2018 under artificial soil in a growth chamber from the National Institute of Crop Science (NICS), Rural Development Administration (RDA), Jeonbuk, Korea. Fifteen oat seeds were washed in water at 20 °C for 18 h, and then germinated at 65% humidity at 25 °C in the dark. The conditions were as follows: temperature, 20 °C; humidity, 60–70%; illumination intensity, 3300–5500 lx; light, 9 h → dark, 15 h (repeated alternatively). The OSs were cultured for 6 different growth times, counting the sowing day as 0 days as follows: 5 days (1st growth), 7 days (2nd growth), 9 days (3rd growth), 11 days (4th growth), 13 days (5th growth), and 15 days (6th growth) (Figure 1B). The harvested seedlings were washed with clean sterile water and freeze-dried at −78 °C. Hexane solvent and *N*-methyl-N-(trimethylysilyl) trifluoroacetamide (MSTFA) were obtained from Sigma-Aldrich (Sigma Co., St. Louis, MO, USA). Policosanol materials including eicosanol (PubChem CID: 12404), heneicosanol (Pub-Chem CID: 85014), docosanol (PubChem CID: 12620), tricosanol (PubChem CID: 18431), tetracosanol (PubChem CID: 10472), hexacosanol (PubChem CID: 68171), heptacosanol (PubChem CID: 74822), octacosanol (PubChem CID: 68406), and triacontanol (PubChem CID: 68972) were also purchased from Sigma-Aldrich. Dulbecco's modified eagle's medium, fetal bovine serum, and antibiotics (streptomycin and penicillin) were acquired from Gibco BRL (Grand Island, NY, USA). Primary (anti-AMPK; antiphospho-AMPK) and secondary (anti-mouse IgG-HRP; anti-rabbit IgG-HRP) antibodies were provided by Santa Cruz Biotechnology Inc. (Dallas, TX, USA). Other solvents and chemical reagents were of analytical grade (Sigma-Aldrich).

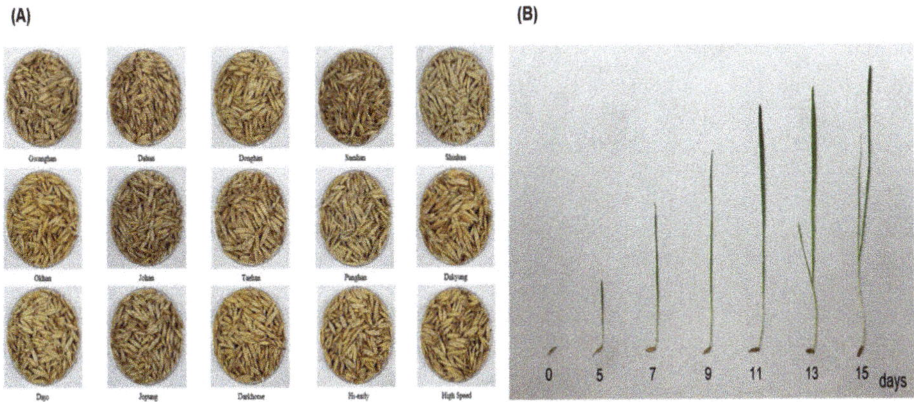

Figure 1. The appearances of oat seedlings (*Avena sativa* L.): (**A**) Oat seeds used in this study; (**B**) Oat seedlings (cv. Gwanghan) at six different growth times.

2.2. Preparation of OSs and Policosanol Materials

The dried powdered OSs were extracted in a shaking incubator with hexane (10 mL) for 12 h at 25 °C. After centrifugation of the crude extract, the supernatant was filtered through a 0.45 μm syringe filter (Whatman Inc., Maidstone, UK). The hexane was removed using an evaporator under reduced pressured and resuspended in MSTFA (250 μL) and chloroform solution (0.5 mL) modified the methods of Choi [10] for the silylation reaction. To analysis of GC-MS, the silylated mixture was stirred with chloroform for 15 min at 60 °C. For quantitative analysis, the policosanol standards were also silylated with MSTFA under the same conditions. To quantify policosanol in OSs, a calibration curve was prepared using 5 concentrations (6.25, 12.5, 25 and 50 μg/mL) of each standard. The quantification of the calibration plot was evaluated using the peak areas of the policosanol standards, and the individual correlation coefficient (r^2) was at least 0.998. The curve regression equation of policosanol and its coefficients are as follows. Eicosanol; $y = 129,494\,x + 447,802$, $r^2 = 0.998$, Heneicosanol; $y = 135,321\,x + 314,175$, $r^2 = 0.998$, Docosanol; $y = 135,968\,x − 262,734$, $r^2 = 0.999$, Tricosanol; $y = 133,917\,x − 635,626$, $r^2 = 0.999$, Tetracosanol; $y = 124,434\,x − 306,413$, $r^2 = 0.998$, Hexacosanol; $y = 144,421\,x − 1,962,403$, $r^2 = 0.999$, Heptacosanol; $y = 135,627\,x − 2,039,018$, $r^2 = 0.999$, Octacosanol; $y = 112,953\,x − 841,945$, $r^2 = 0.998$ and Triacntanol; $y = 114,234\,x − 2,274,380$, $r^2 = 0.999$.

2.3. Instruments

The policosanol components were characterized and examined by an Agilent 7890A GC-MS (Agilent Technologies Inc., PaloAlto, CA, USA). The absorbance results were analyzed using a microplate plate reader (96 well, Molecular Devices, Sunnyvale, CA,

USA). Lysate were transferred to nitrocellulose membranes by electrophoretic transfer cell (Bio-Rad, Hercules, CA, USA).

2.4. GC-MS Conditions for Policosanol Analysis

GC-MS analysis of policosanol was analyzed using previously published methods [12]. The silylated samples were examined by a GC system coupled with a 5977A series mass. Experimental conditions of GC-MS system were as follows: HP-5MS UI (diphenyl 5%-dimethylsiloxane 95% co-polymer) capillary column (30 m $\times$ 0.25 μm $\times$ 0.25 μm film thickness). The helium as the carrier gas with flow rate of at 1.2 mL/min. The ionization energy of MS spectrum was taken at 70 eV and mass range was 40–500 amu. The oven temperature was programmed to rise from 230 to 260 °C with 25 °C/min during 10 min, from 260 to 300 °C with 20 °C/min heating rate and maintained for 7 min. The temperatures of inlet and MS transfer line were 280 °C, and MS source was 230 °C, respectively. Samples were injected 1 μL by auto sampler using split mode injection (1:5). The policosanols were identified through the comparisons of fragmentation patterns of their mass values, retention times of the standards.

2.5. Cell Culture and Measurement of Cell Viability

HepG2 cell line was obtained from a Korean cell line bank (SPL Life Sciences, Pocheon, Korea). Cells were grown in Dulbecco's modified Eagle's medium (DMEM) supplemented with 10% fetal bovine serum (FBS), 1% glutamine, 1% antibiotics (penicillin-streptomycin), and 1.5 g/L sodium bicarbonate [15]. Cells were incubated at 37 °C, in humidified atmosphere containing 5% CO_2, and all the experiments were performed in a clean atmosphere. Cell viability was determined by using the Thiazolyl blue tetrazolium bromide (MTT) method. The cell viability of the hexane extracts of oat seedlings and standard materials in HepG2 cell lines were assessed using the MTT assay as previously reported [12,15]. The HepG2 cells were seeded in 96 well plates at 1×10^4/well and incubated with the different dose of the hexane extract of oat seedlings (concentrations 0, 12.5, 25, 50, 100, and 200 μg/mL) at 37 °C for 24 h in a humidified atmosphere (5% CO_2). After the cells were washed with phosphate-buffered saline (PBS) treated with MTT solution (0.5 mg/mL) to the wells and incubated for 4 h. After incubation, cells were suspended in solubilization buffer of 50% dimethyl sulfoxide (DMSO). Subsequently, the absorbance was measured at 540 nm for absorbance value.

2.6. Western Blot Analysis

Immunoblotting was performed as described with slight modifications as Lee [23]. Brifely, HepG2 cells were lysed in lysis buffer (10 mM Tris-HCl (pH 7.5), 10 mM NaCl, 0.1 M EDTA, 0.1% NP40, 0.1% sodium deoxycholate, and 0.25 M sodium pyrophosphate) containing protease and phosphatase inhibitor cocktails (Sigma P2850) at 4 °C. Lysates of cell supernatants were separated on 10% acrylamide SDS-PAGE and then transferred to 0.2 μm nitrocellulose membranes using electrophoretic transfer cell. The combined membranes were blocked with Tris buffered saline (TBS) containing 5% non-fat milk at 25 °C for 60 min. The membranes were incubated with primary antibody, anti-AMPK, phosphor-AMPK, and anti-β-actin, at 4 °C for 12 h. After incubating with the primary antibody, the membrane is washed with TBS-T for three times. The secondary antibodies, anti-mouse and anti-rabbit immunoglobulin G, incubated with for 60 min at 25 °C and then washed with TBS-T again. The immunoreactive protein bands were detected with super signal picochemilumnescent stain and the protein concentration was determined using the Bradford assay with bovine serum albumin (BSA) as a standard [24]. The band intensities were visualised by a ChemiDoc XRS system (Bio-Rad) and quantified using Gel-Pro Analyser Software (Silk Scientific, Inc., Orem, UT, USA) [15].

2.7. Statistical Analysis

Each experiment of policosanol contents was performed three times, and all data are presented as the mean ± standard derivation (SD) and their differences in growth times were calculated by Duncan's multiple range test by the statistical analysis software (SAS) 9.2 PC package (SAS Institute Inc. Cary, NC, USA). A *p*-value less than 0.05 was considered statistically significant.

3. Results and Discussion

3.1. Identification of Policosanol Compositions in OSs by GC-MS Analysis

In the current research, the policosanol contents in the OSs extracted with *n*-hexane, and analyzed using GC-MS. The policosanol structures and their representative chromatogram are shown in Figure 2. The retention times (t_R) of individual policosanol was in the following order: peak 1 (t_R = 3.45 min), 2 (t_R = 4.09 min), 3 (t_R = 4.75 min), 4 (t_R = 5.73 min), 5 (t_R = 7.05 min), 6 (t_R = 10.63 min), 7 (t_R = 12.51 min), 8 (t_R = 13.60 min), and 9 (t_R = 15.87 min). The mass spectra of the policosanol-Trimethylsilyl (TMS) compositions exhibited distinctive ion peaks of the $[M + 15]^+$ fragmentation pattern owing to the loss of the methyl (CH_3) moiety, facilitating the identification of individual policosanols. The $[M + 15]^+$ ion peaks (m/z) of the individual authentic policosanol-TMS derivatives were 355.3, 369.4, 383.4, 397.4, 411.4, 439.4, 453.5, 467.5, and 495.5, respectively. Furthermore, their mass spectra exhibited the characteristic ions of TMS derivatives of primary alcohols because of fragment ions of $C_4H_9^+$, $OH\text{-}Si(CH_3)^{2+}$, and $CH_2OSi\text{-}(CH3)^{3+}$ with m/z values of 57, 75, and 103, respectively. Therefore, nine peaks were identified by comparison of mass spectra and authentic standards. Eicosanol (**1**) (peak 1, C20-OH), heneicosanol (**2**) (peak 2, C21-OH), docosanol (**3**) (peak 3, C22-OH), tricosanol (**4**) (peak 4, C23-OH), tetracosanol (**5**) (peak 5, C24-OH), hexacosanol (**6**) (peak 6, C26-OH), heptacosanol (**7**) (peak 7, C27-OH), octacosanol (**8**) (peak 8, C28-OH), and triacontanol (**9**) (peak 9, C30-OH).

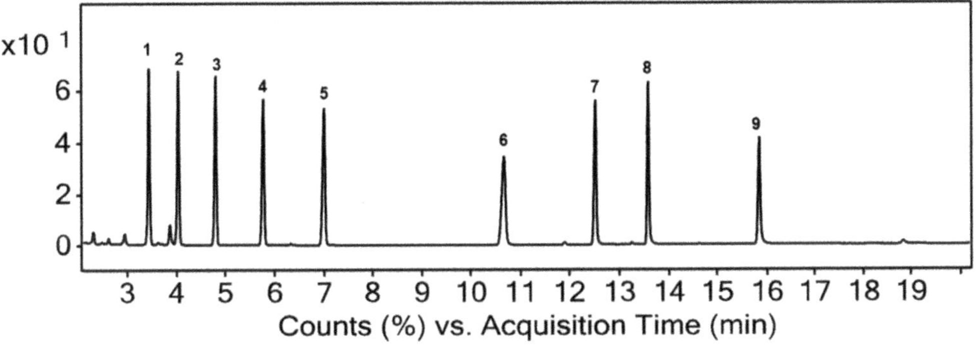

Figure 2. Chemical structures and GC-MS chromatogram of policosanol standards.

3.2. Changes in Policosanol Contents in the Seedlings of Oat Cultivars at Different Growth Times

Many studies have evaluated the policosanol contents in crop seedlings. Numerous researchers have also demonstrated that the metabolite contents showed remarkable differences according to the cultivars and growth periods [25,26]. Unfortunately, to the best of our knowledge, the policosanol contents in the seedlings of oat cultivars have not been investigated at different growth times. Therefore, we examined the policosanol contents of *n*-hexane extracts of OSs at six different growth times using GC-MS and their chromatograms are shown in Figure 3 (cv. Gwanghan). The fragment ions and mass data of policosanols are as follows: eicosanol (**1**) (peak 1, 370.3, 355.3, 103.0, 75.0, 55.1), heneicosanol (**2**) (peak 2, 384.4, 369.4, 103.0, 75.0, 55.1), docosanol (**3**) (peak 3, 398.4, 383.4, 103.0, 75.0, 55.1), tricosanol (**4**) (peak 4, 412.4, 397.4, 103.0, 75.0, 55.1), tetracosanol (**5**) (peak 5, 426.4, 411.4, 103.0, 75.0, 55.1), hexacosanol (**6**) (peak 6, 454.5, 439.4, 129.0, 97.1, 75.0, 55.1), heptacosanol (**7**) (peak 7, 468.5, 453.5, 103.0, 75.0, 55.1), octacosanol (**8**) (peak 8, 482.5, 467.5, 106.0, 75.0, 55.1), and triacontanol (**9**) (peak 9, 510.5, 495.5, 103.0, 75.0, 55.1). In addition, the individual and total policosanol contents of 15 different cultivars at six growth times (Sowing date is counted as 0 days. 5, 7, 9, 11, 13, and 15 days) are presented in Table 1 and their contents are expressed as mg/100 g of OSs. The total policosanols varied widely between cultivars and growth times as follows: 316.0–443.7 mg/100 g (5 days), 413.2–495.2 mg/100 g (7 days), 439.9–611.8 mg/100 g (9 days), 477.1–647.7 mg/100 g (11 days), 462.7–595.3 mg/100 g (13 days), and 367.8–569.8 mg/100 g (15 days), respectively. Especially, peak 6 (hexacosanol) exhibited the highest average values, with 337.8 (5 days, 88%), 416.8 (7 days, 89%), 458.9 (9 days, 90%), 490.0 (11 days, 91%), 479.2 (13 days, 90%), and 427.0 mg/100 g (15 days, 89%), respectively. The second main policosanol, octacosanol contents are as follows according to the growth times: [peak 8; 23.5 (5 days), 23.7 (7 days), 24.5 (9 days), 24.9 (11 days), 24.4 (13 days), and 24.0 mg/100 g (15 days)]. The average contents of other policosanols was as follows with the rank order of increase rates: tetracosanol (**5**) (18.2 mg/100 g) > docosanol (**3**) (7.5 mg/100 g), and the remaining compositions were not detected.

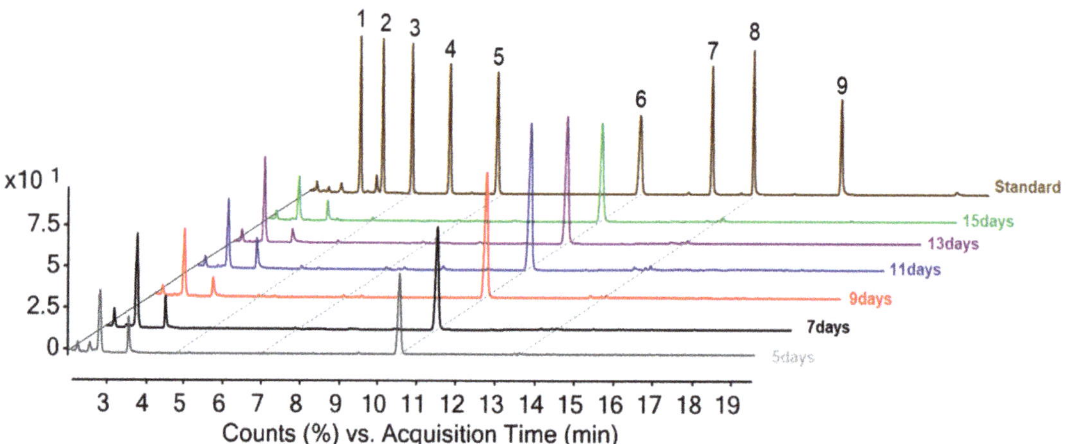

Figure 3. Comparison of GC-MS chromatograms for policosanol standard and oat seedlings at different growth times. (cv. Gwanghan). Eicosanol (**1**), Heneicosanol (**2**), Docosanol (**3**), Tricosanol (**4**), Tetracosanol (**5**), Hexacosanol (**6**), Heptacosanol (**7**), Octacosanol (**8**), Triacontanol (**9**).

The growth times on 11 days through the seedlings of 15 oat cultivars exhibited the predominant average contents with 541.7 mg/100 g, followed by 13 days > 9 days > 15 days > 7 days with 532.2, 508.3, 479.5, and 463.8 mg/100 g, respectively, the lowest average contents were observed at 5 days (383.3 mg/100 g). In other words, the average policosanols increased 463.8 (7 days) → 508.3 (9 days) → 541.7 mg/100 g (11 days) after 7 → 11 days of growth. These observations can be primarily influenced by the most abundant hexacosanol (C26-OH, peak 6), this policosanol, which increased (average contents: 416.8; 7 days → 458.9; 9 days → 490.0 mg/100 g; 11 days) during these growth times. Our results support previous observations that the growth times of crops can strongly affect the policosanol contents. 26 Moreover, extending the growth times from 5 to 7 days considerably increased the total average policosanols in each cultivar (Table 1) and their contents showed 383.3 → 463.8 mg/100 g. In particular, the hexacosanol (**6**) content increased significantly with 337.8 (5 days) → 416.8 mg/100 g (7 days). Interestingly, when the oat seedlings are grown for longer times in 11 → 15 days, the average total policosanols decreased with 541.7 → 532.2 → 479.5 mg/100 g. The above findings have confirmed that the hexacosanol content (average: 490 → 479.2 → 427.0 mg/100 g) can be responsible for the main portion of total policosanols. Based on the considerations, the total policosanol contents in oat seedlings increased mainly during 11 days, while the remaining periods showed reduction phenomena. Therefore, our results suggest that the appropriate growth time regarding policosanol compositions in oat seedlings may be 11 days after sowing. Furthermore, we are confident that the environmental factors including the growth times may be affected with the policosanol contents [25]. The present data were similar to previously reported results concerning variations of phytochemicals at different harvest times [12]. In summarize, our data suggest that the total policosanol content of oat seedlings was closely related to hexacosanol (C-26), the major policosanol. Other components, C22-OH (docosanol peak 3), C24-OH (tetracosanol, peak 5), and C28-OH (octacosanol, peak 8) were observed low average contents with 7.5, 18.2, and 24.2 mg/100 g, respectively, at six different growth times of 15 cultivars. Although many researches have demonstrated that the metabolite contents increased during maturation of crops, our work showed that the policosanol contents were not dependent on the maturation times. Our previous study showed that wheat policosanol content may be affected by factors such as environmental factors and genotypes [27]. The current results support the development of suitable cultivars that can potentially be used as human health foods. Particularly, the oat cultivars such as Gwanghan and Dahan may be recommended as excellent sources owing to high policosanol contents of 640.1 and 647.7 mg/100 g on 11 days. To obtain more information concerning the beneficial effects in oat plant, our research was designed to document the comparison of policosanol in different organs (seeds, roots, and stems) of oat seedlings at 11 days growth times (Figure 4). Interestingly, the stem organ displayed high policosanol contents with 730.9 mg/100 g (docosanol: 70.6, tetracosanol: 70.1, hexacosanol: 515.8, and octacosanol: 74.4 mg/100 g), while other organs were not detected. Therefore, we confirmed that the stems of oat seedlings may be utilized as the best excellent source in terms of functional uses. The present work provided for the first time the policosanol contents in the seedlings of various oat cultivars through different growth times.

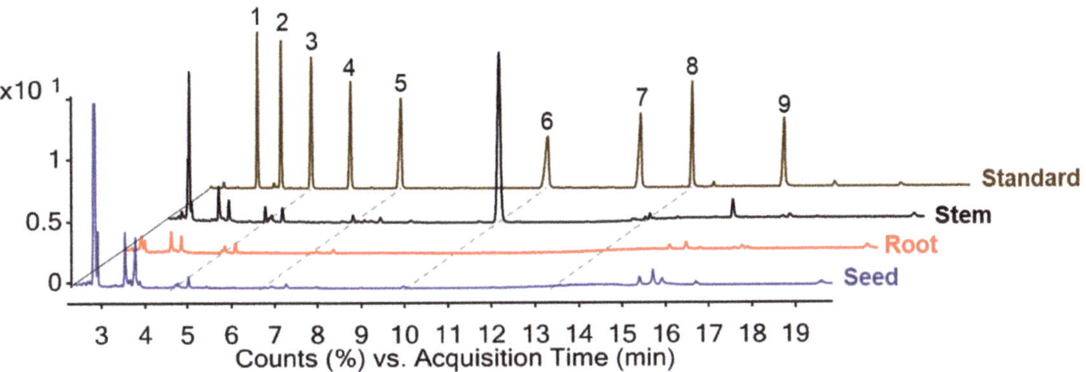

Figure 4. Policosanol in various parts of oat seedlings and standard GC-MS chromatogram (cv. Gwanghan). Eicosanol (**1**), Heneicosanol (**2**), Docosanol (**3**), Tricosanol (**4**), Tetracosanol (**5**), Hexacosanol (**6**), Heptacosanol (**7**), Octacosanol (**8**), Triacontanol (**9**).

Table 1. Changes of policosanol contents in the seedlings of oat cultivars at six different growth times.

Growth Time	Cultivar	PC Content (mg/100 g)									
		C_{20}	C_{21}	C_{22}	C_{23}	C_{24}	C_{26}	C_{27}	C_{28}	C_{30}	Total PC
5 days	Gwanghan	ND	ND	5.3 ± 0.0 [cd]	ND	16.1 ± 0.0 [fg]	339.7 ± 6.5 [cde]	ND	23.1 ± 0.1 [fgh]	ND	384.3 ± 3.2 [cde]
	Dahan	ND	ND	5.3 ± 0.0 [e]	ND	16.5 ± 0.1 [de]	347.3 ± 4.9 [bcd]	ND	23.3 ± 0.1 [efg]	ND	392.4 ± 2.4 [bcd]
	Donghan	ND	ND	5.3 ± 0.0 [cd]	ND	16.3 ± 0.1 [ef]	328.0 ± 12.2 [def]	ND	23.0 ± 0.1 [fg]	ND	372.6 ± 6.1 [def]
	Samhan	ND	ND	5.3 ± 0.0 [cde]	ND	16.0 ± 0.1 [g]	322.5 ± 15.2 [def]	ND	23.5 ± 0.1 [cde]	ND	367.4 ± 7.5 [def]
	Shinhan	ND	ND	5.3 ± 0.0 [de]	ND	16.0 ± 0.1 [g]	323.6 ± 18.6 [def]	ND	23.5 ± 0.3 [cde]	ND	368.4 ± 9.2 [def]
	Okhan	ND	ND	5.3 ± 0.0 [de]	ND	16.5 ± 0.1 [d]	271.9 ± 5.4 [g]	ND	22.9 ± 0.1 [h]	ND	316.7 ± 2.7 [g]
	Johan	ND	ND	5.3 ± 0.0 [bc]	ND	17.5 ± 0.2 [b]	364.5 ± 10.2 [bc]	ND	24.0 ± 0.2 [b]	ND	411.3 ± 5.0 [bc]
	Taehan	ND	ND	5.4 ± 0.0 [ab]	ND	17.3 ± 0.2 [b]	371.7 ± 15.7 [ab]	ND	23.9 ± 0.1 [b]	ND	418.3 ± 7.8 [ab]
	Punghan	ND	ND	5.3 ± 0.0 [cde]	ND	16.7 ± 0.1 [cd]	332.2 ± 10.2 [def]	ND	23.3 ± 0.1 [efg]	ND	377.6 ± 5.1 [def]
	Dakyung	ND	ND	5.4 ± 0.0 [a]	ND	17.8 ± 0.3 [a]	396.6 ± 31.7 [a]	ND	23.8 ± 0.3 [bc]	ND	443.7 ± 15.7 [a]
	Dajo	ND	ND	5.3 ± 0.0 [cde]	ND	16.1 ± 0.2 [fg]	315.7 ± 25. 1 [ef]	ND	23.1 ± 0. 2 [gh]	ND	360.3 ± 12.5 [ef]
	Jopung	ND	ND	5.3 ± 0.0 [bc]	ND	16.6 ± 0.1 [d]	393.1 ± 11.1 [a]	ND	24.3 ± 0.1 [a]	ND	439.4 ± 5.5 [a]
	Darkhorse	ND	ND	5.3 ± 0.0 [ab]	ND	16.8 ± 0.2 [c]	314.5 ± 20.6 [ef]	ND	23.4 ± 0.2 [def]	ND	360.2 ± 10.2 [ef]
	Hi-early	ND	ND	5.3 ± 0.0 [bc]	ND	16.3 ± 0.0 [f]	304.7 ± 5.7 [f]	ND	23.7 ± 0.0 [bcd]	ND	350.0 ± 2.9 [f]
	High Speed	ND	ND	5.3 ± 0.0 [da]	ND	17.5 ± 0.1 [b]	341.1 ± 9.1 [bcd]	ND	23.7 ± 0.2 [bcd]	ND	387.6 ± 4.5 [cde]
7 days	Gwanghan	ND	ND	6.3 ± 0.0 [c]	ND	17.4 ± 0.0 [f]	435.3 ± 2.6 [ab]	ND	23.7 ± 0.1 [cde]	ND	482.7 ± 1.3 [ab]
	Dahan	ND	ND	5.3 ± 0.0 [h]	ND	16.9 ± 0.1 [g]	439.6 ± 9.9 [ab]	ND	23.8 ± 0.1 [cde]	ND	485.5 ± 4.9 [ab]
	Donghan	ND	ND	6.1 ± 0.0 [d]	ND	17.5 ± 0.1 [ef]	386.0 ± 4.7 [cd]	ND	23.4 ± 0.1 [e]	ND	433.0 ± 2.3 [de]
	Samhan	ND	ND	5.4 ± 0.0 [gh]	ND	16.8 ± 0.1 [g]	439.3 ± 18.2 [ab]	ND	24.3 ± 0.1 [ab]	ND	485.8 ± 9.0 [ab]
	Shinhan	ND	ND	5.3 ± 0.0 [h]	ND	16.4 ± 0.1 [h]	419.0 ± 13.8 [abc]	ND	23.8 ± 0.2 [cd]	ND	464.5 ± 6.9 [abcd]
	Okhan	ND	ND	5.3 ± 0.0 [h]	ND	17.4 ± 0.3 [f]	392.6 ± 38.9 [cd]	ND	23.5 ± 0.3 [de]	ND	438.8 ± 19.3 [cde]
	Johan	ND	ND	5.8 ± 0.0 [e]	ND	18.1 ± 0.3 [cd]	423.1 ± 24.4 [abc]	ND	23.6 ± 0.2 [de]	ND	470.7 ± 12.1 [abcd]
	Taehan	ND	ND	5.8 ± 0.0 [e]	ND	17.9 ± 0.2 [de]	440.5 ± 18.8 [ab]	ND	23.9 ± 0.2 [bcd]	ND	488.0 ± 9.3 [ab]
	Punghan	ND	ND	5.4 ± 0.0 [h]	ND	16.8 ± 0.0 [g]	368.1 ± 1.2 [d]	ND	23.0 ± 0.0 [f]	ND	413.2 ± 0.6 [e]
	Dakyung	ND	ND	5.6 ± 0.0 [f]	ND	18.4 ± 0.1 [bc]	447.4 ± 8.4 [a]	ND	23.7 ± 0.0 [cde]	ND	495.2 ± 4.2 [a]
	Dajo	ND	ND	5.5 ± 0.0 [g]	ND	16.7 ± 0.2 [gh]	387.1 ± 24.0 [cd]	ND	23.6 ± 0.3 [de]	ND	432.9 ± 11.9 [de]
	Jopung	ND	ND	5.6 ± 0.0 [f]	ND	16.9 ± 0.2 [g]	433.8 ± 23.6 [ab]	ND	24.4 ± 0.3 [a]	ND	480.8 ± 11.7 [abc]
	Darkhorse	ND	ND	7.0 ± 0.2 [a]	ND	19.4 ± 0.6 [a]	417.4 ± 38.4 [abc]	ND	23.8 ± 0.4 [cde]	ND	467.6 ± 19.0 [abcd]
	Hi-early	ND	ND	5.4 ± 0.0 [gh]	ND	16.8 ± 0.2 [g]	404.2 ± 29.8 [bcd]	ND	24.1 ± 0.3 [abc]	ND	450.5 ± 14.8 [bcde]
	High Speed	ND	ND	6.6 ± 0.1 [b]	ND	18.8 ± 0.3 [b]	418.5 ± 18.9 [abc]	ND	23.9 ± 0.2 [bcd]	ND	467.7 ± 9.3 [abcd]

Table 1. *Cont.*

Growth Time	Cultivar	PC Content (mg/100 g)									
		C_{20}	C_{21}	C_{22}	C_{23}	C_{24}	C_{26}	C_{27}	C_{28}	C_{30}	Total PC
9 days	Gwanghan	ND	ND	8.6 ± 0.0 [b]	ND	19.2 ± 0.1 [c]	559.3 ± 10.9 [a]	ND	24.7 ± 0.2 [abc]	ND	611.8 ± 5.4 [a]
	Dahan	ND	ND	5.4 ± 0.0 [h]	ND	17.4 ± 0.0 [gh]	537.0 ± 7.2 [b]	ND	24.4 ± 0.1 [de]	ND	584.1 ± 3.6 [b]
	Donghan	ND	ND	6.7 ± 0.1 [cd]	ND	18.3 ± 0.2 [de]	495.0 ± 24.0 [c]	ND	24.3 ± 0.2 [de]	ND	544.4 ± 11.9 [c]
	Samhan	ND	ND	5.5 ± 0.0 [h]	ND	16.8 ± 0.0 [i]	443.5 ± 2.8 [def]	ND	24.3 ± 0.1 [de]	ND	490.0 ± 1.4 [ef]
	Shinhan	ND	ND	5.4 ± 0.0 [h]	ND	16.5 ± 0.1 [j]	463.4 ± 13.8 [d]	ND	24.3 ± 0.1 [de]	ND	509.7 ± 6.9 [de]
	Okhan	ND	ND	5.5 ± 0.0 [h]	ND	17.8 ± 0.1 [f]	438.9 ± 5.5 [ef]	ND	24.7 ± 0.0 [bc]	ND	486.9 ± 2.7 [ef]
	Johan	ND	ND	6.5 ± 0.0 [e]	ND	18.2 ± 0.1 [e]	391.6 ± 11.8 [g]	ND	23.5 ± 0.1 [f]	ND	439.9 ± 5.9 [g]
	Taehan	ND	ND	6.9 ± 0.1 [c]	ND	18.5 ± 0.1 [d]	431.0 ± 11.9 [f]	ND	24.7 ± 0.2 [bc]	ND	481.0 ± 5.9 [f]
	Punghan	ND	ND	5.5 ± 0.0 [h]	ND	17.7 ± 0.1 [f]	441.2 ± 9.8 [def]	ND	24.1 ± 0.1 [e]	ND	488.4 ± 4.9 [f]
	Dakyung	ND	ND	6.3 ± 0.0 [f]	ND	19.1 ± 0.1 [c]	400.4 ± 2.9 [g]	ND	24.1 ± 0.2 [e]	ND	450.0 ± 1.4 [g]
	Dajo	ND	ND	5.7 ± 0.0 [g]	ND	17.6 ± 0.1 [fg]	457.7 ± 10.1 [de]	ND	24.4 ± 0.1 [d]	ND	505.5 ± 5.0 [de]
	Jopung	ND	ND	6.6 ± 0.0 [de]	ND	17.5 ± 0.0 [fg]	437.9 ± 5.7 [ef]	ND	24.8 ± 0.1 [ab]	ND	486.9 ± 2.8 [ef]
	Darkhorse	ND	ND	12.2 ± 0.3 [a]	ND	21.8 ± 0.3 [a]	460.0 ± 10.2 [de]	ND	24.5 ± 0.1 [cd]	ND	518.5 ± 5.0 [d]
	Hi-early	ND	ND	5.5 ± 0.0 [h]	ND	17.3 ± 0.1 [g]	462.5 ± 7.6 [d]	ND	25.0 ± 0.1 [a]	ND	510.3 ± 3.8 [de]
	High Speed	ND	ND	8.7 ± 0.2 [b]	ND	20.2 ± 0.4 [b]	463.6 ± 26.5 [d]	ND	25.0 ± 0.3 [a]	ND	517.4 ± 13.1 [d]
11 days	Gwanghan	ND	ND	10.8 ± 0.7 [c]	ND	19.9 ± 0.9 [cd]	583.6 ± 73.9 [a]	ND	25.7 ± 0.8 [a]	ND	640.1 ± 36.5 [a]
	Dahan	ND	ND	5.5 ± 0.0 [j]	ND	18.0 ± 0.2 [ef]	599.0 ± 30.9 [a]	ND	25.2 ± 0.3 [abc]	ND	647.7 ± 15.3 [a]
	Donghan	ND	ND	8.0 ± 0.3 [e]	ND	18.5 ± 0.5 [e]	475.0 ± 53.4 [bc]	ND	24.3 ± 0.5 [de]	ND	525.8 ± 26.5 [bc]
	Samhan	ND	ND	5.6 ± 0.0 [ij]	ND	17.1 ± 0.1 [g]	473.6 ± 8.3 [bc]	ND	24.9 ± 0.1 [bcd]	ND	521.2 ± 4.1 [bc]
	Shinhan	ND	ND	5.7 ± 0.0 [hij]	ND	16.6 ± 0.2 [g]	431.1 ± 27.8 [c]	ND	24.1 ± 0.3 [e]	ND	477.5 ± 13.8 [c]
	Okhan	ND	ND	6.0 ± 0.0 [ghi]	ND	17.9 ± 0.1 [ef]	429.1 ± 9.5 [c]	ND	24.2 ± 0.1 [de]	ND	477.1 ± 4.7 [c]
	Johan	ND	ND	9.6 ± 0.1 [d]	ND	19.7 ± 0.3 [d]	471.7 ± 21.9 [bc]	ND	24.1 ± 0.2 [e]	ND	525.1 ± 10.8 [bc]
	Taehan	ND	ND	8.1 ± 0.1 [e]	ND	19.4 ± 0.0 [d]	493.9 ± 1.6 [bc]	ND	25.5 ± 0.0 [ab]	ND	546.8 ± 0.8 [bc]
	Punghan	ND	ND	5.8 ± 0.1 [hij]	ND	18.3 ± 0.7 [ef]	483.2 ± 75.6 [bc]	ND	24.8 ± 0.8 [bcde]	ND	532.1 ± 37.5 [bc]
	Dakyung	ND	ND	8.2 ± 0.0 [e]	ND	20.5 ± 0.1 [bc]	479.9 ± 9.2 [bc]	ND	25.1 ± 0.1 [abc]	ND	533.8 ± 4.6 [bc]
	Dajo	ND	ND	6.2 ± 0.1 [g]	ND	18.0 ± 0.3 [ef]	485.0 ± 33.3 [bc]	ND	24.6 ± 0.4 [cde]	ND	533.8 ± 16.5 [bc]
	Jopung	ND	ND	7.6 ± 0.1 [f]	ND	18.0 ± 0.2 [ef]	496.2 ± 21.4 [bc]	ND	25.4 ± 0.3 [ab]	ND	520.2 ± 10.6 [bc]
	Darkhorse	ND	ND	15.0 ± 0.1 [a]	ND	23.3 ± 0.3 [a]	514.4 ± 17.1 [b]	ND	25.2 ± 0.2 [abc]	ND	578.0 ± 8.4 [b]
	Hi-early	ND	ND	6.1 ± 0.1 [gh]	ND	17.8 ± 0.3 [f]	485.8 ± 31.3 [bc]	ND	24.7 ± 0.3 [cde]	ND	524.4 ± 15.6 [bc]
	High Speed	ND	ND	11.3 ± 0.2 [b]	ND	20.9 ± 0.2 [b]	485.0 ± 14.7 [bc]	ND	25.2 ± 0.2 [abc]	ND	542.4 ± 7.3 [bc]

Table 1. *Cont.*

Growth Time	Cultivar	C_{20}	C_{21}	C_{22}	C_{23}	C_{24}	C_{26}	C_{27}	C_{28}	C_{30}	Total PC
							PC Content (mg/100 g)				
13 days	Gwanghan	ND	ND	10.4 ± 0.0 [d]	ND	18.6 ± 0.1 [e]	518.8 ± 26.7 [ab]	ND	24.8 ± 0.2 [abc]	ND	572.5 ± 13.3 [ab]
	Dahan	ND	ND	5.9 ± 0.0 [gh]	ND	17.8 ± 0.2 [fghi]	547.1 ± 11.9 [a]	ND	24.4 ± 0.3 [bcde]	ND	595.3 ± 5.9 [a]
	Donghan	ND	ND	11.6 ± 0.8 [c]	ND	19.2 ± 0.7 [d]	511.5 ± 60.0 [abc]	ND	24.6 ± 0.6 [abcd]	ND	566.9 ± 29.7 [abc]
	Samhan	ND	ND	5.9 ± 0.0 [h]	ND	17.5 ± 0.1 [ghi]	487.5 ± 12.9 [bcd]	ND	25.0 ± 0.1 [a]	ND	535.8 ± 6.4 [bc]
	Shinhan	ND	ND	6.6 ± 0.1 [fg]	ND	17.3 ± 0.2 [i]	484.6 ± 28.7 [bcd]	ND	24.5 ± 0.2 [abcde]	ND	533.0 ± 14.2 [bc]
	Okhan	ND	ND	6.7 ± 0.1 [f]	ND	18.1 ± 0.4 [efg]	422.2 ± 31.0 [bcd]	ND	23.7 ± 0.2 [f]	ND	470.7 ± 15.4 [de]
	Johan	ND	ND	17.0 ± 0.5 [b]	ND	20.6 ± 0.3 [c]	458.2 ± 16.7 [ef]	ND	24.1 ± 0.2 [ef]	ND	520.0 ± 8.2 [bcd]
	Taehan	ND	ND	11.8 ± 0.1 [c]	ND	19.7 ± 0.2 [d]	463.4 ± 18.9 [cdef]	ND	24.7 ± 0.2 [abc]	ND	519.6 ± 9.3 [bcd]
	Punghan	ND	ND	5.8 ± 0.0 [h]	ND	18.3 ± 0.1 [ef]	471.0 ± 10.8 [bcde]	ND	24.0 ± 0.2 [ef]	ND	519.2 ± 5.3 [bcd]
	Dakyung	ND	ND	6.3 ± 0.1 [fgh]	ND	19.2 ± 0.2 [d]	463.1 ± 1.3 [cdef]	ND	24.3 ± 0.1 [cde]	ND	512.9 ± 0.6 [cd]
	Dajo	ND	ND	6.1 ± 0.0 [fgh]	ND	18.1 ± 0.1 [efg]	478.5 ± 7.0 [bcd]	ND	24.1 ± 0.1 [def]	ND	526.8 ± 3.4 [bc]
	Jopung	ND	ND	7.6 ± 0.1 [e]	ND	18.0 ± 0.2 [efgh]	511.7 ± 19.8 [abc]	ND	24.8 ± 0.2 [ab]	ND	562.1 ± 9.8 [abc]
	Darkhorse	ND	ND	18.2 ± 0.9 [a]	ND	23.1 ± 0.6 [a]	490.7 ± 31.1 [bcd]	ND	24.7 ± 0.4 [abc]	ND	556.7 ± 15.2 [abc]
	Hi-early	ND	ND	6.7 ± 0.1 [f]	ND	17.4 ± 0.2 [hi]	414.9 ± 32.2 [f]	ND	23.7 ± 0.4 [f]	ND	462.7 ± 16.0 [e]
	High Speed	ND	ND	17.4 ± 0.8 [b]	ND	22.2 ± 0.7 [b]	465.2 ± 38.9 [cdef]	ND	24.6 ± 0.3 [abcd]	ND	529.4 ± 19.2 [bc]
15 days	Gwanghan	ND	ND	11.9 ± 0.2 [c]	ND	18.3 ± 0.1 [cd]	406.0 ± 6.3 [ef]	ND	23.4 ± 0.2 [fgh]	ND	459.6 ± 3.1 [de]
	Dahan	ND	ND	6.0 ± 0.0 [h]	ND	17.0 ± 0.0 [e]	436.2 ± 8.7 [bcde]	ND	23.3 ± 0.1 [ghi]	ND	482.5 ± 4.3 [cde]
	Donghan	ND	ND	13.6 ± 0.3 [b]	ND	18.6 ± 0.3 [c]	414.6 ± 21.8 [def]	ND	23.8 ± 0.5 [defg]	ND	470.7 ± 10.7 [cde]
	Samhan	ND	ND	6.4 ± 0.1 [gh]	ND	18.2 ± 0.3 [cd]	439.4 ± 23.4 [bcde]	ND	24.7 ± 0.3 [bc]	ND	488.7 ± 11.6 [bcd]
	Shinhan	ND	ND	7.2 ± 0.1 [ef]	ND	17.8 ± 0.1 [d]	470.4 ± 14.9 [abc]	ND	24.8 ± 0.2 [b]	ND	520.1 ± 7.4 [bc]
	Okhan	ND	ND	7.9 ± 0.2 [d]	ND	18.3 ± 0.4 [cd]	386.5 ± 37.0 [f]	ND	23.5 ± 0.1 [efgh]	ND	436.2 ± 18.4 [e]
	Johan	ND	ND	15.4 ± 0.2 [a]	ND	20.0 ± 0.1 [b]	434.2 ± 8.8 [bcde]	ND	23.8 ± 0.1 [defg]	ND	493.4 ± 4.3 [bcd]
	Taehan	ND	ND	13.5 ± 1.0 [b]	ND	19.9 ± 0.9 [b]	445.7 ± 58.5 [bcde]	ND	24.8 ± 0.7 [b]	ND	504.0 ± 28.8 [bcd]
	Punghan	ND	ND	6.1 ± 0.1 [h]	ND	18.4 ± 0.2 [cd]	453.1 ± 20.2 [bcde]	ND	24.0 ± 0.3 [de]	ND	501.5 ± 10.0 [bcd]
	Dakyung	ND	ND	6.7 ± 0.1 [fg]	ND	19.5 ± 0.4 [b]	460.1 ± 29.9 [abcd]	ND	24.2 ± 0.2 [cd]	ND	510.5 ± 14.8 [bcd]
	Dajo	ND	ND	6.0 ± 0.1 [h]	ND	16.6 ± 0.2 [e]	322.3 ± 26.1 [g]	ND	22.8 ± 0.2 [i]	ND	367.8 ± 13.0 [f]
	Jopung	ND	ND	7.5 ± 0.1 [de]	ND	16.6 ± 0.1 [e]	331.0 ± 19.1 [g]	ND	23.1 ± 0.1 [hi]	ND	378.3 ± 9.5 [f]
	Darkhorse	ND	ND	15.8 ± 0.8 [a]	ND	22.4 ± 0.6 [a]	506.2 ± 30.9 [a]	ND	25.4 ± 0.4 [a]	ND	569.8 ± 15.2 [a]
	Hi-early	ND	ND	7.4 ± 0.1 [de]	ND	18.0 ± 0.2 [d]	421.5 ± 14.1 [cdef]	ND	24.0 ± 0.2 [def]	ND	470.8 ± 7.0 [cbd]
	High Speed	ND	ND	14.1 ± 0.2 [b]	ND	21.9 ± 0.2 [a]	477.4 ± 12.9 [ab]	ND	24.8 ± 0.2 [b]	ND	538.1 ± 6.4 [ab]

All values are the mean ± SD derivation of three experiments. C_{20} = Eicosanol; C_{21} = Heneicosanol; C_{22} = Docosanol; C_{23} = Tricosanol; C_{24} = Tetracosanol; C_{26} = Hexacosanol; C_{27} = Heptacosanol; C_{28} = Octacosanol; C_{30} = Triacontanol; total PC, = Total policosanol content; ND = not detected; Data with different superscript letters differed significantly by date and cultivar with respect to each policosanol row. (Duncan's multiple range test $p < 0.05$).

3.3. Properties of AMPK Phosphorylation in Oat Seedlings

AMPK regulates oxidative stress to control fat metabolism and the AMPK activation is best known to improve insulin resistance [12,15]. Previously, the policosanol components displayed considerable AMPK-activating abilities [15]. In addition, we demonstrated that the wheat and barley seedlings were detected potential properties of AMPK activation [26]. However, to the best of our knowledge, AMPK activation in oat seedlings has not been reported yet. Therefore, we investigated the activation of AMPK in the hexane extract of this crop. The OS extracts for AMPK phosphorylation in HepG2 cells were examined, and cell viability was determined using the MTT assay. To observe the changes in AMPK regulation, various concentrations (0, 12.5, 25, 20, 100, and 200 µg/mL, cv. Gwanghan) were used to measure cell viability. Cell viability was approximately 98% at a concentration of 200 µg/mL (Figure 5A). We also examined the relative effectiveness of each extract in different growth times. The extracts of oat seedlings of 9 and 11 days showed high ratios with 205 and 210% in comparison with positive control (100%) (Figure 5B). The previous study reported that hexacosanol was related to increased AMPK activation. 15 In this regard, we focused on the growth time of 11 days and evaluated the effects of different doses of the tested samples (25, 50, and 100 µg/mL) on rates of AMPK expression and phosphorylation in HepG2 cells using western blotting to stimulate AMPK phosphorylation by using a positive control (β-actin) (Figure 5C). As a result, the ratio of AMPK to phosphorylated AMPK increased in a dose-dependent manner. Our results indicate that the effects of OSs on AMPK activation can be correlated with high policosanol content. Consequently, the major policosanol of oat seedlings, hexacosanol (**6**) may be an excellent factor for AMPK activation. We believe that the oat seedlings on 11 days growth time may be considered a potential AMPK activation source for developing a new health functional food.

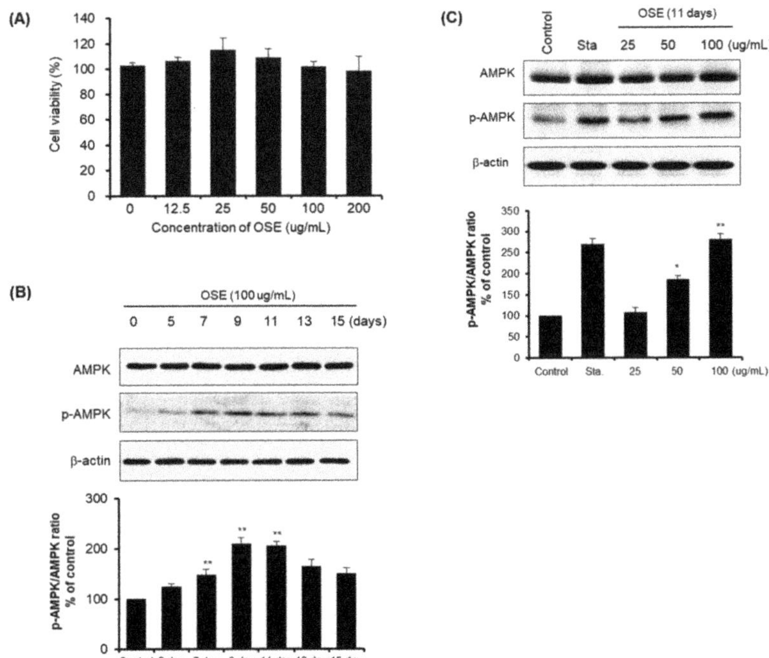

Figure 5. Effects of policosanol on HepG2 cells induced p-AMPK expression via activating AMPK. (**A**) Cell viability of each concentration (0, 12.5, 25, 50, 100, and 200 µg/mL) measured by MTT on

HepG2 cells. (**B**) HepG2 cells were treated with OS extracts from different growth stages for different growth times (5, 7, 9, 11, 13, and 15 days) and p-AMPK expression was measured by western blot analysis. β-Actin was used as the internal standard. (**C**) HepG2 cells were treated with variable concentrations (25, 50, and 100 μg/mL) of OS extracts under 11 days. Under the same 11 days growth time, p-AMPK expression was measured by western blot analysis. Data represent the mean ± standard deviation of three independent experiments. (*) $p < 0.05$, (**) $p < 0.01$ when compared to corresponding control cells.

4. Conclusions

The present work was elucidated for the first time nine policosanol compositions by GC-MS analysis from the hexane extract of oat seedlings. We have also proven the variations of individual and total policosanols in Korean cultivars during different growth times of 5, 7, 9, 11, 13, and 15 days. Policosanols exhibited significant differences between cultivars and growth times, especially, hexacosanol (**6**) showed the predominant component with 88–91% of the total average content. In addition, this policosanol displayed remarkable variations according to the growth times as follows: 337.8 → 416.8 → 458.9 → 490.0 → 479.2 → 427.0 mg/100 g. Interestingly, the harvested seedlings on 11 days exhibited the highest average policosanols with 541.7 mg/100 g in 15 oat cultivars, and the remaining sources were in decreasing order: 532.2 (13 days) > 508.3 (9 days) > 479.5 (15 days) > 463.8 (7 days) > 383.3 mg/100 g (5 days). The hexane extract of 11 days was observed the highest effects of AMPK activation in HepG2 cells, and the main policosanol, hexacosanol (**6**) may be considered as excellent AMPK activator. Based on the above findings, the optimum growth time can be in 11 days due to the policosanol contents and AMPK activations, specifically, Gwanghan and Dahan cultivars may be considered as potential materials for functional agents using oat seedlings. We believe that the oat seedlings can be employed as an excellent material for improving human nutrition and health.

Author Contributions: H.-G.L. and S.-Y.W. contributed to the writing of the manuscript and performed the major experiments. H.-J.A., J.-Y.Y., M.-J.L., H.-Y.K. and S.-Y.S. performed minor experiments and prepared raw materials. J.-H.L. and W.-D.S. planned and led this research. The authors have read and approval the final manuscript.

Funding: This research was funded by Rural Development Administration (RDA), Korea. Project title: Enhancement of secondary metabolites from crop sprouts and their improved effects against atopy and alopecia disease, Project No. PJ01421201.

Institutional Review Board Statement: Not applicable.

Informed Consent Statement: Not applicable.

Data Availability Statement: The datasets used and/or analyzed during the current study are available from the corresponding author Woo-Duck Seo (swd2002@korea.kr) on reasonable request.

Acknowledgments: This work was conducted with the support of the "Cooperative Research Program for Agriculture Science and Technology Development".

Conflicts of Interest: The authors declare that they have no known competing financial interests or personal relationships that could have appeared to influence the work reported in this paper.

Abbreviations

AMPK	adenosine 5′-monophosphate-activated protein kinase
MSTFA	N-methyl-N-(trimethylysilyl)trifluoroacetamide
MTT	thiazolyl blue tetrazolium bromide
GC-MS	gas chromatography-mass spectrometry

References

1. Arruzazabala, M.L.; Noa, M.; Menéndez, R.; Más, R.; Carbajal, D.; Valdés, S.; Molina, V. Protective effect of policosanol on atherosclerotic lesions in rabbits with exogenous hypercholesterolemia. *Braz. J. Med. Biol. Res.* **2000**, *33*, 835–840. [CrossRef] [PubMed]
2. Jang, Y.; Kim, D.; Han, E.; Jung, J. Physiological activities of policosanol extracted from sugarcane wax. *Nat. Prod. Sci.* **2019**, *25*, 293–297. [CrossRef]
3. Aneiros, E.; Mas, R.; Calderon, B.; Illnait, J.; Fernandez, L.; Castano, G.; Cesar, J.C. Effect of policosanol in lowering cholesterol levels in patients with type II hypercholesterolemia. *Curr. Ther. Res. Clin. Exp.* **1995**, *56*, 176–182. [CrossRef]
4. Arruzazabala, M.L.; Molina, V.; Mas, R.; Fernández, L.; Carbajal, D.; Valdés, S.; Castaño, G. Antiplatelet effects of policosanol (20 and 40 mg/day) in healthy volunteers and dyslipidaemic patients. *Clin. Exp. Pharmacol. Physiol.* **2002**, *29*, 891–897. [CrossRef]
5. Arruzazabala, M.L.; Carbajal, D.; Mas, R.; Garcia, M.; Fraga, V. Effects of policosanol on platelet aggregation in rats. *Thromb. Res.* **1993**, *69*, 321–327. [CrossRef]
6. Carbajal, D.; Molina, V.; Valdés, S.; Arruzazabala, L.; Más, R. Anti-ulcer activity of higher primary alcohols of beeswax. *J. Pharm. Pharmacol.* **1995**, *47*, 731–733. [CrossRef]
7. Fernández-Arche, A.; Marquez-Martín, A.; de la Puerta Vazquez, R.; Perona, J.S.; Terencio, C.; Perez-Camino, C.; Ruiz-Gutierrez, V. Long-chain fatty alcohols from pomace olive oil modulate the release of proinflammatory mediators. *J. Nutr. Biochem.* **2009**, *20*, 155–162. [CrossRef]
8. Ravelo, Y.; Molina, V.; Carbajal, D.; Fernández, L.; Fernández, J.C.; Arruzazabala, M.L.; Más, R. Evaluation of anti-inflammatory and antinociceptive effects of D-002 (beewax alcohols). *J. Nat. Med.* **2011**, *65*, 330–335. [CrossRef]
9. Chen, Y.; Dunford, N.T.; Goad, C. Tangential abrasive dehulling of wheat grain to obtain policosanol and phytosterol enriched fractions. *J. Cereal Sci.* **2013**, *57*, 258–260. [CrossRef]
10. Choi, S.J.; Park, S.Y.; Park, J.S.; Park, S.K.; Jung, M.Y. Contents and compositions of policosanols in green tea (*Camellia sinensis*) leaves. *Food Chem.* **2016**, *204*, 94–101. [CrossRef]
11. Harrabi, S.; Boukhchina, S.; Mayer, P.M.; Kallel, H. Policosanol distribution and accumulation in developing corn kernels. *Food Chem.* **2009**, *115*, 918–923. [CrossRef]
12. Seo, W.D.; Yuk, H.J.; Curtis-Long, M.J.; Jang, K.C.; Lee, J.H.; Han, S.I.; Kang, H.W.; Nam, M.H.; Lee, S.J.; Lee, J.H.; et al. Effect of the growth stage and cultivar on policosanol profiles of barley sprouts and their adenosine 5′-monophosphate-activated protein kinase activation. *J. Agric. Food Chem.* **2013**, *61*, 1117–1123. [CrossRef] [PubMed]
13. Irmak, S.; Dunford, N.T. Policosanol contents and compositions of wheat varieties. *J. Agric. Food Chem.* **2005**, *53*, 5583–5586. [CrossRef]
14. Irmak, S.; Dunford, N.T.; Milligan, J. Policosanol contents of beeswax, sugar cane and wheat extracts. *Food Chem.* **2006**, *95*, 312–318. [CrossRef]
15. Lee, J.H.; Jia, Y.; Thach, T.T.; Han, Y.; Kim, B.; Wu, C.; Kim, Y.; Seo, W.D.; Lee, S.J. Hexacosanol reduces plasma and hepatic cholesterol by activation of AMP-activated protein kinase and suppression of sterol regulatory element-binding protein-2 in HepG2 and C57BL/6J mice. *Nutr. Res.* **2017**, *43*, 89–99. [CrossRef] [PubMed]
16. Oliaro-Bosso, S.; Calcio Gaudino, E.; Mantegna, S.; Giraudo, E.; Meda, C.; Viola, F.; Cravotto, G. Regulation of HMGCoA Reductase Activity by Policosanol and Octacosadienol, a New Synthetic Analogue of Octacosanol. *Lipids* **2009**, *44*, 907. [CrossRef] [PubMed]
17. Singh, D.K.; Li, L.; Porter, T.D. Policosanol inhibits cholesterol synthesis in hepatoma cells by activation of AMP-kinase. *J. Pharmacol. Exp. Ther.* **2006**, *318*, 1020–1026. [CrossRef]
18. Darzian Rostami, Z.; Asghari, A.; Jahandideh, A.; Mortazavi, P.; Akbarzadeh, A. Effect of oat (*Avena sativa* L.) extract on experimental sciatic nerve injury in rats. *Arch. Razi Inst.* **2020**, *75*, 249–256.
19. Liu, B.; Yang, T.; Luo, Y.; Zeng, L.; Shi, L.; Wei, C.; Nie, Y.; Cheng, Y.; Lin, Q.; Luo, F. Oat β-glucan inhibits adipogenesis and hepatic steatosis in high fat dietinduced hyperlipidemic mice via AMPK signaling. *J. Funct. Foods* **2018**, *41*, 72–82. [CrossRef]
20. Ahmed, S.; Gul, S.; Gul, H.; Bangash, M.H. Anti-inflammatory and anti-platelet activities of Avena sativa are mediated through the inhibition of cyclooxygenase and lipoxygenase enzymes. *Int. J. Endorsing Health Sci. Res.* **2013**, *1*, 62–65. [CrossRef]
21. Feng, B.; Ma, L.J.; Yao, J.J.; Fang, Y.; Mei, Y.A.; Wei, S.M. Protective effect of oat bran extracts on human dermal fibroblast injury induced by hydrogen peroxide. *J. Zhejiang Univ. Sci. B* **2013**, *14*, 97–105. [CrossRef]
22. Woo, S.Y.; Lee, K.S.; Shin, H.L.; Kim, S.H.; Lee, M.J.; Kim, H.Y.; Ham, H.; Lee, D.J.; Choi, S.W.; Seo, W.D. Two new secondary metabolites isolated from *Avena sativa* L. (Oat) seedlings and their effects on osteoblast differentiation. *Bioorg. Med. Chem. Lett.* **2020**, *30*, 127250. [CrossRef] [PubMed]
23. Lee, J.H.; Lee, S.Y.; Kim, B.; Seo, W.D.; Jia, Y.; Wu, C.; Jun, H.J.; Lee, S.J. Barley sprout extract containing policosanols and polyphenols regulate AMPK, SREBP2 and ACAT2 activity and cholesterol and glucose metabolism in vitro and in vivo. *Food Res. Int.* **2015**, *72*, 174–183. [CrossRef]
24. Bradford, M.M. A rapid and sensitive method for the quantitation of microgram quantities of protein utilizing the principle of protein-dye binding. *Anal. Biochem.* **1976**, *72*, 248–254. [CrossRef]
25. Lee, J.H.; Park, M.J.; Ryu, H.W.; Yuk, H.J.; Choi, S.-W.; Lee, K.-S.; Kim, S.-L.; Seo, W.D. Elucidation o phenolic antioxidants in barley seedlings (*Hordeum vulgare* L.) by UPLC-PDA-ESI/MS and screening for their contents at different harvest times. *J. Funct. Foods* **2016**, *26*, 667–680. [CrossRef]

26. Ra, J.E.; Woo, S.Y.; Lee, K.S.; Lee, M.J.; Kim, H.Y.; Ham, H.M.; Chung, I.M.; Kim, D.H.; Lee, J.H.; Seo, W.D. Policosanol profiles and adenosine 5′-monophosphate-activated protein kinase (AMPK) activation potential of Korean wheat seedling extracts according to cultivar and growth time. *Food Chem.* **2020**, *317*, 126388. [CrossRef]
27. Irmak, S.; Jonnala, R.S.; MacRitchie, F. Effect of genetic variation on phenolic acid and policosanol contents of Pegaso wheat lines. *J. Cereal Sci.* **2008**, *48*, 20–26. [CrossRef]

MDPI
St. Alban-Anlage 66
4052 Basel
Switzerland
Tel. +41 61 683 77 34
Fax +41 61 302 89 18
www.mdpi.com

Plants Editorial Office
E-mail: plants@mdpi.com
www.mdpi.com/journal/plants